International Series in Operations Research & Management Science

Volume 312

The book series *International Series in Operations Research and Management Science* encompasses the various areas of operations research and management science. Both theoretical and applied books are included. It describes current advances anywhere in the world that are at the cutting edge of the field. The series is aimed especially at researchers, doctoral students, and sophisticated practitioners. The series features three types of books:

- *Advanced expository books* that extend and unify our understanding of particular areas.
- *Research monographs* that make substantial contributions to knowledge.
- *Handbooks* that define the new state of the art in particular areas. They will be entitled *Recent Advances in (name of the area)*. Each handbook will be edited by a leading authority in the area who will organize a team of experts on various aspects of the topic to write individual chapters. A handbook may emphasize expository surveys or completely new advances (either research or applications) or a combination of both.

The series emphasizes the following four areas: *Mathematical Programming:* Including linear programming, integer programming, nonlinear programming, interior point methods, game theory, network optimization models, combinatorics, equilibrium programming, complementarity theory, multiobjective optimization, dynamic programming, stochastic programming, complexity theory, etc.

Applied Probability: Including queuing theory, simulation, renewal theory, Brownian motion and diffusion processes, decision analysis, Markov decision processes, reliability theory, forecasting, other stochastic processes motivated by applications, etc. *Production and Operations Management:* Including inventory theory, production scheduling, capacity planning, facility location, supply chain management, distribution systems, materials requirements planning, just-in-time systems, flexible manufacturing systems, design of production lines, logistical planning, strategic issues, etc. *Applications of Operations Research and Management Science:* Including telecommunications, health care, capital budgeting and finance, marketing, public policy, military operations research, service operations, transportation systems, etc.

More information about this series at http://www.springer.com/series/6161

Joe Zhu • Vincent Charles
Editors

Data-Enabled Analytics

DEA for Big Data

Springer

Editors
Joe Zhu
Business School
Worcester Polytechnic Institute
Worcester, MA, USA

Vincent Charles
University of Wales
Trinity Saint David
Birmingham, UK

ISSN 0884-8289 ISSN 2214-7934 (electronic)
International Series in Operations Research & Management Science
ISBN 978-3-030-75161-6 ISBN 978-3-030-75162-3 (eBook)
https://doi.org/10.1007/978-3-030-75162-3

This Springer imprint is published by the registered company Springer Nature Switzerland AG
The registered company address is: Gewerbestrasse 11, 6330 Cham, Switzerland

Preface

Data envelopment analysis (DEA) has been and continues to be a widely used technique both in performance and productivity measurement, having covered a plethora of challenges and debates within the modelling framework. Over the past four decades, DEA models have been applied in almost every major field of study. Despite this, however, DEA has not been used to its fullest extent. As the inter- and intra-disciplinary research grows, DEA could be used in potentially many other ways. DEA could be viewed as a data-oriented data science tool for data-enabled analytics, benchmarking, performance evaluation, and developing composite indexes, among other new uses, in addition to the traditional uses, such as production efficiency and productivity measurement. One opportunity is brought by the existence of big data. Although big data have existed for a while now, gaining popularity among insight seekers, we are still in incipient stages when it comes to taking full advantage of their potential. As the amount of (big) data keeps growing in an exponential manner, so does its complexity; in this sense, various types of data are surfacing, whose study and examination could shed new light on phenomena of interest.

A quick review of existing literature shows that big data is a new entrant within the DEA framework. Recently, there has been an increasing interest in bringing the two concepts together, with research studies aiming to integrate DEA and big data concepts within a single framework. Despite this, however, more work is needed to fully explore the value of their intersection. It is thus time to view DEA considering its potential usage in new fields or new usage within the existing fields, under the big data umbrella. Otherwise stated, it is time to view DEA models beyond their present scope to mine new insights for better data-driven decision-making. This book seeks new DEA developments that are tailored for big data research and data-enabled analytics.

In the chapter "Data Envelopment Analysis and Big Data: A Systematic Literature Review with Repeated Bibliometric Analysis", Vincent Charles, Tatiana Gherman, and Joe Zhu aim to identify the current avenues of research for studies integrating DEA with big data. The analysis performed shows that big data is a new entrant within the DEA literature, with the recent body of work in the field being indicative of an increasing interest in bringing the two concepts together under a single framework.

In the chapter "Acceleration of Large-Scale DEA Computations Using Random Forest Classification", Anyu Yu, Yu Shi, and Joe Zhu propose a novel approach to accelerate DEA computations involving voluminous data. The proposed method uses random forest (RF) classification to predict and search for the best-practice decision-making units (DMUs) within the large-scale observations. The effectiveness of the proposed method is tested using numerical cases involving large-scale data. The authors find that the proposed DEA-RF method can decrease computation time significantly, while ensuring an acceptable level of accuracy.

In the chapter "The Estimation of Productive Efficiency Through Machine Learning Techniques: Efficiency Analysis Trees", Juan Aparicio, Miriam Esteve, Jesus J. Rodriguez-Sala, and Jose L. Zofio revise the fundamentals of a new technique recently proposed in the literature for estimating production frontiers based on decision trees, called efficiency analysis trees (EAT), and extend it to the context of measuring productive efficiency under convexification, using the directional distance function. The authors further illustrate how the different methods work by resorting to two real datasets.

In the chapter "Hybrid Data Science and Reinforcement Learning in Data Envelopment Analysis", Chia-Yen Lee, Yu-Hsin Hung, and Yen-Wen Chen propose a hybrid data science (DS) framework and reinforcement learning (RL) in DEA to complement efficiency analysis, which they validate via an empirical study of the US coal-fired and oil-fired power plants operating from 2004 through 2019. The authors find that the hybrid DS framework and RL can enhance the interpretation of the production frontier and identify the optimal resource policy, thus guiding productivity improvement strategy.

Motivated by the increasing attention to dimension reduction in the context of DEA with large dimensions for inputs and outputs, in the chapter "Aggregation of Outputs and Inputs for DEA Analysis of Hospital Efficiency: Economics, Operations Research and Data Science Perspectives", Bao Hoang Nguyen and Valentin Zelenyuk investigate the two most popular dimension reduction approaches: PCA-based aggregation and price-based aggregation for hospital efficiency analysis. Using data on public hospitals in Queensland, Australia, the authors find, among others, that the PCA-based aggregation can be viewed as a viable alternative for DEA practitioners who are unable to/or unwilling to use the price-based approach, for example, due to unavailable or unreliable price information.

In the chapter "Parallel Processing and Large-Scale Datasets in Data Envelopment Analysis", Dariush Khezrimotlagh illustrates the main existing methods to decrease the elapsed time of applying a DEA model to evaluate a large-scale dataset. Using different datasets, it is shown that the strengths of the existing methods are affected when cardinality, dimension, and density are changed. Then, the author proposes a new methodology using the combination of two existing methods. In general, the proposed method is faster than all existing methods regardless of cardinalities, dimensions, and densities.

In the chapter "Network DEA and Big Data with an Application to the Coronavirus Pandemic", Hirofumi Fukuyama and William L. Weber examine how NDEA models can accommodate big data and further estimate a dynamic network model of the coronavirus pandemic in the United States. The model assumes that states seek to simultaneously maximise real gross domestic product and minimise deaths from Covid-19 given inputs. Additionally, the authors investigate whether intertemporal reallocations of Covid tests could have helped reduce Covid-19 deaths.

In the chapter "Hierarchical Data Envelopment Analysis for Classification of High-Dimensional Data", Ming-Miin Yu, Kok Fong See, and Bo Hsiao provide an application of big data, data science, and data analytics methods in the hierarchical DEA (H-DEA) framework for the classification of high-dimensional data. The authors examine global food security performance using an H-DEA model and then use a multi-level K means clustering approach to cluster the 110 sampled countries into homogeneous and distinct groups. Under the scoring clustering approach, the results can help relevant policymakers to understand the benchmarking process and the learning path so as to design relevant policies.

In the chapter "Dominance Network Analysis: Hybridizing DEA and Complex Networks for Data Analytics", Laura Calzada-Infante and Sebastian Lozano advocate for the hybridisation of DEA and complex networks considering the advantages such hybridisation brings in terms of the multidimensional benchmarking prowess of DEA and the versatility, computational efficiency, and modelling capabilities of the network paradigm. The methodology presented is based on dominance network (DN) analysis and is further illustrated with data on how the COVID-19 pandemic has affected the different countries.

In the chapter "Value extracting in relative performance appraisal with network DEA: an application to US equity mutual funds", Hirofumi Fukuyama and Don U.A. Galagedera discuss the contribution of network DEA in mutual funds (MF) performance appraisal and highlight that when MF management process is conceptualised as a network structure, it is possible to extract valuable information from MF specific data analogous to data mining in the case of big data. The information extracted via network DEA is practical and valuable to all stakeholders involved.

In the chapter "Measuring Chinese Bank Performance with Undesirable Outputs: A Slack-Based Two-Stage Network DEA Approach", Ya Chen, Mengyuan Wang, and Jingyu Yang propose a slack-based two-stage DEA model with undesirable outputs under variable returns-to-scale (called the UVSBM model) to measure both overall and sub-stage efficiencies of banks. Among others, by considering the internal production process in bank efficiency evaluation, the results help to identify the source of inefficiency in bank operations.

In the chapter "Using Network DEA and Grey Prediction Model for Big Data Analysis: An Application in the Global Airline Efficiency", Wen-Min Lu, Qian Long Kweh, Mohammad Nourani, and Hsiu-Fei Wang illustrate how to use network DEA integrated with multiplicative efficiency aggregation and grey prediction model to uncover valuable information in a big data context, with an application to global airlines. In essence, the study advances an approach to transform large volumes of data into multiple pieces of useful information, helping to extract value from big data.

The many academics and researchers who contributed chapters and the experts within the field who reviewed the chapters made this book possible – we thank you! The chapters contributed to this book should be of considerable interest and provide our readers with informative reading.

Worcester, MA, USA Joe Zhu

Birmingham, UK Vincent Charles

Contents

Data Envelopment Analysis and Big Data: A Systematic Literature Review with Bibliometric Analysis

Vincent Charles, Tatiana Gherman, and Joe Zhu

Abstract Data envelopment analysis (DEA) is a powerful data-enabled, big data science tool for performance measurement and management, which over time has been applied across a myriad of domains. Over the past years, various advancements in big data have captured the attention of DEA scholars, which in turn, has translated into the emergence of new research strands. In the present work, we perform a systematic literature review with bibliometric analysis of studies integrating DEA with big data, in an attempt to answer the question: *what are the current avenues of research for such studies?* The results obtained are further complemented with a thematic analysis. Among others, findings indicate that big data is still a new entrant within the DEA literature, that most of the studies have focused on developing faster and more accurate computational techniques to handle problems with a large number of decision-making units (DMUs), and that most of the studies have been carried out in the area of environmental efficiency evaluation. This work should contribute to the construction of an overview of the existing literature on DEA-big data studies, as well as stimulate the interest in the topic.

Keywords Data envelopment analysis · Data-enabled analytics · Big data · Systematic literature review · Bibliometric analysis

V. Charles (✉)
University of Wales Trinity Saint David, Birmingham, UK
e-mail: c.vincent@uwtsd.ac.uk

T. Gherman
Faculty of Business and Law, University of Northampton, Northampton, UK
e-mail: tatiana.gherman@northampton.ac.uk

J. Zhu
Business School, Worcester Polytechnic Institute, Worcester, MA, USA
e-mail: jzhu@wpi.edu

J. Zhu, V. Charles (eds.), *Data-Enabled Analytics*, International Series in Operations Research & Management Science 312, https://doi.org/10.1007/978-3-030-75162-3_1

1 Introduction

Data envelopment analysis (DEA) is a non-parametric mathematical programming approach for performance evaluation and identification of best practices when multiple performance metrics or measures are present. Although with strong links in economic and production theory (Farrell, 1957), DEA is used for benchmarking in operations management, wherein the efficient decision-making units (DMUs) may not necessarily form a "production frontier", but rather lead to a "best-practice frontier" (Cook et al., 2014). The initial DEA model was introduced by Charnes et al. (1978) and assumes constant returns-to-scale, *i.e.*, that outputs are increased proportionally to inputs. The model was later adapted by Banker et al. (1984) to accommodate variable returns-to-scale. In time, the DEA literature has seen a great variety of applications across a plethora of domains, with DEA becoming a powerful data-driven management science tool (Charles et al., 2018), more recently coined as a tool for data-oriented or data-enabled analytics (Zhu, 2020).

In recent years, the emergence of big data has created opportunities for pursuing new avenues of research. Without much doubt, big data have developed into a big phenomenon. Currently, there is no precise and uniform definition of big data, although it is commonly agreed that big data are "datasets that are too large for traditional data-processing systems and that therefore require new technologies" (Provost & Fawcett, 2013). Likewise, it is more common for the literature to refer to the four dimensions of big data defined by Laney (2001): volume, velocity, variety, and veracity, although there are other Vs, *e.g.*, value. In a big data context, the data are massive and generated continuously, are of different types (structured, semi-structured, and unstructured), and are characterised by uncertainty. The above highlights the computational complexities and technical requirements associated with big data. Coupled with ethical challenges (Charles et al., 2015), analysing big data confronts researchers with many difficulties (Bizer et al., 2012). Charles and Gherman (2013) emphasised that in order to create value and competitive advantage, big data should be further considered in view of the dimensions of context, connectedness, and complexity.

Studies have already shown that big data can be used to improve company productivity. For instance, see the works by Brynjolfsson et al. (2011) and Müller et al. (2018), which proved that big data are associated with improvement in firm productivity by using empirical study methods. Likewise, the research by Manyika et al. (2011) showed that big data will be essential for enterprises to grow and achieve competitive advantage. Over the years, big data have drawn the attention of researchers across a variety of domains, revolutionising business, scientific research, and public administration alike (Chen & Zhang, 2014). For example, Wu et al. (2014) indicated that big data have rapidly expanded in all science and engineering domains, such as biological, biomedical, and physical sciences. In the process, organisations of every kind have learned to take advantage of big data-driven strategies to innovate, compete, and capture value from big data, although we are yet to see big data being fully translated into societal value (Charles & Gherman, 2018).

Big data have transformed our data paradigms, opening new opportunities and improving established analytic techniques. Big data analytics can be defined as the process of extracting useful information (*e.g.*, finding patterns in the data or deriving decision-models) from a pre-processed dataset. The boom of big data analytics has brought in a significant revolution in thinking and behaviour in all sectors of modern societies (Zhang et al., 2019). Michael and Miller (2013) remarked that the development of big data and big data analytics can help with comprehensive data analyses, supporting improved policy- and decision-making.

Zhu (2020) noted that "as big data research becomes an important area of operations analytics, DEA is evolving into data enabled analytics. DEA can be viewed as a data-oriented data science tool for productivity analytics, benchmarking, performance evaluation, and composite index construction, among other new uses, in addition to the traditional uses such as, production efficiency and productivity measurement" (p. 2). When it comes to big data and DEA, big data have brought many challenges. For example, the larger scale number of DMUs is one among these, as it may take an impractical amount of time to finish the efficiency evaluation of all the DMUs. Therefore, researchers have been generally concerned with developing methods for reducing the solution time for DEA problems under a big data environment. Another challenge associated with big data is extracting *value* from the same, with Zhu (2020), however, demonstrating that network DEA (NDEA) can be used to deal with this dimension.

In this work, we aim to provide an overview of the literature on DEA-big data by performing a systematic literature review and analysing an extensive range of bibliometric indicators and employing software for bibliographic mapping, in an attempt to answer the question: *what are the current avenues of research for such studies?* In essence, the systematic literature review involves a well thought out search strategy that helps to identify and synthesise the scholarly research on the topic and bibliometric analysis embodies both the statics and dynamics of the literature set, emphasising important trends and patterns in the topic.

The remainder of this work is organised as follows. Section 2 details the methodology. Section 3 outlines the initial document results with regard to the literature on DEA and big data, and further presents a bibliometric analysis of the studies integrating DEA with big data. Section 4 delves deeper into the studies that have attempted to integrate DEA with big data and presents the findings resulted from the systematic literature review performed. The section also offers a blend of bibliometric analysis with thematic analysis. Section 5 presents a discussion of main results and conclusions.

2 Methodology

Our intention is to provide an overview of the current avenues of research for the studies aimed at integrating DEA with big data. To achieve this aim, we have performed a blend of systematic literature review, bibliometric analysis, and

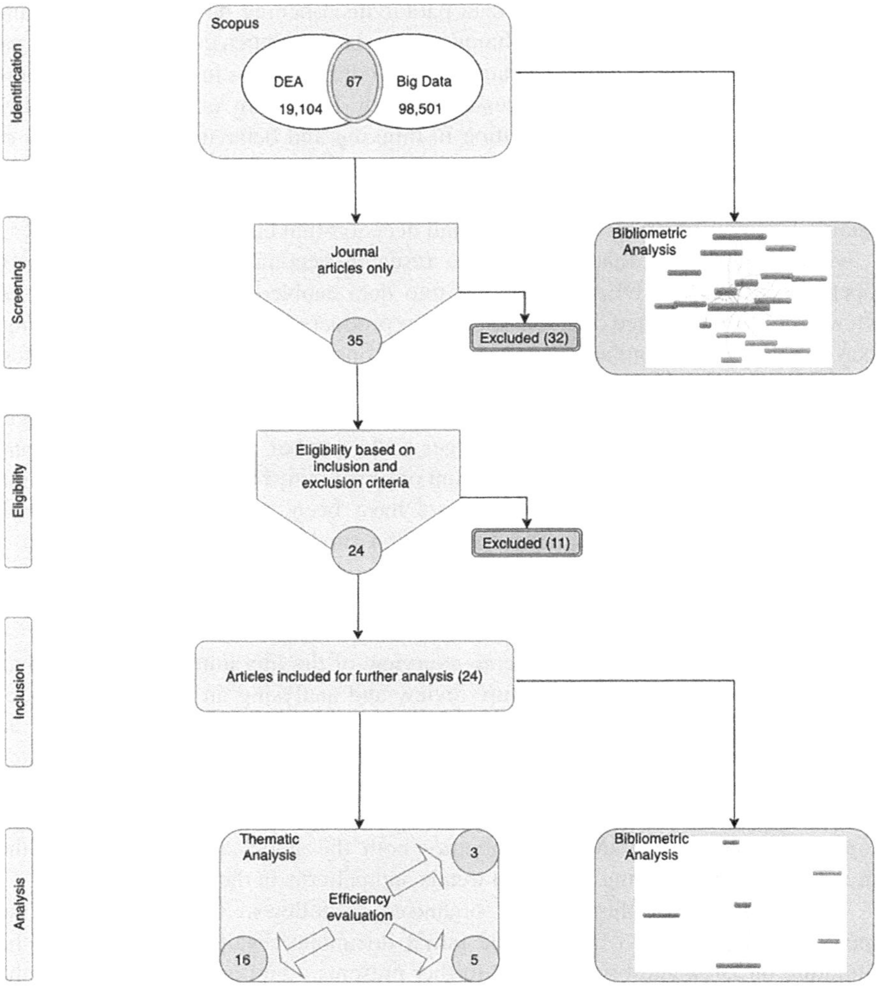

Fig. 1 Flowchart of the systematic literature review with bibliometric analysis and thematic analysis

thematic analysis on the Scopus database. The Scopus database was chosen in view of the fact that it is the largest database of peer-reviewed literature. Figure 1 depicts the flowchart of the approach followed.

The first phase consisted in identifying and mapping the current literature on the topics of DEA and big data. The literature search through the Scopus database was first conducted using two keywords: "data envelopment analysis" and "big data", respectively. This is because, in general, if a document uses DEA and/or big data, it is expected that such document will mention these specific terms in the title, abstract, and/or keywords; hence, it was deemed that there was no need to also search for related or alternative keywords. The search for the term "data envelopment analysis" on the 28th February 2021 in the title, abstract, and keywords of the material deposited in the Scopus database yielded 19,104 document results, published during the period 1980–2021. A similar search for the term "big data" on the 28th February 2021 in the title, abstract, and keywords of the material deposited in the Scopus database yielded 98,501 document results, published during the period 1957–2021. Interestingly enough, the simultaneous search for the terms "data envelopment analysis" and "big data" in the title, abstract, and keywords of the material deposited in the Scopus database yielded only 67 document results, published between 2013 and 2021; these results were further subjected to a co-occurrence analysis using the VOSviewer software for bibliometric analysis. Details with regards to the above can be appreciated in Sect. 3.1 (for DEA document results), Sect. 3.2 (for big data document results), and Sect. 3.3 (for DEA-big data document results).

In the second phase, we screened the 67 document results and selected only the journal articles for further analysis. This decision was taken in view of the controversy surrounding the issue of whether to include conference papers, which generally do not provide enough information about the research conducted, as we encounter in full papers. Additionally, conference papers are normally written to present preliminary results, constituting works in progress rather than full papers (Mubin et al., 2018). Book chapters, conference reviews, and reviews do different work than journal articles; hence, these were also excluded from the pool. Moreover, these types of publications only constituted 11.1% of the total number of publications, indicating a marginal impact, if any, on the overall analysis. This screening led to the consideration of 35 research articles for further processing, constituting 52.2% of the publications.

In the next phase, we have proceeded with a systematic literature review of the 35 research articles. In this sense, a set of inclusion and exclusion criteria were established, and a manual checking of the articles was performed to identify those articles that complied with the criteria. Such endeavour resulted in 24 eligible articles, which were then passed through a bibliometric analysis and thematic analysis. Section 4 contains the details of these analyses and the results obtained.

3 An Overview of the DEA and Big Data Literatures

3.1 "DEA" Document Results (19,104 Documents)

The 19,104 document results show the great interest that DEA has accumulated over time, with a markedly upward trend at the beginning of the twenty-first century. Figure 2 shows the evolution of the number of publications over the period mentioned (1980–2021). To be noted that the number of publications in 2021 is currently 349, but this is, of course, caused by the limited time frame covered in the search, as only publications up to 28th February were considered (Note: Such consideration is to be exercised for the remaining visualisations).

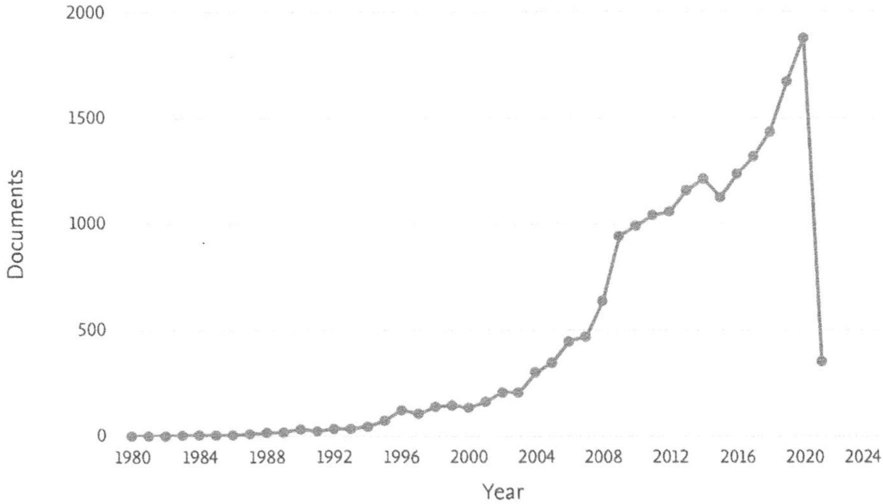

Fig. 2 DEA – Annual scientific production. (Source: Scopus 2021)

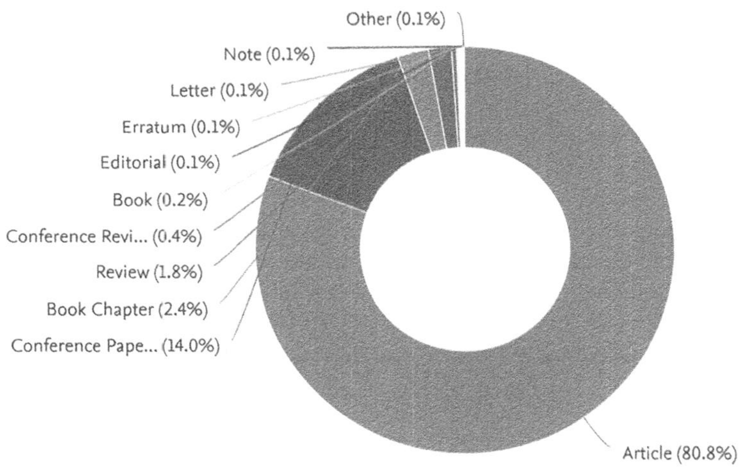

Fig. 3 DEA – Descriptive summary: documents by type. (Source: Scopus 2021)

Furthermore, by looking at Fig. 3, we can also notice that the DEA literature is dominated by research articles (which constitute 80.8% of the number of publications), followed by conference papers (14%). Figure 4 shows the documents by subject area. Here, we can observe that the area of "business, management, and accounting" has received the most interest, with 15% of the publications. This is closely followed by "engineering" (12.6%), "computer science" (12.2%), and "decision sciences" (11.6%), respectively.

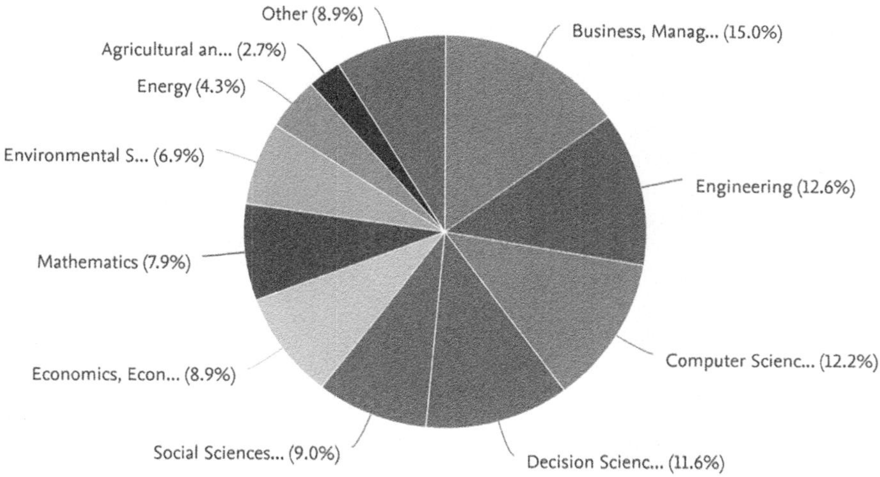

Fig. 4 DEA – Descriptive summary: documents by subject area. (Source: Scopus 2021)

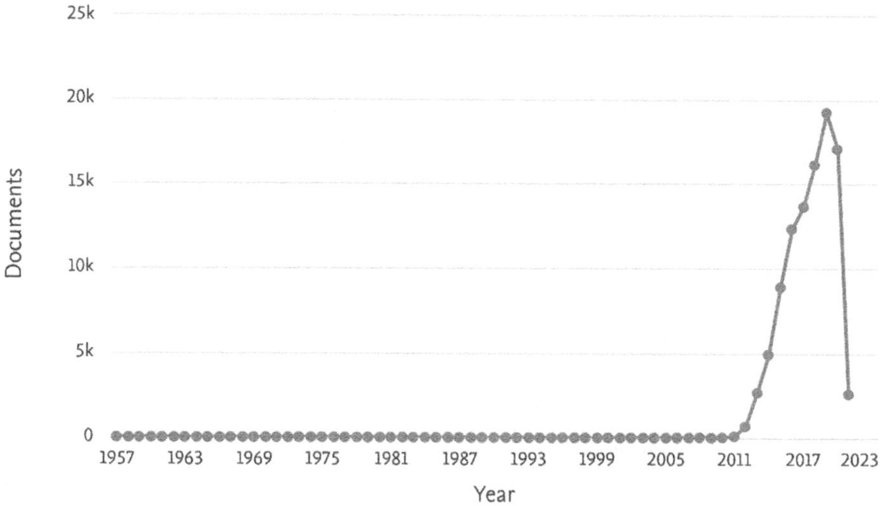

Fig. 5 Big data – Annual scientific production. (Source: Scopus 2021)

3.2 "Big Data" Document Results (98,501 Documents)

Figure 5 shows that studies on big data took off after the year 2011 in an exponential manner, with the peak in 2019, with 19,266 publications. In 2020, the number decreased slightly to 17,085 publications, but this may have been caused by the COVID-19 pandemic, which saw many conferences, for example, being cancelled. Moreover, in view of the fact that conference papers represent the biggest percentage

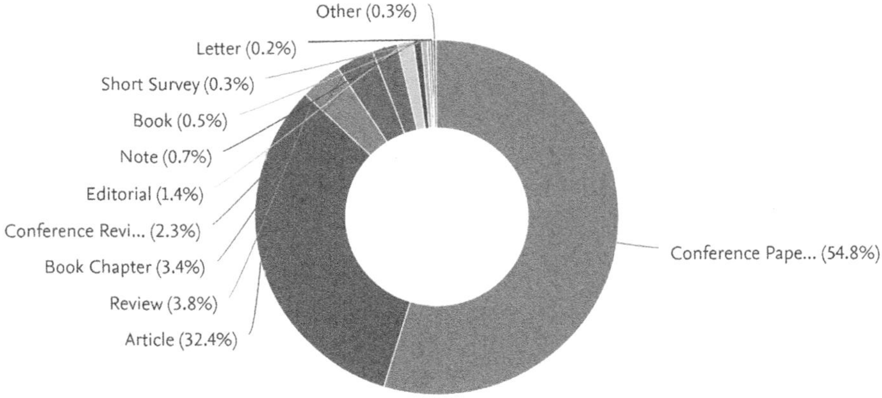

Fig. 6 Big data – Descriptive summary: documents by type. (Source: Scopus 2021)

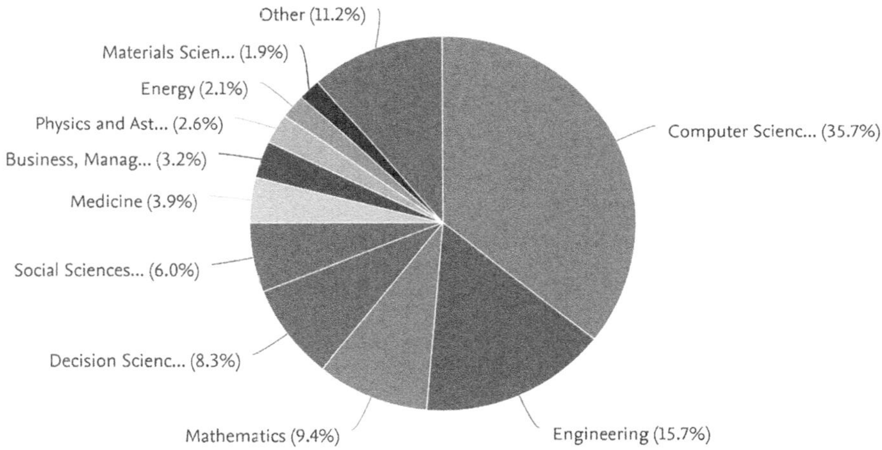

Fig. 7 Big data – Descriptive summary: documents by subject area. (Source: Scopus 2021)

(54.8%) of the publications on big data (Fig. 6), it does make sense to exercise caution in interpreting the 2020 publications. By contrast to DEA publications, the area of "computer science" concentrates most of the publications (35.7%), followed by "engineering" (15.7%). Interestingly, the area of "business, management, and accounting" captures only 3.2% of the publications on big data, perhaps indicating that this is still a young area when it comes to capitalising on the benefits brought by big data (Fig. 7).

3.3 A Brief Bibliometric Analysis of the DEA-Big Data Literature (67 Studies)

We have performed a simultaneous search for the terms "data envelopment analysis" and "big data" in the article title, abstract, and keywords of the Scopus database, which, as mentioned, yielded 67 document results, all published between 2013 and 2021.

In this section, we graphically analyse the bibliographic material on DEA-big data using the VOSviewer software. The software considers the co-occurrence of all keywords, with full counting. The co-occurrence of keywords measures the most common keywords and those that appear more frequently in the same papers. Table 1 provides a summary of the keywords whose co-occurrence is at least 5 times. Table 1 is further visually depicted in Figs. 8 and 9.

A total of 18 keywords were identified (Table 1). Keywords are labelled with coloured frames (Fig. 8), wherein the size of the frames is positively correlated with the occurrence of the keyword in the publication. Therefore, the size of the label and the frame of a keyword is determined by the weight of the item, with a greater weight being associated with a larger label and frame. The results identified "data envelopment analysis" (with a total link strength of 177) and "big data" (with a total link strength of 176) as the most common keywords, followed by "efficiency" (with

Table 1 Co-occurrence of all keywords

Keywords	Occurrences	Total link strength
data envelopment analysis	54	177
big data	53	176
efficiency	28	110
decision making	16	73
efficiency evaluation	8	43
dea	9	39
decision making unit	7	34
sustainable development	6	31
environmental management	6	26
relative efficiency	5	26
data mining	7	25
economics	6	25
advanced analytics	5	24
energy efficiency	7	24
data envelopment analysis (dea)	8	22
technical efficiency	5	21
artificial intelligence	5	20
information management	5	20

Note. "Total link strength" refers to the total strength of the links of an item with other items.
Source: Scopus 2021

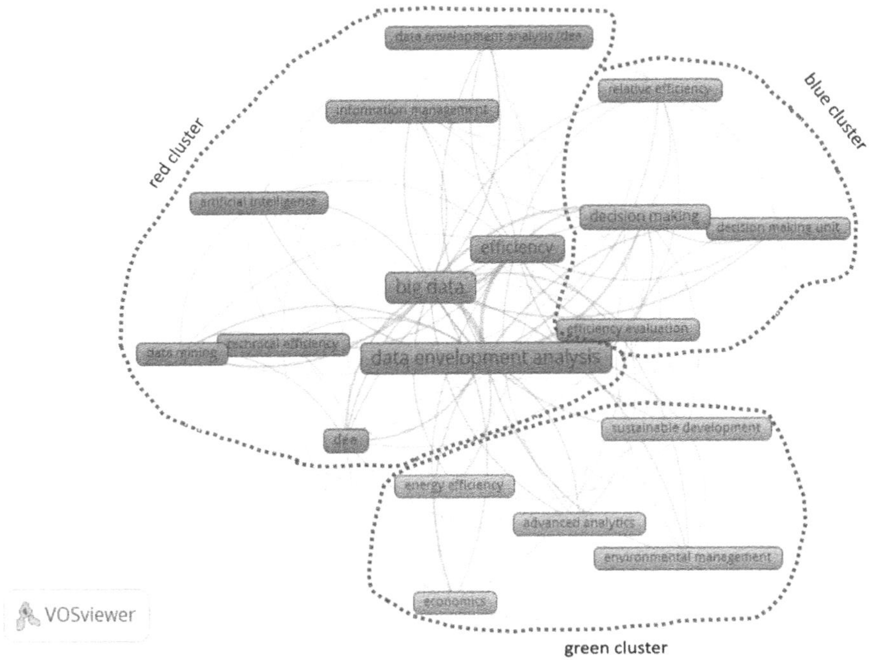

Fig. 8 Network map showing the relations between various topics in the DEA-big data field (based on the pool of 67 studies)

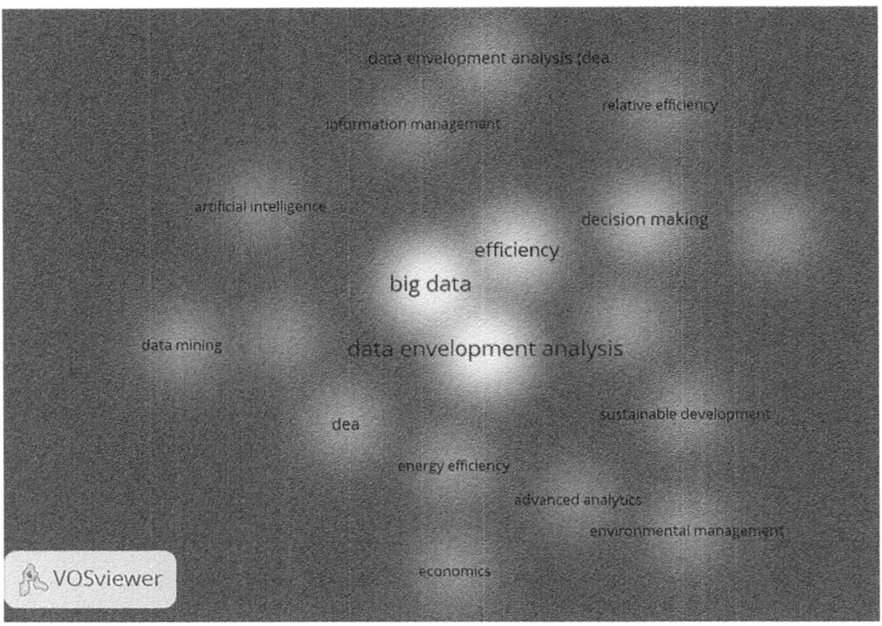

Fig. 9 Density visualisation

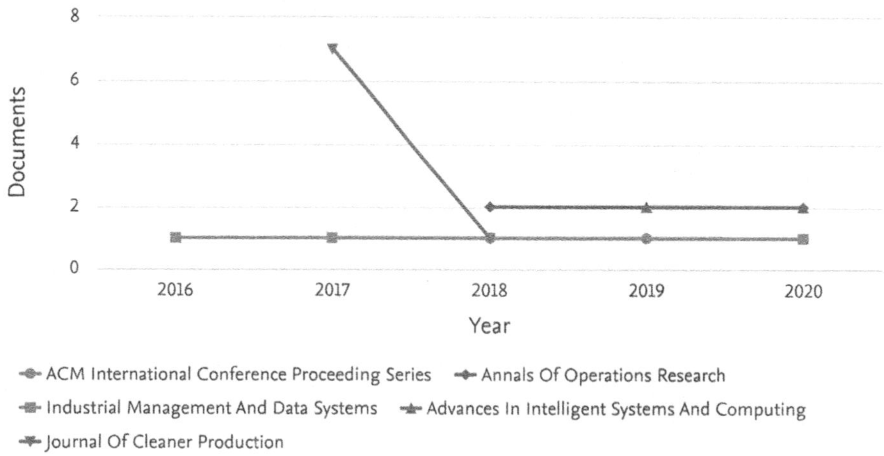

Fig. 10 DEA-big data – Descriptive summary: documents per year by source. (Source: Scopus 2021)

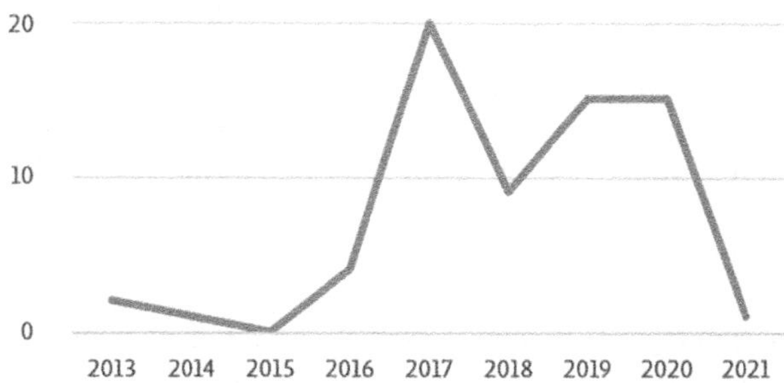

Fig. 11 DEA-big data – Descriptive summary: documents by year. (Source: Scopus 2021)

a total link strength of 110) and "decision-making" (with a total link strength of 73). These keywords were further classified by the software into three large clusters (Fig. 8) that seem to assume a prominent role vis-à-vis "computational paradigms" (nine items, red cluster), "measures and decision-making" (four items, blue cluster), and "areas of application" (five items, green cluster).

Figure 10 shows the most relevant sources in the Scopus literature collection. The graph displays the five journals that have published most of the material on DEA and big data. These journals are Journal of Cleaner Production (8 publications), ACM International Conference Proceeding Series (4 publications), Annals of Operations Research (4 publications), Advances in Intelligent Systems and Computing (4 publications), and Industrial Management and Data Systems (3 publications). Together, these journals account for 23 documents out of the 67 results.

Figure 11 further indicates that there has been a generally increasing interest in researching DEA under a big data environment, particularly in the last five

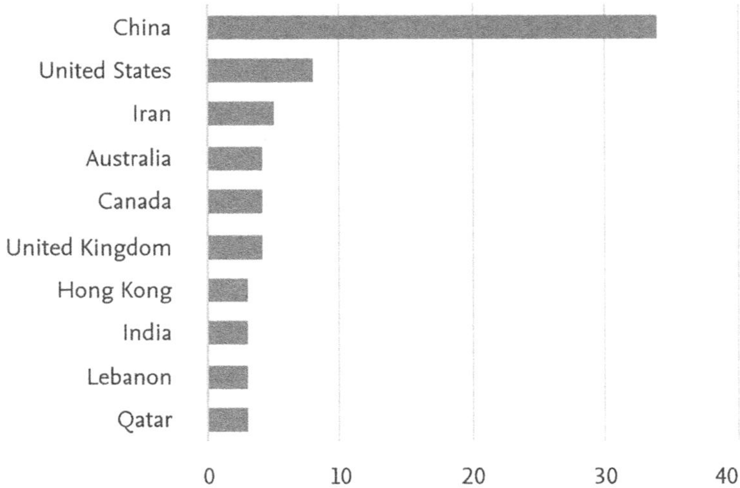

Fig. 12 DEA-big data – Descriptive summary: countries of the publications. (Source: Scopus 2021)

years. Another interesting observation to make is that while searching for relevant literature, we placed no constraints regarding the year of publication; yet, we were not able to find any studies on DEA-big data before the year 2013 in the Scopus database, revealing that the field is still in its incipient stages.

Figure 12 visually depicts the top 10 countries with the highest number of publications. The countries of origin for the 67 documents were determined by considering the country of the corresponding author. It is to be noted that China ranks first with 34 publications, followed by the United States with 8 publications. It is nice to note that interest in the topic is spread across a mix of developed and developing countries.

Figure 13 shows that the DEA-big data literature is dominated by articles (which constitute more than half of the publications, 52.2%), followed by conference papers (31.3%).

Lastly, Fig. 14 displays the documents by subject areas. Here, we can observe that the area of "computer science" has received the most interest, with 22.6% of the publications. This is followed by "engineering" (19.5%), "business, management, and accounting" (13.8%), and "decision sciences" (11.9%). Such results are not surprising, especially considering that the field of computer science has been redefined by the exponential growth of new computing technologies in view of big data, cloud computing and machine learning, and so on. It is also known that, due to the rapid growth of these new technologies, more experts in modern computer science are needed to analyse and solve data-driven problems.

It should be noted that while this list of 67 studies may not be comprehensive or fully accurate due to possible errors arisen during the filtering of the thousands of studies (for example, a publication on DEA can mention "big data" without actually

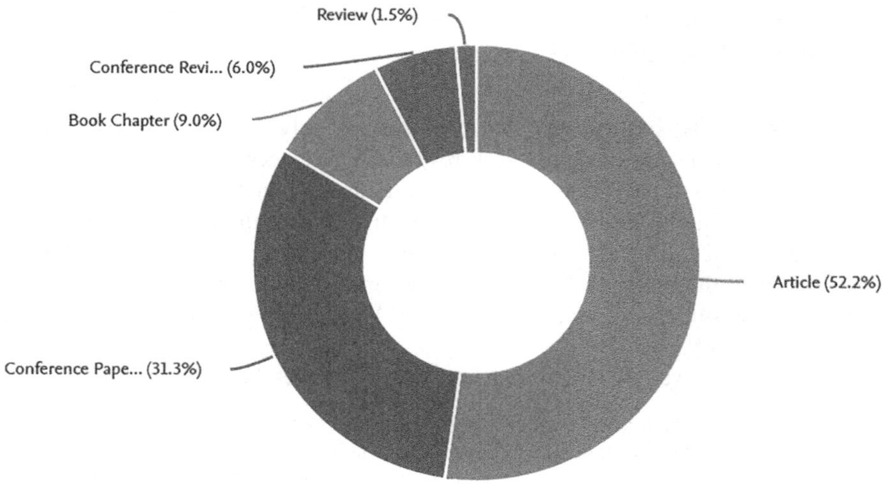

Fig. 13 DEA-big data – Descriptive summary: publications by type. (Source: Scopus 2021)

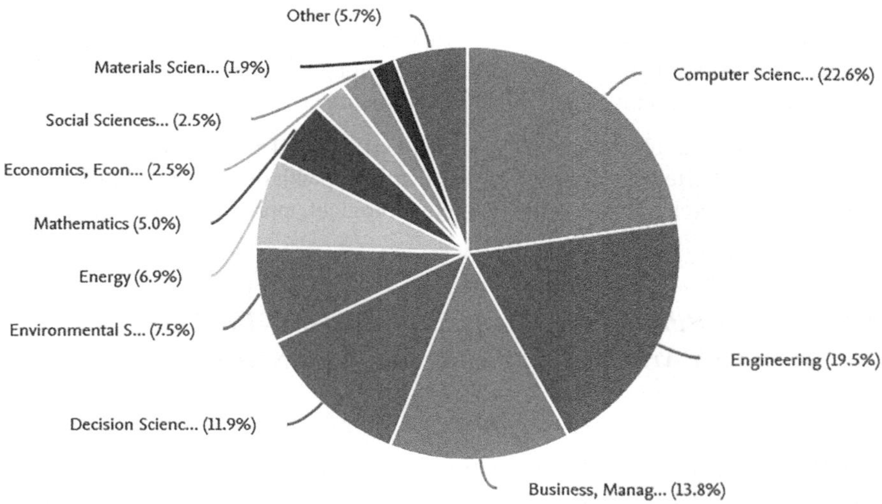

Fig. 14 DEA-big data – Descriptive summary: documents by subject area. (Source: Scopus 2021)

employing big data in any way), it does nonetheless provide a generally good picture of "what is out there" and of what the research interests are. For example, immediate observations point to the fact that the computer sciences and engineering fields concentrate most of these studies. Also, that research has, nonetheless, been conducted in a variety of institutional settings, as will further be appreciated in the following section.

4 A Systematic Literature Review of the DEA-Big Data Research Articles (35 Articles)

In this section, we further restricted our analysis only to analysing the articles on DEA-big data. Although no year restriction has been applied, interestingly enough, all the search results filtered by article type in Scopus (a total of 35 peer-reviewed journal articles) have been published over the past five years only, during 2016–2021. Considering the low number of articles yielded, we have further complemented the search with a manual checking of the referred articles, to make sure that these studies did indeed consider DEA under a big data environment as a core development.

Hence, studies were included in the review if they met the following criteria:

1. Big data were treated as a core question in the study, along with DEA. Such treatment could be both empirical and theoretical.
2. The articles involved research published in English, irrespective of year of publication.

Studies were specifically excluded from the review when:

1. They were conference papers, book chapters, conference reviews, and reviews.
2. The topic of big data was only casually mentioned in the DEA studies, without receiving any real empirical or theoretical treatment.

Results after the above criteria were applied resulted in a total of 24 relevant articles. Table 2 offers an overview of these articles, with details regarding authors, article title, research aim, data source, methodological approach, and article type.

4.1 A Brief Bibliometric Analysis of the DEA-Big Data Research Articles Composing the Final Sample (24 Articles)

A brief bibliometric analysis of the 24 research articles composing the final sample of studies integrating DEA with big data identified six keywords as the most common keywords (whose co-occurrence is at least three times) (Figs. 15 and 16). These keywords were further classified by the software into two clusters (Fig. 15) that seem to assume a prominent role vis-à-vis "computational paradigms for environmental efficiency" (four items, red cluster) and "decision-making" (two items, green cluster).

Table 2 Characteristics of the DEA-big data studies reviewed

Authors	Journal	Article title	Research aim	Data source	Methodological approach	Article type
Herranz et al. (2017)	Journal of Business Economics and Management	Leveraging financial management performance of the Spanish aerospace manufacturing value chain	To assess the financial management performance during 2008–2013 for the Spanish aerospace manufacturing value chain and the links with managerial decisions.	Company financial statements (over the period 2008–2013)	Principal Component Analysis, DEA, Artificial Neural Network	Methodological/Application (Financial efficiency evaluation – Aerospace)
Zhan et al. (2020)	Annals of Operations Research	Evaluation of food security based on DEA method: a case study of Heihe River Basin	To analyse the agricultural production efficiency in the Heihe River Basin and identify what was the role played by big data in the assessment of food security.	Socio-economic data on 11 counties of the Heihe River Basin over the period 1990–2012	DEA, Malmquist total factor productivity index	Methodological/Application (Agricultural efficiency evaluation – Food security)
Chen and Jia (2017)	Journal of Cleaner Production	Environmental efficiency analysis of China's regional industry: a data envelopment analysis (DEA) based approach	To assess the dynamic environmental efficiency of China's regional industry.	China Statistical Yearbook (2008–2012)	SBM-DEA	Methodological/Application (Environmental efficiency evaluation – Regional industry)

(continued)

Table 2 (continued)

Authors	Journal	Article title	Research aim	Data source	Methodological approach	Article type
An et al. (2017)	Journal of Cleaner Production	Allocation of carbon dioxide emission permits with the minimum cost for Chinese provinces in big data environment	To propose a new DEA approach to evaluate the efficiency of DMUs in a big data environment and set the carbon dioxide emission permits for each DMU with the minimum costs, with an application to 29 Chinese provinces.	China Statistical Yearbook, 2013; and China Energy Statistical Yearbook, 2013	DEA	Methodological/Application (Environmental efficiency evaluation – Carbon dioxide emissions)
Li et al. (2017)	Journal of Cleaner Production	Evaluation on China's forestry resources efficiency based on big data	To evaluate the forestry resources efficiency of China's 31 inland provinces and municipalities.	Annual statistical data from 2005 to 2013 of China's forestry resource	DEA, Malmquist total factor productivity index	Methodological/Application (Environmental efficiency evaluation – Forestry)
Liu et al. (2017)	Journal of Cleaner Production	DEA cross-efficiency evaluation considering undesirable output and ranking priority: a case study of eco-efficiency analysis of coal-fired power plants	Aims at incorporating undesirable outputs into DEA cross-efficiency evaluation and solve the well-known problem of the non-uniqueness of optimal weights.	China Statistical Yearbook, China Energy Statistical Yearbook, China Electric Power Yearbook, China Environmental Statistical Yearbook	DEA	Methodological/Application (Environmental efficiency evaluation – Coal-fired power plants)

Gong et al. (2017)	Journal of Cleaner Production	An approach for evaluating cleaner production performance in iron and steel enterprises involving competitive relationships	Aims at proposing an approach for evaluating the performance of iron and steel enterprises' cleaner production technologies.	Hypothetical numerical example	DEA	Methodological/Application (Environmental efficiency evaluation – Iron and steel enterprises cleaner production technologies)
Zhu et al. (2017)	Journal of Cleaner Production	China's regional natural resource allocation and utilization: a DEA-based approach in a big data environment	Aims at proposing a DEA-based approach for China's regional natural resource allocation and utilisation.	China Statistical Yearbook, China City Statistical Yearbook, China Energy Statistical Yearbook (2005-2012)	SBM-DEA	Methodological/Application (Environmental efficiency evaluation – Natural resource allocation and utilisation)
Chu et al. (2018)	Annals of Operations Research	An SBM-DEA model with parallel computing design for environmental efficiency evaluation in the big data context: a transportation system application	Aims at using SBM-DEA, also, at proposing an approach comprised of two algorithms for environmental efficiency evaluation in a big data context (i.e., for concurrently computing the environmental efficiencies of a massive number of DMUs).	30 actual DMUs (transportation systems) from Chang et al. (2013) and 2100 simulated DMUs	SBM-DEA	Methodological/Application (Environmental efficiency evaluation – transportation systems)

(continued)

Table 2 (continued)

Authors	Journal	Article title	Research aim	Data source	Methodological approach	Article type
Kiani Mavi et al. (2019)	Technological Forecasting and Social Change	Joint analysis of eco-efficiency and eco-innovation with common weights in two-stage network DEA: A big data approach	To propose a novel approach to measure the eco-efficiency and eco-innovation in the form of two-stage process in the context of big data.	Data on eco-efficiency and eco-innovation of OECD countries	NDEA	Methodological/Application (Environmental efficiency evaluation – Eco-efficiency and eco-innovation)
Khezrimotlagh et al. (2019)	European Journal of Operational Research	Data envelopment analysis and big data	To propose a new framework to significantly decrease the required DEA calculation time in comparison with the existing methodologies when a large set of DMUs (e.g., 20,000 DMUs or more) is present.	Real data set consisting of 30,099 electric power plants in the United States from 1996 to 2016	DEA	Methodological/Application (Environmental efficiency evaluation – Electric power plants)
Fan et al. (2019)	Energy	Comprehensive method of natural gas pipeline efficiency evaluation based on energy and big data analysis	To develop a natural gas pipeline efficiency evaluation method focusing on the pipeline energy input-output by monitoring the energy and transmission amount changes along the pipeline.	Operating data of a main natural gas transmission pipeline as collected by the China Petroleum Corporation	DEA, AHP	Methodological/Application (Environmental efficiency evaluation – Energy)

Tayal et al. (2020)	Sustainable Cities and Society	Integrated frame work for identifying sustainable manufacturing layouts based on big data, machine learning, meta-heuristic and data envelopment analysis	To propose a novel four-stage methodology using Big Data Analytics, Machine Learning, Hybrid Meta-heuristic, DEA, and K-mean clustering for designing an energy-efficient sustainable sub-optimal manufacturing layout under uncertain (stochastic) demand over multiple periods.	Hypothetical data from the literature	Big Data Analytics, Machine Learning, Hybrid Meta-Heuristic, DEA, K-mean clustering	Methodological/Application (Environmental efficiency evaluation – Facility layout design)
Taboada and Han (2020)	Electronics	Exploratory data analysis and data envelopment analysis of urban rail transit	To characterise the efficiency and sustainability of urban rail transit using a proposed methodology based on exploratory data analysis and DEA.	Open data from Transport for London and online services	Exploratory Data Analysis, DEA	Methodological/Application (Environmental efficiency evaluation – Urban rail transit)

(continued)

Table 2 (continued)

Authors	Journal	Article title	Research aim	Data source	Methodological approach	Article type
Zhu et al. (2020)	Science of The Total Environment	The potential for energy saving and carbon emission reduction in China's regional industrial sectors	To propose a new DEA model to analyse the energy and environmental efficiency of industrial sectors from China's 30 provincial-level regions in order to determine the potential and route for energy saving and carbon emission reduction.	Regional industrial dynamic dataset of China	DEA	Methodological/Application (Environmental efficiency evaluation – Energy saving and carbon emission reduction)
Kiani Mavi and Kiani Mavi (2021)	Technological Forecasting and Social Change	National eco-innovation analysis with big data: A common-weights model for dynamic DEA	To analyse the eco-innovation efficiency over time via a novel technique based on goal programming to find a common set of weights in relational dynamic DEA.	Eco-innovation data of 27 members of the European Union (EU-27), during the period 2011–2013. Data of eco-patents from www.stats.oecd.org, data of energy productivity from http://ec.europa.eu, and other data from www.worldbank.org	Dynamic DEA	Methodological/Application (Environmental efficiency evaluation – Eco-innovation)
Chen et al. (2017)	Transportation Research Part A: Policy and Practice	Balancing equity and cost in rural transportation management with multi-objective utility analysis and data envelopment analysis: A case of Quinte West	To develop a new methodology for rural transportation management which takes into consideration both the equity and cost factors under multiple objectives.	Case study (Quinte West, a municipality in Southeastern Ontario, Canada)	MCDM, DEA, Heuristics	Methodological/Application (Rural transportation management)

Badiezadeh et al. (2018)	Computers & Operations Research	Assessing sustainability of supply chains by double frontier network DEA: A big data approach	To develop an NDEA model for calculating optimistic and pessimistic efficiency.	Data obtained via surveys from companies which produce tomato paste	NDEA	Methodological/Application (Supply chain management)
He et al. (2019)	IEEE Access	Big Data-Oriented Product Infant Failure Intelligent Root Cause Identification Using Associated Tree and Fuzzy DEA	To propose a new big data-oriented root cause identification approach based on the associated tree and fuzzy DEA.	Data on washing machines in batch manufacturing	Fuzzy DEA, Associated Tree	Methodological/Application (Infant failure of the vibration and noise of a washing machine)
Song et al. (2017)	Production Planning & Control	A theoretical method of environmental performance evaluation in the context of big data	To present a set of scientific and axiomatised methods for environmental performance evaluation based on the big data environment.	N/A	DEA	Methodological/Theoretical treatment of environmental efficiency evaluation
Zhu (2020)	Annals of Operations Research	DEA under big data: data enabled analytics and network data envelopment analysis	To position DEA as data-enabled analytics and propose NDEA as an approach to deal with the value dimension of big data.	N/A	DEA, NDEA	Methodological

(continued)

Table 2 (continued)

Authors	Journal	Article title	Research aim	Data source	Methodological approach	Article type
Zhu et al. (2018)	Computers & Operations Research	Efficiency evaluation based on data envelopment analysis in the big data context	To propose novel algorithms to accelerate the computation process in the big data environment.	Simulated cases from Chen and Cho (2009), Dulá and López (2009), Dulá (2011), and Chen and Lai (2015)	DEA	Methodological
Zelenyuk (2020)	European Journal of Operational Research	Aggregation of inputs and outputs prior to Data Envelopment Analysis under big data	To explore the possible solutions to a 'big data' problem related to the very large dimensions of input-output data.	Simulated data	DEA	Methodological
Song et al. (2018)	Annals of Operations Research	Environmental performance evaluation with big data: theories and methods.	To present the theories and technologies regarding big data, along with the opportunities, applications, and challenges in the context of environmental management.	N/A	DEA	Theoretical treatment in the context of environmental efficiency evaluation

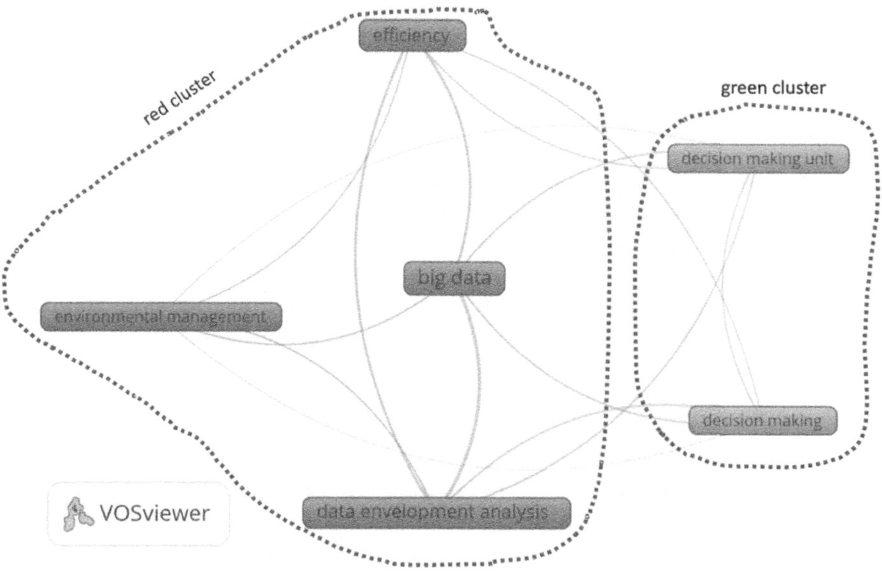

Fig. 15 Network map showing the relations between various topics in the DEA-big data field (based on the pool of 24 research articles)

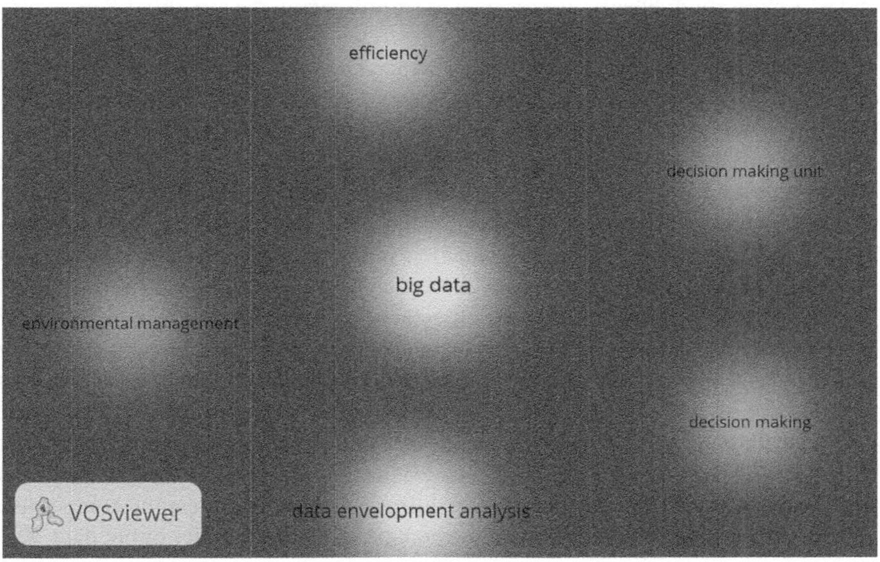

Fig. 16 Density visualisation

4.2 Thematic Analysis of the DEA-Big Data Research Articles (24 Articles)

4.2.1 Purely Methodological Articles (3 Articles)

The literature reviewed revealed the existence of three articles that make purely methodological contributions. Zhu (2018) proposed that DEA should be viewed as a method (or tool) for data-enabled analytics in performance evaluation and benchmarking and further advocated NDEA as an approach to deal with the value dimension of big data. Zhu et al. (2018) proposed novel algorithms to accelerate the computation process in the big data environment. Zelenyuk (2020) discussed possible solutions to one of the major challenges of the 'big data' related to the very large dimensions in the context of DEA.

4.2.2 Environmental Efficiency Evaluation (16 Articles)

As it can be observed, most of the research efforts have been dedicated to integrating big data with DEA for environmental efficiency evaluations, with applications to a wide range of domains: regional industry, carbon-dioxide emissions, forestry, coal-fired power plants, iron and steel enterprises cleaner production technologies, natural resource allocation and utilisation, transportation systems, eco-efficiency and eco-innovation, electric power plants, facility layout design, urban rail transit, and energy saving and carbon emission reduction.

Chen and Jia (2017) considered big data for DEA to perform an environmental efficiency analysis of China's regional industry. An et al. (2017) proposed a new DEA approach to evaluate the efficiency of DMUs in a big data environment and solve the carbon emission permits allocation issue. Li et al. (2017) used DEA in conjunction with big data theory to evaluate the forestry resources efficiency of China's 31 inland provinces and municipalities based on big data. They predominantly considered the numerous evaluation indexes, as well as the huge amount of data available, when performing the efficiency evaluation. Liu et al. (2017) introduced a new DEA-based cross-efficiency approach, which they applied for eco-efficiency analysis of coal-fired power plants in a big data environment. The proposed approach accommodates undesirable output and the ranking preferences of the DMUs, and the authors further incorporate big data theory to handle the large amount of data and the numerous input and output indicators. Gong et al. (2017) proposed an approach for evaluating the performance of iron and steel enterprises' cleaner production technologies, which considers the competitive relationship among the enterprises in the context of the availability of big data. Zhu et al. (2017) proposed a DEA-based approach in a big data environment to assess China's regional natural resource allocation and utilisation. The authors incorporate big data technology to support the characterisation of the production technology for each region.

Chu et al. (2018) used an SBM-DEA model with parallel computing for environmental efficiency evaluation in the big data context. Kiani Mavi et al. (2019) proposed a novel approach to find the common set of weights in a two-stage NDEA based on goal programming to analyse the joint effects of eco-efficiency and eco-innovation, considering the undesirable inputs, intermediate products, and the outputs in the context of big data. Khezrimotlagh et al. (2019) proposed a new framework to deal with large-scale DEA; more specifically, the technique decreases the computational time to measure the performance scores of big data sets. Fan et al. (2019) developed a novel natural gas pipeline efficiency evaluation method focusing on the pipeline energy input-output by monitoring the energy and transmission amount changes along the pipeline. The authors noted that the application of big data to pipeline energy monitoring had not been studied before. Tayal et al. (2020) proposed a novel 4-stage methodology using Big Data Analytics, Machine Learning, Hybrid Meta-heuristic, DEA, and K-mean clustering for designing an energy-efficient sustainable sub-optimal manufacturing layout under uncertain (stochastic) demand over multiple periods. In this paper, Big Data-Machine Learning (ML) is used to reduce and derive the sustainable criteria for sustainability. Taboada and Han (2020) assessed the efficiency and sustainability of urban rail transit (URT) using exploratory data analytics and DEA, under a big data context. Zhu et al. (2020) proposed a new DEA model to analyse the energy and environmental efficiency of industrial sectors from China's 30 provincial-level regions in order to determine the potential and route for energy saving and carbon emission reduction. The new DEA model considers dynamic data under a big data environment. More recently, Kiani Mavi and Kiani Mavi (2021) assessed the environmental performance of organisations, regions, and countries to analyse eco-innovation in a big data context. To this aim, the authors proposed a novel technique based on goal programming to find a common set of weights (CSW) in relational dynamic DEA.

Lastly, Song et al. (2017) presented a set of scientific and axiomatised methods and proposed approaches to evaluate environmental efficiency in the context of big data. And Song et al. (2018) presented the theories and technologies regarding big data, along with a discussion of challenges, opportunities, and applications in the context of environmental management. Unlike the other articles in this thematical category, these last two studies represent a theoretical treatment of environmental efficiency evaluation.

4.2.3 Other Types of Efficiency Evaluation (5 Articles)

Herranz et al. (2017) used Principal Component Analysis, DEA, and Artificial Neural Network to study the financial management performance during 2008–2013 of the Spanish aerospace manufacturing value chain using data from company financial statements. Among others, the study contributes by employing a big data sample that closely represents the population. Taking the Heihe River Basin (HRB) as a case study area and using DEA and the Malmquist index, Zhan et al. (2020) analysed the agricultural production efficiency in the HRB and further aimed to identify what was the role played by big data in the assessment of food security. Chen et al. (2017) developed a new methodology for rural transportation management which takes into consideration both the equity and cost factors under multiple objectives. The authors utilised the Geographic Information System as a big data platform to develop a decision support system for compiling, exporting, importing, and synchronising data and analytical results. Badiezadeh et al. (2018) proposed a new NDEA model to assess the optimistic and pessimistic efficiency of sustainable supply chain management given undesirable outputs, under a big data environment. Last but not least, He et al. (2019) proposed a novel big data-oriented root cause identification approach based on fuzzy DEA with the help of an established failure associated tree to study the infant failure of the vibration and noise of a washing machine.

5 Discussion and Conclusion

It has been the endeavour of the current study to perform a systematic literature review with bibliometric analysis (with software for bibliographic mapping) and thematic analysis of studies integrating DEA with big data, in an attempt to answer the question: *what are the current avenues of research for such studies?* All in all, the analysis performed shows that big data is a new entrant within the DEA literature, with the recent body of work in the field being indicative of an increasing interest in bringing the two concepts together under a single framework.

At the outset, it can be noted that, generally, the articles reviewed aimed at making methodological contributions, either purely or partially. Interestingly enough, in terms of methodological approaches adopted, it can be observed that in their attempts to integrate DEA with big data, the DEA analyses have been complemented with techniques such as: Multi-Objective Decision-making, Principal Component Analysis, Artificial Neural Network, Malmquist total factor productivity index, Machine Learning, Hybrid Meta-Heuristic, and K-mean clustering. As for DEA, the variants most commonly used in a big data environment are NDEA, dynamic DEA, SBM-DEA, and fuzzy DEA, with a significant body of DEA research focusing on NDEA (Zhu, 2020).

In terms of applications, scholars have deployed big data for DEA studies to measure efficiency in a variety of settings, such as the environmental efficiency of regional industry (Chen & Jia, 2017), energy saving and carbon dioxide emissions (An et al., 2017; Zhu et al., 2020), forestry resources (Li et al., 2017), coal-fired power plants, iron and steel enterprises cleaner production technologies (Gong et al., 2017), natural resource allocation and utilisation (Zhu et al., 2017), transportation systems (Chu et al., 2018), eco-efficiency and eco-innovation (Kiani Mavi et al., 2019; Kiani Mavi & Kiani Mavi, 2021), electric power plants (Khezrimotlagh et al., 2019), facility layout design (Tayal et al., 2020), urban rail transit (Taboada & Han,

2020), supply chain management (Badiezageh et al., 2018), and infant failure of the vibration and noise of a washing machine (He et al., 2019), among others.

A closer look at the articles reviewed shows that one of the biggest challenges in applying big data in DEA is posed by the large number of DMUs (*e.g.*, Chu et al., 2018; Khezrimotlagh et al., 2019; Liu et al., 2017; Song et al., 2017, 2018; Zhu et al., 2018). Therefore, it comes as no surprise that most of the studies on the topic of DEA-big data have focused on developing faster and more accurate computational techniques to handle problems with a large number of DMUs (*e.g.*, Zelenyuk, 2020; Zhu et al., 2018). Challenges also arise from the complicated interrelations and interactions among the DMUs, inputs, and outputs (*e.g.*, Zhu et al., 2017).

This piece of research provided an insight into the development of the DEA-big data literature. Although clearly expanding, the number of relevant DEA-big data studies was identified as being only 24, limiting thus the number of contributions that could be analysed. This is indicative, nonetheless, of the nascent nature of the DEA-big data research area. In terms of further research avenues, in this study, we have employed the Scopus database; therefore, future studies could leverage other databases, which may yield complementary insights.

Acknowledgement The authors are thankful to the reviewers for their valuable feedback on the previous version of this research.

References

An, Q., Wen, Y., Xiong, B., Yang, M., & Chen, X. (2017). Allocation of carbon dioxide emission permits with the minimum cost for Chinese provinces in big data environment. *Journal of Cleaner Production, 142*, 886–893.

Badiezadeh, T., Saen, R. F., & Samavati, T. (2018). Assessing sustainability of supply chains by double frontier network DEA: A big data approach. *Computers and Operations Research, 98*, 284–290.

Banker, R. D., Charnes, A., & Cooper, W. W. (1984). Some models for estimating technical and scale inefficiencies in Data Envelopment Analysis. *Management Science, 30*, 1078–1092.

Bizer, C., Boncz, P., Brodie, M. L., & Erling, O. (2012). The meaningful use of big data: Four perspectives-four challenges. *ACM SIGMOD Record, 40*(4), 56–60.

Brynjolfsson, E., Hitt, L. M., & Kim, H. H. (2011). *Strength in numbers: How does data-driven decision making affect firm performance?* Social Science Electronic Publishing.

Chang, Y. T., Zhang, N., Danao, D., & Zhang, N. (2013). Environmental efficiency analysis of transportation system in China: A non-radial DEA approach. *Energy Policy, 58*, 277–283.

Charles, V., & Gherman, T. (2013). Achieving competitive advantage through big data. Strategic implications. *Middle-East Journal of Scientific Research, 16*(8), 1069–1074.

Charles, V., & Gherman, T. (2018). Big data and ethnography: Together for the greater good. In A. Emrouznejad & V. Charles (Eds.), *Big data for the greater good* (pp. 19–34). Springer.

Charles, V., Tavana, M., & Gherman, T. (2015). The right to be forgotten – Is privacy sold out in the big data age? International Journal of Society Systems Science, 7(4), 283-298.

Charles, V., Tsolas, I. E., & Gherman, T. (2018). Satisficing data envelopment analysis: A Bayesian approach for peer mining in the banking sector. *Annals of Operations Research, 269*(1–2), 81–102.

Charnes, A., Cooper, W. W., & Rhodes, E. (1978). Measuring the efficiency of decision making units. *European Journal of Operational Research, 2*(6), 429–444.
Chen, C., Achtari, G., Majkut, K., & Sheu, J.-B. (2017). Balancing equity and cost in rural transportation management with multi-objective utility analysis and data envelopment analysis: A case of Quinte West. *Transportation Research Part A, 95*, 148–165.
Chen, C. P., & Zhang, C. Y. (2014). Data-intensive applications, challenges, techniques and technologies: A survey on Big Data. *Information Sciences, 275*, 314–347.
Chen, L., & Jia, G. (2017). Environmental efficiency analysis of China's regional industry: A data envelopment analysis (DEA) based approach. *Journal of Cleaner Production, 142*, 846–853.
Chen, W. C., & Cho, W. J. (2009). A procedure for large-scale DEA computations. *Computers & Operations Research, 36*(6), 1813–1824.
Chen, W. C., & Lai, S. Y. (2015). Determining radial efficiency with a large data set by solving small-size linear programs. *Annals of Operations Research, 250*, 147–166.
Chu, J.-F., Wu, J., & Song, M.-L. (2018). An SBM-DEA model with parallel computing design for environmental efficiency evaluation in the big data context: A transportation system application. *Annals of Operations Research, 270*(1-2), 105–124.
Cook, W. D., Tone, K., & Zhu, J. (2014). Data envelopment analysis: Prior to choosing a model. *Omega, 44*, 1–4.
Dulá, J. H. (2011). An algorithm for data envelopment analysis. *INFORMS Journal on Computing, 23*(2), 284–296.
Dulá, J. H., & López, F. J. (2009). Preprocessing DEA. *Computers & Operations Research, 36*(4), 1204–1220.
Fan, M.-W., Ao, C.-C., & Wang, X.-R. (2019). Comprehensive method of natural gas pipeline efficiency evaluation based on energy and big data analysis. *Energy, 188*, 116069.
Farrell, M. J. (1957). The measurement of productive efficiency. *Journal of the Royal Statistical Society: Series A, 120*(3), 253–281.
Gong, B., Guo, D., Zhang, X., & Cheng, J. (2017). An approach for evaluating cleaner production performance in iron and steel enterprises involving competitive relationships. *Journal of Cleaner Production, 142*, 739–748.
He, Z., He, Y., Liu, F., & Zhao, Y. (2019). Big data-oriented product infant failure intelligent root cause identification using Associated tree and fuzzy DEA. *IEEE Access, 7*(8667817), 34687–34698.
Herranz, R. E., Estévez, P. G., Oliva, M. A. D. V. Y., & Dé, R. (2017). Leveraging financial management performance of the Spanish aerospace manufacturing value chain. *Journal of Business Economics and Management, 18*(5), 1005–1022.
Khezrimotlagh, D., Zhu, J., Cook, W. D., & Toloo, M. (2019). Data envelopment analysis and big data. *European Journal of Operational Research, 274*(3), 1047–1054.
Kiani Mavi, R., & Kiani Mavi, N. (2021). National eco-innovation analysis with big data: A common-weights model for dynamic DEA. *Technological Forecasting and Social Change, 162*, 120369.
Kiani Mavi, R., Saen, R. F., & Goh, M. (2019). Joint analysis of eco-efficiency and eco-innovation with common weights in two-stage network DEA: A big data approach. *Technological Forecasting and Social Change, 144*, 553–562.
Laney, D. (2001). *3D data management: Controlling data volume, velocity and variety. Applications delivery strategies*. META Group (now Gartner) [online] http://blogs.gartner.com/doug-laney/files/2012/01/ad949-3D-Data-Management-Controlling-Data-Volume-Velocity-and-Variety.pdf.
Li, L., Hao, T., & Chi, T. (2017). Evaluation on China's forestry resources efficiency based on big data. *Journal of Cleaner Production, 142*, 513–523.
Liu, X., Chu, J., Yin, P., & Sun, J. (2017). DEA cross-efficiency evaluation considering undesirable output and ranking priority: A case study of eco-efficiency analysis of coal-fired power plants. *Journal of Cleaner Production, 142*, 877–885.
Manyika, J., Chui, M., Brown, B., Bughin, J., Dobbs, R., Roxburgh, C., & Hung Byers, A. (2011). Big data: The next frontier for innovation, competition and

productivity. *McKinsey Quarterly*. Retrieved on 13 January 2021 from https://www.mckinsey.com/~/media/McKinsey/Business%20Functions/McKinsey%20Digital/Our%20Insights/Big%20data%20The%20next%20frontier%20for%20innovation/MGI_big_data_exec_summary.pdf.

Michael, K., & Miller, K. W. (2013). Big data: New opportunities and new challenges [guest editors' introduction]. *Computer, 46*(6), 22–24.

Mubin, O., Arsalan, M., & Al Mahmud, A. (2018). Tracking the follow-up of work in progress papers. *Scientometrics, 114*, 1159–1174.

Müller, O., Fay, M., & Vom Brocke, J. (2018). The effect of big data and analytics on firm performance: An econometric analysis considering industry characteristics. *Journal of Management Information Systems, 35*(2), 488–509.

Provost, F., & Fawcett, T. (2013). *Data science for business: What you need to know about data mining and data-analytic thinking*. O'Reilly.

Song, M.-L., Fisher, R., Wang, J.-L., & Cui, L.-B. (2018). Environmental performance evaluation with big data: Theories and methods. *Annals of Operations Research, 270*(1–2), 459–472.

Song, M., Du, Q., & Zhu, Q. (2017). A theoretical method of environmental performance evaluation in the context of big data. *Production Planning and Control, 28*(11–12), 976–984.

Taboada, G. L., & Han, L. (2020). Exploratory data analysis and data envelopment analysis of urban rail transit. *Electronics, 9*(8), 1–29.

Tayal, A., Solanki, A., & Singh, S. P. (2020). Integrated frame work for identifying sustainable manufacturing layouts based on big data, machine learning, meta-heuristic and data envelopment analysis. *Sustainable Cities and Society, 62*, 102383.

Wu, X., Zhu, X., Wu, G. Q., & Ding, W. (2014). Data mining with big data. *IEEE Transactions on Knowledge and Data Engineering, 26*(1), 97–107.

Zelenyuk, V. (2020). Aggregation of inputs and outputs prior to Data Envelopment Analysis under big data. *European Journal of Operational Research, 282*(1), 172–187.

Zhan, J., Zhang, F., Li, Z., Zhang, Y., & Qi, W. (2020). Evaluation of food security based on DEA method: A case study of Heihe River Basin. *Annals of Operations Research, 290*(1–2), 697–706.

Zhang, Y., Huang, Y., Porter, A. L., Zhang, G., & Lu, J. (2019). Discovering and forecasting interactions in big data research: A learning-enhanced bibliometric study. *Technological Forecasting and Social Change, 146*, 795–807.

Zhu, J. (2020). DEA under big data: Data enabled analytics and network data envelopment analysis. *Annals of Operations Research*, 1–23.

Zhu, Q., Li, X., Li, F., & Zhou, D. (2020). The potential for energy saving and carbon emission reduction in China's regional industrial sectors. *Science of the Total Environment, 716*, 135009.

Zhu, Q., Wu, J., Li, X., & Xiong, B. (2017). China's regional natural resource allocation and utilization: A DEA-based approach in a big data environment. *Journal of Cleaner Production, 142*, 809–818.

Zhu, Q., Wu, J., & Song, M. (2018). Efficiency evaluation based on data envelopment analysis in the big data context. *Computers and Operations Research, 98*, 291–300.

Acceleration of Large-Scale DEA Computations Using Random Forest Classification

Anyu Yu, Yu Shi, and Joe Zhu

1 Introduction

"Big data" can be defined as high volume, high velocity, high variety, high veracity, and high value (5V) information (Chang et al., 2014). Since its emergence in the 1980s, big data has become more influential, and in recent years, the term is pervasive thanks to the constant technological advancements of social media, including various network platforms and communication channels. The emergence and prevalence of big data has galvanized the field of data science by inciting the development of more advanced decision-making tools that can handle the growing data size.

Many existing classical decision-making methods struggle to deal with the high volume of big data efficiently, in the sense that the computation time would increase significantly, given the large size of data. This phenomenon is specifically common in performance evaluation studies with data envelopment analysis (DEA) (Charnes et al., 1978). The term DEA encompasses a dual-concept of data envelopment analysis and data enabled analytics (Zhu, 2020). The traditional technique of data envelopment analysis is a popular data-driven tool for the performance evaluation of decision-making units (DMUs), and data enabled analytics is an expansion of the definition of DEA that accentuates the data-oriented characteristic of performance evaluation and the pertinence of data envelopment analysis to the value dimension of big data. A DEA evaluation measures the performance of a DMU based on how well the DMU converts the resources or inputs consumed to output products. A DEA

A. Yu (✉)
International Business School, Zhejiang Gongshang University, Hangzhou, People's Republic of China

Y. Shi · J. Zhu
Business School, Worcester Polytechnic Institute, Worcester, MA, USA
e-mail: yshi2@wpi.edu; jzhu@wpi.edu

J. Zhu, V. Charles (eds.), *Data-Enabled Analytics*, International Series in Operations Research & Management Science 312, https://doi.org/10.1007/978-3-030-75162-3_2

model is typically solved with linear programming. Although linear programming is polynomial-time solvable, when the programming size increases due to the growing number of observations, the limited computer memory causes the run time to grow exponentially (Chen & Cho, 2009).

Therefore, currently, DEA suffers the disadvantage of taking too long to compute with observations of massive sizes. This problem is not commonly observed in previous studies as it is not often addressed, but with the advancement of technology in the big-data context, it is now more feasible to obtain and collect large data sets, and the problem of high computational cost is becoming more imminent than ever. For example, the performances of employees, product lines, and products, may have been hard to gauge in the past can now be monitored and collected with the help of technology. In linear programming, it is well-known that greater amounts of DMUs complexify the DEA computation, resulting in increased computation time. Additionally, the computation time grows at a faster rate than the number of observations does.

To this end, important insights into reducing programming complexity and cutting down computation time have been developed in the DEA field. One of the most efficient solutions is to identify the DMUs which perform the best (termed as best-practice DMUs) and exclude the remaining DMUs. This is based on the idea that DEA performance results are determined only by the best-practice DMUs (Khezrimotlagh et al., 2019; Chen and Cho, 2009). When the sample density decreases, the computation time to find the best-practice DMUs also decreases (Khezrimotlagh et al., 2019). In other words, only the best-practice DMUs are used to evaluate the performance and the remaining DMUs are discarded, hence the DEA programming size and the shorter computation time are reduced.

In the literature, there are a few studies in this line of research. Specifically, Ali (1993) proposed a restricted basis entry (RBE) method to decrease the computation time of DEA computations. Restricted basis entry method first computes the conventional CCR DEA model for each DMU, and then removes the instrument variables lambda of DMUs which do not perform the best one by one, to decrease the sample size of DEA computations. Barr and Durchholz (1997) created a hierarchal decomposition (HD) method that decomposes the DMU sample into several subgroups and finds the best-practice DMUs in each subgroup. The computation process can be repeated to reduce the block numbers and to integrate the best-practice DMU sets into one block and then identify the final best-practice DMUs. This subgroup classification concept is also adopted in other studies. Korhonen and Siitari (2009) used lexicographic parametric programming to reduce the number of inputs and outputs of a DEA model. This method first partitions the original sample into subsamples dimensionally and then identifies the best-practice DMUs of the subsamples. Second, the best-practice DMUs in subsamples are used as an initial approximation, and finally, the best-practice DMUs are compiled to identify the final best-practice DMUs. The subgroup classification concept which accelerates DEA computation is also discussed in Zhu et al. (2018). Moreover, Dulá and López (2009) proposed a build-hull (BH) method, which first proposes the adjusted dual DEAs starting with only one DMU. Second, the method assumes that a DMU with the largest value of the first output lies on the best-practice frontier. Third, the method uses the dual multiplier DEA models to identify the best-practice DMU, and the best-practice DMU's lambda are added to the envelopment model to evaluate the final DEA scores. Lastly, all the remaining DMUs are added and evaluated, and all the best-practice DMUs are located. Chen and Cho (2009) proposed a new method to deal with large-scale DEA computations. This method firstly transforms data into a polar coordinate system and classifies the DMU into groups with different possibilities of being best-practice. Due to the potential performance category, indexes of mix and magnitude aspects are used to find the neighboring sets, and then the DEA scores are solved for and the remaining sets are checked. This process is repeated until solutions based on neighbor and peer group sets are identical, and the best-practice DMUs are then obtained. Chen and Lai (2017) proposed an algorithm to control the size of subsamples and compute the individual linear programming for each subsample. Then the DEA computations are processed across DMU subsamples in iterations through adding or dropping DMUs. This computation uses small-size linear programming to reduce computation time. Khezrimotlagh et al. (2019) proposed a method to divide the DMU sample into subsets, in which best-practice DMUs are identified and continually added to the exterior subset. This method ensures the entire DMU sample is checked and all the best-practice DMUs are identified step by step.

The aforementioned methods mainly achieve computation time reduction by searching for the best-practice DMUs (Ali, 1993; Barr & Durchholz, 1997; Khezrimotlagh et al., 2019). Although much effort has been launched into overcoming the problem of complex computation and long computation time, the problem still exists in the DEA filed. This is because the aforementioned approaches find the best-practice DMUs by searching over the entire DMU sample, which requires computing the DEA results with subsamples for all the DMUs first, and in a big data context, the computation of a massive observation size with millions or billions of observations, already takes substantial time.

Noteworthily, in an environment with big data and enormous observations, machine learning techniques have been widely adopted instead of traditional statistical analyses, to aid decision-making. These techniques include but are not limited to classification, correlation tests, clustering, and causal analyses. Among all the machine learning methods, random forest is a very useful and practical tool, especially for the classification of big data. Random forest (RF) is a model that uses ensemble decision trees for classification (Breiman, 2001). The mechanism of RF is to build multiple decision trees at training time and processing their results to obtain a stable prediction result (Herce-Zelaya et al., 2020). RF encapsulates the core capabilities of decision trees and uses them to conduct the classification. The class is selected based on the mode of the classes output by the individual trees (Herce-Zelaya et al., 2020). Trees are built each using the randomly selected attributes per node (Singh et al., 2014). Each decision tree in the model is built using bootstrap sampling, or sampling with replacement. RF combines trees grown on bootstrap samples of data and a random subset bagging of predictor variables (Breiman, 2001). It constitutes a novel way of combining information, at each node, an individual decision tree determines the split based on a smaller and random selection of contextual variables, but not on all the contextual variables (Wanke & Barros, 2016). This approach adds an element of randomness to the modeling process and allows for a broad search of the decision space, without explicitly needing to calculate it in its entirety (Jazar & Dai, 2000).

Random forest algorithm is used in this study because of its high prediction accuracy, capability of handling data characterized by a very large number and diverse types of descriptors, ease of training, computational efficiency (Singh et al., 2014), and robustness to outliers and noise (Yeh et al., 2014). Moreover, RF is capable of predicting results that are not overfitting, because of the law of large numbers (Herce-Zelaya et al., 2020). In literature, RF algorithm has also been used jointly with DEA. A notable case is Wanke and Barros (2016), in which RF is used to mine the heterogeneity impacts on performance for ranking the insurance sector. It is used to obtain the final ranking of the contextual variables with the consideration of their importance in a classification task. RF is also used as a classification method to rank journals from across the globe and DEA is used to aggregate the ratings (Tüselmann et al., 2015). The RF here estimates the individual rank probabilities when making a prediction.

Noteworthily, the searching process for the best-practice DMUs is also a classification process. Machine learning methods, which are suitable for the classification of big data, can also be used to accelerate the classification of best-practice DMUs. The DMUs that are not best-practice can then be discarded and DEA programming size can be reduced. To the best of our knowledge, in the literature, there is no research pertaining to the application of machine learning methods in searching for best-practice DMUs of DEA models, leading to a significant research gap that needs to be filled.

Therefore, this study proposes a novel method and framework that incorporates DEA and random forest (termed as DEA-RF) to overcome the computation issue of large-scale data. The proposed method aims to reduce computation time by identifying the best-practice DMUs, in a big data context. This is achieved by incorporating DEA evaluation with a machine learning approach. We also use gigantic observed and simulated samples of DMU observations to test the accuracy and computation speed of the DEA-RF method. The random forest method is also compared with the other machine learning methods and its advantage in accelerating large-scale DEA computations is testified.

The following section introduces the new method incorporating DEA and random forest methods. The numerical case and the discussions are provided in Sect. 3. The conclusion follows in Sect. 4.

2 Methodology

This section introduces the DEA-RF algorithm to reduce the complexity in DEA computations. Because the DEA performance measure for each DMU is always determined by the best-practice DMUs, the DEA-RF algorithm postulates that the best-practice DMUs in the DMU sample should be identified. We first propose a basic variable returns to scale (VRS) DEA model to illustrate the algorithm. The algorithm can also be adopted in the constant returns to scale (CRS) DEA model.

Notably, DEA models have two basic forms, the envelopment model, and the multiplier model. The former optimizes the levels of input or output measures to reach the best-practice performance on the DEA frontier for each DMU, while the latter optimizes the weights of the inputs or outputs to determine the best-practice DMUs. Both models are dual. More details of basic DEA models can be seen in Cooper et al. (2011). We use the envelopment DEA model instead of the multiplier model to conduct the DEA computations, because the practical computation is less time-consuming than that of the envelopment model since fewer constraints and a smaller basis inverse are maintained in the envelopment model (Barr & Durchholz, 1997). The envelopment model is shown in model (1).

In model (1), there are n DMUs, denoted DMU_j, $j = 1, 2, \ldots, n$. k means the DMU under evaluation. x and y denote the input and output elements, and there are m inputs and s outputs. The input-to-output transformational performance is defined as the DEA performance, and λ is the instrumental weights attached to the participate DMUs. $\sum_{j=1}^{n} \lambda_j = 1$ is the constraint which reflects the variable returns to scale (VRS) model setting. The objective function ensures the performance scores are within the range of (0, 1). A larger score indicates better performance, therefore, a score of one indicates this observation performs the best.

$$
\begin{aligned}
\min \quad & \theta_k \\
\text{s.t.} \quad & \sum_{j=1}^{n} \lambda_j x_{ij} \leq \theta_0 x_{ik}, \ j = 1, 2, \ldots, m, \\
& \sum_{j=1}^{n} \lambda_j y_{rj} \leq y_{rk}, \ r = 1, 2, \ldots, s, \\
& \sum_{j=1}^{n} \lambda_j = 1, \\
& \lambda_j \geq 0.
\end{aligned}
\tag{1}
$$

Because a primary way of accelerating DEA computations is finding the best-practice DMUs, we provide the simplified DEA model as model (2). Noteworthily, in model (2), if the DMU under evaluation is not included in the DMU set that forms the best-practice frontier, the linear program can possibly become infeasible under a VRS setting. However, if we identify all the best-practice DMUs in the entire dataset, the program will always be feasible for all the DMUs.

$$
\begin{aligned}
\min\ & \theta_k \\
\text{s.t.}\ & \sum_{j=1}^{n'} \lambda_j x_{ij}^e \le \theta_0 x_{ik},\ j = 1, 2, \ldots, m, \\
& \sum_{j=1}^{n'} \lambda_j y_{rj}^e \le y_{rk},\ r = 1, 2, \ldots, s, \\
& \sum_{j=1}^{n'} \lambda_j = 1, \\
& \lambda_j \ge 0.
\end{aligned} \tag{2}
$$

In model (2), x_{ij}^e and y_{rj}^e, $j = 1, 2, \ldots, n'$ are the input and output elements of best-practice DMUs obtained by solving model (1). Other variables in model (2) have the same meanings as in model (1). n in model (1) is significantly more than n' in model (2), thus the variable size of model (2) is significantly less than in model (1). Model (2) is the simplified model that is ultimately utilized by the DEA-RF algorithm to accelerate the DEA computations.

Definition 1 DMU_A dominates DMU_B if and only if, $x_A \le x_B$ and $y_A \ge y_B$.

Definition 2 DMU_A is termed as a best-practice DMU if and only if it is dominated neither by any other DMUs nor by any virtual DMUs generated by linear combinations of other DMUs.

The main idea of DEA-RF algorithm is to use RF classification to reduce the searching range of the best-practice DMUs. This algorithm uses partial DMU observations to compute the DEA model and then classifies the observations into high-score and low-score groups. The selected DMU sample is trained, and its classification is then used to predict the classification of the remaining DMUs. The classification puts DMUs that perform well into a small-size group, and this downsized group can simplify the DEA programming size and used to quick in search of the best-practice DMUs.

The main steps of DEA-RF algorithm are summarized below. The illustration of the main steps is followed by the details of the steps.

Step 1. Select two sets of DMUs as a training set Θ and a test set Ω from the whole set ψ. There is no overlapping of DMUs in sets Θ and Ω.

Step 2. Solve model (1) for the DMUs within Θ and Ω, and obtain their DEA scores (termed as machine learning DEA scores). Locate the best-practice DMUs, and denote the sets containing the best-practice DMUs within Θ and Ω as Θ^b and Ω^b.

Step 3. Group classification determination. We classify DMUs in Θ into two groups. One group, denoted as "high-score" group, contains DMUs with high DEA scores (including the best-practice DMUs), and the other group includes the remaining DMUs.

Step 4. Use the sample Θ to predict the two-group classification of DMUs in Ω using random forest. This training based upon Θ predicts the relationship between the input and output data of each DMU and the group classification of each DMU.

Step 5. Test the accuracy of the group classification in Ω, based on the machine learning DEA scores obtained from step 2. Accuracy ratio is computed as the number of best-practice DMUs in the "high-score" group to the actual number of best-practice DMUs in Ω. This study aims for a 100% accuracy rate and deems a ratio of 100% acceptable. If the accuracy rate is below 100% (unacceptable), go back to step 4 and lower the threshold of "high-score" to increase the number of DMUs in the "high-score" group. Otherwise, proceed to step 6.

Step 6. Use the trained group classification to predict the classification of the remaining DMUs in set Ψ using RF.

Step 7. Find the "high-score" group DMUs in the remaining DMUs in Ψ and denote them as Ψ^h.

Step 8. Merge Θ^b, Ω^b, and Ψ^h into a new DMU set Φ.

Step 9. Use model (1) to identify the best-practice DMUs within Φ as Φ^b.

Step 10. Use Φ^b to compute model (2) and obtain the actual performance results (termed as actual performance).

Step 11. End.

The rule on the selection of the training set and the test set is empirical, and it depends on the specific data set. In the case of less than 15,000 DMUs, the combined size of the training and the test set is recommended to be less than 30% of the entire DMU set. In our larger sample of more than 100,000 DMUs, the summed set is recommended to be less than 5% of the entire DMU set. This ensures the size of the training and test sets combined is about 5,000 DMUs observations. Besides, in this study, the size of the training sample is set to 2–2.5 times the size of the test sample. Machine learning DEA scores computed for the training set Θ are used to determine the "high-score" group DMUs, and machine learning DEA scores computed for the test set Ω are used to verify the accuracy of the "high-score" group classification in step 4.

In step 3, a set of "high-score" DMUs is determined. Initially, we select the top 100 DMUs with the highest scores as the "high-score" DMUs. In this step, the objective is to make sure that all the best-practice DMUs of set Θ are included in the "high-score" group. Other reasonable rules for classifying "high-score" DMUs include selecting the DMUs with scores in or above the 95th percentile.

In step 4, a random forest (RF) algorithm predicts the "high-score" DMUs in set ψ using the training based on Θ. The purpose of the training process is to establish a relationship between the input and output data of each DMU and the DMU's classification into the high-score or low-score group. RF is proposed by Breiman (2001) for the purposes of classification as well as regression. This model splits the data at each node by randomly selecting a predictor, and creates a wide array of diverse decision trees that contribute to the voting process. In RF modeling, only two parameters should be defined: the total number of trees in the forest and the number of variables to be considered for random selection when splitting at each node. Its prediction function for the group category is given as:

$$P\left(g\left|\upsilon\right.\right)=\sum_{t=1}^{T}P_{t}\left(g_{i}\left|\upsilon\right.\right) \tag{3}$$

In Eq. (3), t denotes the index of the tree, T is the total number of trees, and g_i is the predicted group. υstands for the variables used in prediction, which are inputs and outputs from the DEA model. RF model has only two parameters, can reduce overfitting and yield high model performance (Jazar & Dai, 2000). More illustrations of RF can be found in Jazar and Dai (2000).

In step 5, the prediction accuracy is computed as the ratio between the number of best-practice DMUs within the "high-score" group and the number of best-practice DMUs in Ω. The best-practice DMUs are determined by the machine learning DEA scores in step 2. To ensure a satisfactory prediction accuracy, we only consider the accuracy ratio acceptable if it is 100%. We discuss the model settings more in detail in our numerical cases. The accuracy ratio can be increased in two ways. One way is to increase the size of the training sample, and the other is to increase the size of the "high-score" group. We choose the second way, because the computation using RF can be rapid and easily repeated. In practice, RF method can predict the classification of 400,000 DMUs within 4.5 s, and it is quite a short time given the large size of the DMUs. However, a change in the training sample size will cost additional time to re-compute the machine learning DEA scores, for example, the computation of machine learning DEA scores for 3000 DMUs would cost nearly 1 min, which is much greater than what RF costs.

The generation of Φ and the relationships among Θ^b, Ω^b, and Ψ^h within ψ are shown in Fig. 1. Φ represents the downsized sample of the complete sample ψ, and Φ contains the best-practice DMUs. The set $\Phi^b$is the set of best-practice DMUs searched out from sample Φ in step 10, and this search would be less time-consuming than the search for best-practice DMUs in the complete set ψ.

Note that the best-practice DMUs identified by the machine learning DEA scores from both the training test sets always include the actual best-practice DMUs of the complete sample ψ. Since the complete sample ψ includes both the training and the test samples Θ and Ω, the actual performance scores of sample ψ from model (1) will be lower than the machine learning DEA scores of sample Θ and Ω (because the model with ψ has more constraints and smaller feasible regions of variables). Therefore, the best-practice DMUs based on the actual performance results of ψ are included in the best-practice DMUs based on the machine learning DEA scores. Following this logic, we can also use the test best-practice sets of Θ^b and Ω^b, respectively, instead of the high-score groups of Θ and Ω, to reduce the size of sample Φ.

With the best-practice DMUs Φ^b from step 10, we can calculate model (2) with less time, because the programming size of model (2) is less than that of model (1). The acceleration of computations is achieved.

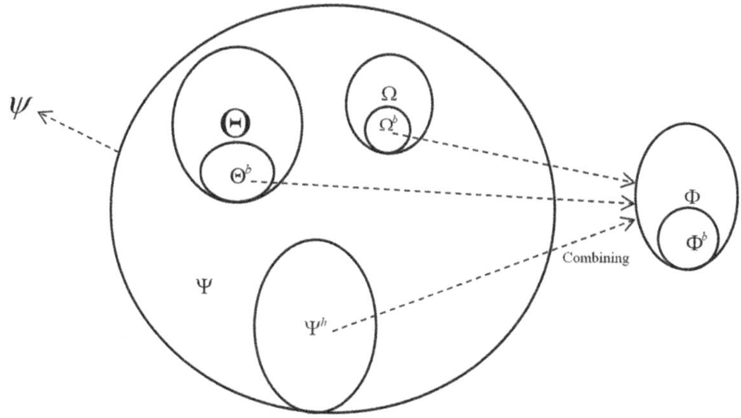

Fig. 1 Sizing down the entire sample to obtain a new sample

The proposed DEA-RF algorithm identifies the best-practice DMUs from the observed DMUs. This algorithm transforms the traditional searching process of best-practice DMUs into a classification process. RF is applied to implement the classification of large-scale DMUs, due to its capability of handling large datasets more efficiently. Figure 2 demonstrates the identified best-practice DMUs, the "high-score" group, and the "low-score" group derived using the algorithm. In general, Φ^b is always a small set within Φ, and Φ is also a small set within ψ. Therefore, the number of DMUs in Φ^b would be less than the number of DMUs in Φ. Furthermore, the number of DMUs in both sets are less than the number of DMUs in sample $\psi \backslash \Phi$. From ψ to Φ^b, the computation size of the DEA model has been significantly reduced.

Overall, the rationale of the above algorithm is to classify a sample set of DMUs in the entire DMU set into high-score and low-score groups according to their DEA scores and train them to conduct the group classification in the prediction. The purpose of training is to determine the high-score group for the remaining DMUs. Then, only the high-score DMUs are subsequently searched for the best-practice DMUs. This way, the searching time for best-practice DMUs is significantly reduced with the help of RF. Noteworthily, the proposed algorithm can also be adopted for accelerating the DEA computations of the multiplier model.

We acknowledge that the prediction of the group classification can still have errors, and the DEA-RF algorithm can only provide approximate results. However, the prediction accuracy can be increased by adjusting the high-score group classification in Θ. The effect of the DEA-RF method is a tradeoff between the accuracy and the computation time of large-scale DEA computations. Figure 3 demonstrates our DEA-RF approach in a conceptual framework.

Note that if the number of DMUs in the whole set is huge (i.e., more than one million DMUs) and the resulting high-score DMU group is also of a huge size, the search for the best-practice DMUs in the high-score group would also need

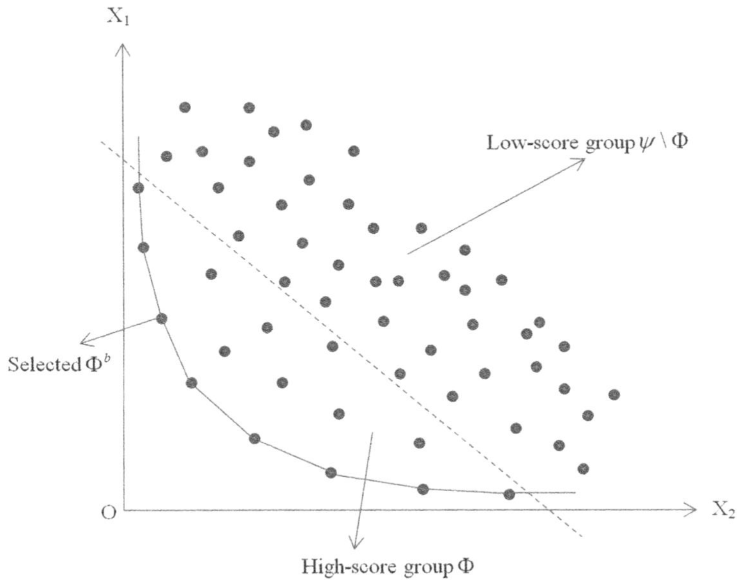

Fig. 2 Classifying DMUs in the high-score group Φ using random forest

a substantial amount of time. One solution to this problem is using the algorithm again in the high-score group. If the whole data set has more than one million DMUs, the proposed algorithm can also be adopted to the sample of high-score group. Additionally, the classification and prediction processes from step 1 to 5 in Fig. 3 can also be repeated, until the high-score group is downsized sufficiently.

3 Numerical Examples

3.1 Computer and Data Sample

To test the effectiveness of our DEA-RF algorithm, we use a laptop with a hexa-core CPU, Intel (R) Core (TM) i7-9750H CPU@ 2.60GHz, 16 gigabytes memory, and a 64-bit Windows 10 operating system. We use Matlab2017a to perform the computation. The RF method is assisted by a package "random forest" created in R. The package is based upon Breiman's random forest algorithm (based on Breiman and Cutler's original Fortran code) which is used to predict the group classification.

Our observed data sample is based on the registered high-tech firms located in Shanghai of mainland China. Data are collected from the database of the Chinese regulatory agency's project, "Shanghai Soft Science Research Project". The observed period is 2014–2017. The duration of four years is selected for the balance

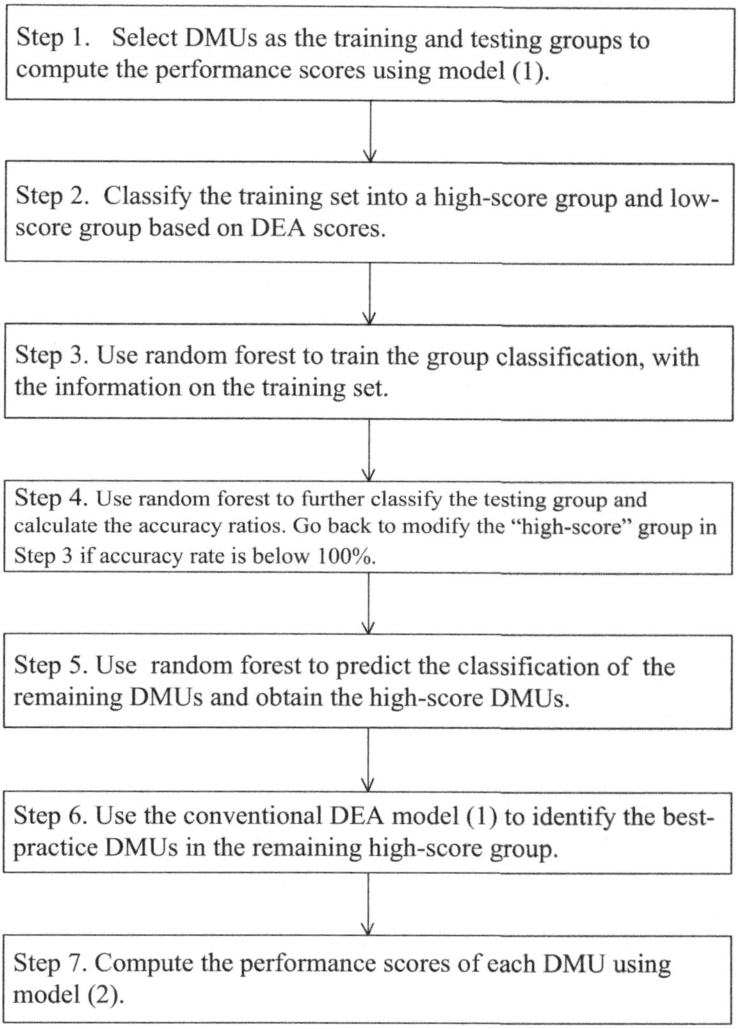

Fig. 3 Flowchart of the DEA-RF framework

between informativeness and stability of performance evaluation (Charnes et al., 1994). The observed sample includes 13,162 observations. To construct the DEA model, three inputs: labor, capital stock, and R&D expense, and two outputs: high-tech revenue and other (non-high-tech) revenues, are selected. These performance metrics are used to evaluate the performance of the 13,162 high-tech firms. All currency data are deflated to the base year of 2015. The descriptive statistics of the observed data sample are shown in Table 1.

Table 1 Descriptive statistics

	Capital stock (Thousand Yuan)	Labor (People)	R&D expense (Thousand Yuan)	Other revenue (Thousand Yuan)	High-tech revenue (Thousand Yuan)
Min	0.3374	1.0000	0.9372	0.0094	0.9372
Max	9,539,894	13,746	6,774,452	35,986,487	72,877,472
Ave.	72,720	246	25,155	92,724	289,101
Std.	346,395	598	157,244	642,764	1,408,254

Furthermore, we provide two simulated data samples. These data are simulated based on the normal distribution of inputs and outputs in the observed data sample. The simulated data samples include 50,000 observations and 400,000 observations, respectively, and they are numbered data samples 1 and 2. All the data in both samples are selected nonnegative.

3.2 Numerical Case

Next, we test the effectiveness of the proposed DEA-RF method on the numerical cases with large-scale data. Two key indicators of effectiveness, computation time and the predicted accuracy ratio (i.e., the number of DMUs in the classified high-score group divided by the number of best-practice DMUs) are considered. As mentioned above, we use the envelopment DEA model to complete the DEA-RF computation.

First, we use the proposed method to accelerate the computation of the observed data sample. The observed data sample includes 13,162 DMUs. We randomly choose 2500 DMUs as the training set, and 1000 DMUs as the test set. Following the steps of the DEA-RF method, we compute the machine learning DEA scores of the 3500 DMUs of both the training and the test sets and classify the DMUs in the training set into a high-score group and a low-score group. Then, RF is used to train the correlation between inputs & outputs and the group classification of each DMU in the training set, and to predict the group classification of DMUs in the test set. We set the number of trees in RF to 500 and the number of categories at each node to two.

After predicting the high-score DMUs in the test set using RF, the accuracy ratio of finding the best-practice DMUs in the high-score group of the test set is shown in Table 2. In order to define "high-score", we first set "high-score" as the first 100, 200, 300, 400, 500, 600, 700, 800, 900, and 1000 highest DEA scores, and run the classification 10 times. We find that the optimal accuracy rate of 100% is achieved when we include 300 DMUs with the highest DEA scores in the high-score group of the training set. In other words, when the high-score group of the training set consists of 300 DMUs with the highest DEA scores, all the best-practice DMUs are found in the high-score group of the test set. This result is shown in Table 2. Therefore, we define "high-score" as the first 300 DMUs with

Table 2 Accuracy ratios of classification to the observed data sample

	Accuracy	
High scores	Test set	Remaining set
Top-100	0.8571	0.8889
Top-200	0.8571	0.9630
Top-300	1.0000	0.9630
Top-400	1.0000	0.9630
Top-500	1.0000	0.9630
Top-600	1.0000	0.9630
Top-700	1.0000	0.9630
Top-800	1.0000	1.0000
Top-900	1.0000	1.0000
Top-1000	1.0000	1.0000

the highest DEA scores. Note that when more DMUs are added into the high-score group of the training set, the group size is enlarged and the chance of the DMUs in the training high-score group being classified into the high-score group for the test set will become higher, resulting in a higher accuracy ratio. However, if there are more DMUs in the high-score group of the test set, the computation time would be increased, because the best-practice DMUs will be searched for in a larger dataset. There is a trade-off between achieving the highest accuracy and reducing the computation time to the highest extent, and the decision should be made circumstantially.

Moreover, we use the training set to predict the classification of the remaining 9662 DMUs (named the remaining set). The actual DEA scores of the remaining set are computed using a conventional DEA model including 13,162 DMUs. The accuracy ratios are shown in Table 2. Again, we set the high-score DMUs as the first 100, 200, 300, ... 1000 highest scores in the training set. It is found that when the high-score group includes more than 800 DMUs with the highest DEA scores in the training set, the prediction of the remaining set achieves a 100% accuracy. This indicates that when the number of observations is massive, a larger training set is required, or alternatively, the high-score group of the training set should be larger. Note that the actual DEA scores of the remaining set are also computed, using the 13,162 complete DMU set complete. This is to examine the accuracy of the RF prediction. However, practically, users only need to compute the accuracy ratio of the test set and compare it with a predetermined satisfactory accuracy level.

Finally, we use the proposed DEA-RF method in the DEA programming of the simulated data sample 1. The training set and the test set include 7000 and 3000 randomly chosen DMUs, respectively. The remaining 40,000 DMUs are termed as the remaining set. Similarly, we compute the machine learning DEA scores of the total 10,000 DMUs in the training and test sets, and the actual scores of the remaining set are computed with the DEA model including all the 50,000 DMUs. The accuracy results are shown in Table 3. It is noticeable that when the number of DMUs in the high-score group of the training set reaches 500, the testing and the remaining sets can both achieve a 100% accuracy ratio. This result corroborates that

Table 3 Accuracy ratios of classification to the simulated data

	Accuracy	
High scores	Test set	Remaining set
Top-100	0.4138	0.8000
Top-200	0.9310	0.9275
Top-300	0.9655	0.9615
Top-400	0.9655	0.9846
Top-500	1.0000	1.0000
Top-600	1.0000	1.0000
Top-700	1.0000	1.0000
Top-800	1.0000	1.0000
Top-900	1.0000	1.0000
Top-1000	1.0000	1.0000

RF is capable of predicting an accurate classification for the remaining set when the training set is large enough (i.e., 7000). Moreover, the larger the size of the training set, the higher the prediction accuracy of the RF method. In summary, all the above results verify that RF can reduce the computation time of the DEA model while ensuring computation accuracy.

After obtaining the accuracy ratio of the observed data sample and simulated data sample 1, we run the DEA-RF algorithm and record the corresponding computation time. We use the DMUs in the high-score group to continue the steps 6 and 7 in Fig. 2. This time, the high-score group of the observed training set is comprised of 800 DMUs with the highest DEA scores, and the high-score group of the simulated training set contains 500 DMUs with the highest scores. These two high-score sets are named ob_800 and sim_500, respectively. We also use a training set high-score group including 300 DMUs (ob_300). It is observed that the accuracy ratio of predicting the test set is 100% when there are 300 DEAs in the high-score group of the observed dataset, therefore, the ob_300 case is also adopted.

Furthermore, we compute the DEA scores of data sets of ob_300, ob_800, and sim_500, and find 43, 44, and 155 best-practice DMUs, respectively. If we compute the entire observed and simulated samples combined, with the basic DEA model, there are 44 and 155 best-practice DMUs, respectively. These results once again prove that RF can identify the best-practice DMUs accurately. For the ob_300 case, one best-practice DMU is missing from the high-score group, implying that an insufficient size of training set, or an insufficient size of the high-score group, could harm the accuracy of predicting the classification of the remaining set. In this case, the accuracy ratio is 96%, as shown in Table 2. Since one best-practice DMU is missing, we also find two infeasible DMUs. Their DEA scores are greater than one and both 1.000002.

With the obtained best-practice DMUs in the cases of ob_300, ob_800, and sim_500, we use model (2) to perform the simplified DEA computation. At this point, the proposed DEA-RF process has been completed, and the corresponding time for computing the sets ob_300, ob_800, and sim_500 with DEA-RF are provided in Fig. 4. For comparison, we also provide the time to complete conventional

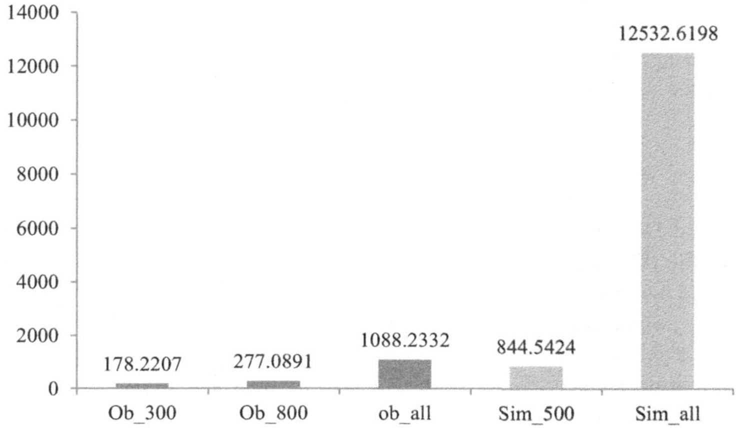

Fig. 4 Comparison of computation time (seconds) of different classified data samples

DEA computations of the whole observed and the whole simulated samples in Fig. 4.

As shown in Fig. 4, the computation time of DEA-RF for the ob_300 case (i.e., 178.2207s, 178.2207 s) is less than that for the ob_800 case (i.e., 277.0891s), but both are significantly less than the time cost by conventional DEA to compute the entire observed dataset (i.e., 1088.2332s). Moreover, we find that the computation time of sim_500 (i.e., 844.5424s) is also significantly less than the computation time of the conventional whole simulated sample (i.e., 12532.6198s). The findings from Fig. 4 show the rapidness of our DEA-RF algorithm, when used in large-scale DEA computations. These findings are also consistent with the belief that RF is more suitable for classifying larger to massive DMU sets. Because more time was saved in the sim_500 case, we conclude that the larger the data sample is, the more time the DEA-RF method saves. Unlike RBE, HD, and BH, RF can identify best-practice DMUs without requiring the DEA scores for all the DMUs to be computed. When the size of the observed data sample is sufficiently large, the computation time can be saved substantially with the DEA-RF method.

To determine the accuracy of our DEA scores solved with the DEA-RF method, we also compare the DEA scores of ob_300, ob_800, and sim_500 with the conventional DEA results. The descriptive statistics of the differences between the DEA-RF results and the conventional DEA results are shown in Table 4. It is observed that the results of the ob_800 case are the same as the complete observed sample, and this is because all the best-practice DMUs are located in the ob_800 dataset. However, even though all the best-practice DMUs are located in sim_500, the sim_500 results and the sim_all results are not entirely the same. The sim_500 case has a 99.99% accuracy ratio of the DEA results, which is satisfactory. Moreover, for the ob_300 case, 43 best-practice DMUs are identified from the total number of 44 DMUs. There is also a discrepancy between the ob_300 results and the

Table 4 Differences in DEA scores using different machine learning classifications

	Ob_300	Ob_800	Sim_500
Max	0.0513	0.0000	0.3221
Min	0.0000	0.0000	0.0000
Ave.	0.0000	0.0000	0.0000
Std.	0.0007	0.0000	0.0015

conventional DEA results, particularly, seven DMUs have differences greater than 0.01, and the differences are 0.0154, 0.0173, 0.0179, 0.0255, 0.0340, 0.0401, and 0.0513. The accuracy of the ob_300 case is 99.95%, this is considered acceptable, because the computation time of ob_300 case is less than the ob_800 case.

The DEA-RF results show reasonable prediction accuracy ratios and uses shorter computation time. The above findings indicate that RF classification could be subject to minor prediction errors, but it is still worth using, because it saves a significant amount of time on the identification of best-practice DMUs, especially when big data of millions or billions of observations are present.

Furthermore, we use a larger sample, simulated sample 2, to investigate the computation speed of the DEA-RF method. Because simulated sample 2 is huge, we only compute the DEA scores of some of the DMUs. The computation of the entire simulated sample 2 would be very time-consuming, with 400,000 DMUs. We randomly choose 10000 DMUs as the training set and 5000 other DMUs as the test set. Firstly, we calculate the machine learning DEA scores of the training and the test sets of 15,000 DMUs in total. This process takes 1972.9393s. Then, we train the RF method using the training set to predict the classification of the test set. We find that if the high-score group includes 600 DMUs with the highest DEA scores, we obtain the predicted classification of the test set with a 100% accuracy. Therefore, we use 600 DMUs in the high-score group to classify the remaining 385,000 DMUs and acquire 17,659 DMUs in the high-scores group. We find 255 best-practice DMUs after computing the DEA scores of the DMUs in the high-score group. This process takes 2020.3988s. We also provide the total computation time in computing the DEA scores of 100 DMUs within the 400,000 sample, by using the conventional DEA envelopment model, the multiplier model, and the DEA-RF method. The comparison of computation time is shown in Fig. 5.

From Fig. 5, we find that the multiplier model incurs the longest computation time. The DEA-RF model only takes 12.8172 s to compute 100 DMUs. This shows that the proposed model cuts down more than 95% of the time used by the envelopment model, when the data sample is large. Notably, the additional running-time cost by the RF-classification is also reflected in the DEA-RF result in Fig.4. The additional computation adds an extra 0.9983 s to the total computation time of 100 DMUs (i.e., $(1972.9393 + 2020.3988)/400000*100 = 0.9983$).

Overall, the time spent on large-scale DMU computations is proved to be significantly reduced by the proposed DEA-RF method. The effectiveness of the DEA-RF method in saving the computation time of big data is verified.

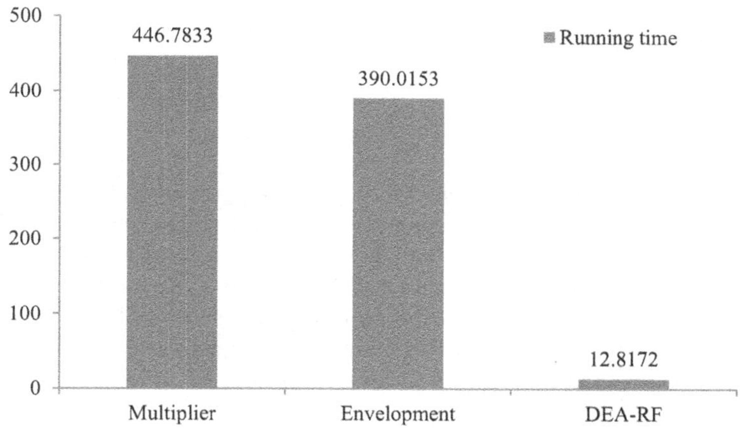

Fig. 5 Comparison of computation time (seconds) for 100 DMUs within the 400,000-DMU set

3.3 *Comparisons of Different Machine Learning Methods*

We discuss the advantage of RF in large-scale DEA computations, compared with other machine learning methods. We use other machine learning methods, including support vector machine (SVM), logistics regression, and artificial neural network (ANN)[1] to predict the classification of the selected 3000 DMUs in the test sets. More details of these machine learning methods can be seen in Jazar and Dai (2000). First, we use these methods to predict the high-score group and examine the accuracy ratios of the classification. Then, we use these methods to find the best-practice DMUs in step 6 of the proposed algorithm. The numerical case used is the simulated data sample 1, and the corresponding training and test sets include 7000 and 3000 DMUs respectively. We include 500 DMUs with the highest DEA scores in the high-score group of the training set.

We use the Matlab software packages to conduct the classification and prediction of the simulated data sample 1. We use the basic settings of the software packages for all these other machine learning methods. The machine learning DEA scores are also computed with the combined set of the training and test sets. The predicted accuracy ratios of the best-practice DMUs in the high-score group are shown in Table 5.

In Table 5, RF method is observed to have the best prediction accuracy ratio, 100%, followed by SVM, logistics regression, and artificial neural network. Moreover, SVM has the high-score group of the largest size (708 DMUs), which is

[1]Some basic parameter settings of ANN. The network has one layer, 10 hidden neurons, and 2 output neurons. The training method is "gradient descent method with adaptive LR momentum". Maximized training times are 500; the goal of training error is 0.01; learning rate is 0.01

Table 5 Comparison between performance results of different machine learning methods

Accuracy ratio	SVM	Logistics	RF	ANN
Test set	89.66%	55.17%	100%	13.79%
High-score group size	708	89	162	12

far larger than the high-score group size of logistics regression (89), RF (162), and ANN (12). Thus, SVM's searching process for the best-practice DMUs in the high-score group would take the longest time. In fact, among the actual DEA scores of the 3000 DMUs in the test set, only eight DMUs are best-practice; among the machine learning DEA scores, only 29 DMUs are best-practice. It can be concluded that classifications or predictions by RF is more accurate than those by other machine learning methods, and its high-score group size is also acceptable. And, most conspicuously, RF's prediction results are not prone to overfitting (Herce-Zelaya et al., 2020; Jazar & Dai, 2000).

4 Conclusion

This study proposes a DEA-RF method to handle large-scale DEA computations under a big data environment. The proposed method incorporates DEA with a machine learning method, random forest classification. The resulting combination is first used to reduce the computation time of DEA computations with enormous DMUs. The DEA-RF method is then applied in numerical cases including observed and simulated data samples. It is observed that the DEA-RF method can significantly accelerate large-scale DEA computations while ensuring a satisfactory prediction for the best-practice DMUs based on a classification of high-score groups. The greater the number of DMU observations, the more time the algorithm saves. The numerical results validate the effectiveness of our proposed DEA-RF method in the computations of big data.

Notably, modified and more improved techniques of RF, SVM and Artificial neural networks (ANNs) can be implemented in future studies to further accelerate the computations of large-scale DEA and to increase prediction accuracy. Moreover, previous algorithms such as RBE, HD, and BH are suggested to be integrated with our method to achieve an even greater reduction in DEA computation time.

Acknowledgement The authors are sincerely grateful to Dr. Dariush Khezrimotlagh for his valuable comments and suggestions on an earlier version of the paper.

References

Ali, A. I. (1993). Streamlined computation for data envelopment analysis. *European Journal of Operational Research, 64*(1), 61–67.

Barr, R. S., & Durchholz, M. L. (1997). Parallel and hierarchical decomposition approaches for solving large-scale data envelopment analysis models. *Annals of Operations Research, 73*, 339–372.

Breiman, L. (2001). Random forests. *Machine Learning, 45*(1), 5–32.

Charnes, A., Cooper, W. W., Lewin, A. Y., & Seifor, L. M. (1994). *Data envelopment analysis: Theory, methodology, and application*. Kluwer Academic Publishers.

Charnes, A., Cooper, W. W., & Rhodes, E. (1978). Measuring the inefficiency of decision making units. *European Journal of Operational Research, 2*(6), 429–444.

Chang, R. M., Kauffman, R. J., & Kwon, Y. (2014). Understanding the paradigm shift to computational social science in the presence of big data. *Decision Support Systems, 63*, 67–80.

Chen, W. C., & Cho, W. J. (2009). A procedure for large-scale DEA computations. *Computers & Operations Research, 36*, 1813–1824.

Chen, W. C., & Lai, S. Y. (2017). Determining radial efficiency with a large data set by solving small-size linear programs. *Annals of Operations Research, 250*, 147–166.

Cooper, W. W., Seiford, L. M., & Zhu, J. (2011). *Handbook on data envelopment analysis*. Springer.

Dulá, J. H., & López, F. J. (2009). Preprocessing DEA. *Computers & Operations Research, 36*(4), 1204–1220.

Herce-Zelaya, J., Porcel, C., Bernabé-Moreno, J., Tejeda-Lorente, A., & Herrera-Viedma, E. (2020). New technique to alleviate the cold start problem in recommender systems using information from social media and random decision forests. *Information Sciences, 536*, 156–170.

Jazar, R. N., & Dai, L. (2000). *Nonlinear approaches in engineering applications automotive applications of engineering problems*. Springer.

Khezrimotlagh, D., Zhu, J., Cook, W. D., & Toloo, M. (2019). Data envelopment analysis and big data. *European Journal of Operational Research, 274*, 1047–1054.

Korhonen, P. J., & Siitari, P. A. (2009). A dimensional decomposition approach to identifying efficient units in large-scale DEA models. *Computers and Operations Research, 36*(1), 234–244.

Singh, K., Guntuku, S. C., Thakur, A., & Hota, C. (2014). Big data analytics framework for peer-to-peer botnet detection using random forests. *Information Sciences, 278*, 488–497.

Tüselmann, H., Sinkovics, R. R., & Pishchulov, G. (2015). Towards a consolidation of worldwide journal rankings – A classification using random forests and aggregate rating via data envelopment analysis. *Omega, 51*, 11–23.

Wanke, P., & Barros, C. P. (2016). Efficiency drivers in Brazilian insurance: A two-stage DEA meta frontier-data mining approach. *Economic Modelling, 53*, 8–22.

Yeh, C. C., Chi, D. J., & Lin, Y. R. (2014). Going-concern prediction using hybrid random forests and rough set approach. *Information Sciences, 254*, 98–110.

Zhu, J. (2020). DEA under big data: Data enabled analytics and network data envelopment analysis. *Annals of Operations Research*. (in press).

Zhu, Q., Wu, J., & Song, M. (2018). Efficiency evaluation based on data envelopment analysis in the big data context. *Computers and Operations Research, 98*, 291–300.

The Estimation of Productive Efficiency Through Machine Learning Techniques: Efficiency Analysis Trees

Juan Aparicio, Miriam Esteve, Jesus J. Rodriguez-Sala, and Jose L. Zofio

1 Introduction

Machine learning (ML) is one of the ways of referring to the set of modern techniques, based on computer algorithms, which automatically improve from the information provided by a 'training' dataset, with the aim of making predictions or decisions without being explicitly supervised to do so. Nowadays, these techniques are usually linked to the quantitative analysis of big databases. ML has been utilized with success in a wide range of real applications, such as spam email filtering and pattern recognition in medicine. Recently, ML has been also introduced for the estimation of production frontiers in the field of Economics and Industrial Engineering. In particular, Esteve et al. (2020) have defined a new methodology based on decision trees for estimating production frontiers satisfying essential postulates of microeconomics, such as free disposability. This new approach, baptized as Efficiency Analysis Trees (EAT), shares some similarities with the Free Disposal Hull (FDH) technique. However, and in contrast to FDH, EAT overcomes the problem of overfitting by using cross-validation to prune back the deep tree obtained in a first stage. Esteve et al. (2020) also show that the performance of EAT, measured through Monte Carlo simulations, clearly outperforms the FDH

J. Aparicio (✉) · M. Esteve · J. J. Rodriguez-Sala
Center of Operations Research (CIO), University Miguel Hernandez of Elche (UMH), Elche (Alicante), Spain
e-mail: j.aparicio@umh.es

J. L. Zofio
Department of Economics, Universidad Autónoma de Madrid, Madrid, Spain

Erasmus Research Institute of Management, Erasmus University, Rotterdam, The Netherlands

J. Zhu, V. Charles (eds.), *Data-Enabled Analytics*, International Series in Operations Research & Management Science 312, https://doi.org/10.1007/978-3-030-75162-3_3

technique, showing that the new approach can reduce the mean squared error associated with the estimation of the true frontier by 13–70% and the bias by 10–49%.

The Free Disposal Hull technique was introduced by Deprins et al. (1984) and relies only on a few microeconomic postulates. Free disposability is one of them and states that if a certain input and output bundle is producible, then any input and output bundle that presents a greater value for inputs and a lower value for outputs is also producible. In some respects, it means that doing it worse is always feasible. Graphically, in the case of producing only one output, this property is related to the estimation of non-decreasing production functions, ruling out negative marginal factor productivities. Additionally, FDH is assumed to be deterministic. This means that the observations (examples) that belong to the data sample (i.e., the learning sample following the ML terminology) have been observed without noise. In other words, the production possibility set determined by FDH always contains all these observations and, graphically, the production frontier envelops all the data cloud from above. Finally, the third postulate that FDH satisfies is that known as minimal extrapolation and is associated with the typical problem-solving principle of Occam's razor. A lot of possible estimators can satisfy free disposability and 'deterministicness'. Accordingly, additional requirements are needed to select the suitable one. In this respect, if one abides by Occam's razor, the most conservative estimation of the production frontier would be that related to a surface enveloping the data from above and, at the same time, as close as possible to the data cloud. Therefore, from a non-parametric perspective, FDH is very attractive since it does not rely on limiting hypothesis on the Data Generating Process (DGP). This is a feature shared with usual machine learning techniques, which are clearly data-driven approaches. However, standard FDH cannot be considered a proper ML technique. The main objective of FDH seems to be the description of the extreme behavior of a set of observations, satisfying the aforementioned microeconomic axioms. In contrast, most ML techniques (as for example, Classification and Regression Trees, Bagging, Random Forest, Boosting, Support Vector Machines, to name a few) additionally present other objectives beyond descriptive statistics. They deal with the notion of generalization or prediction error, which is a measure of how accurately a method can predict outcome values for previously unseen data. This objective is inherited from the theory of Statistical Learning (see, for example, Vapnik, 2013). In practice, the way of evaluating the generalization error of any approach is not using the same data utilized in the fitting process of the model, but resorting to a different test data sample (for example, by applying cross-validation). The idea behind generalization in statistical learning is as follows: if the right balance is struck between the accuracy obtained on a particular training set, and the ability of the approach to 'learn' any training set without error, then the best generalization performance will be achieved. In contrast, FDH is not based on these fundamentals. In particular, the minimal extrapolation principle does not allow FDH to leave the field of descriptive statistics to enter the more interesting and modern field of statistical learning. A consequence of this feature of FDH is its well-known overfitting problem. The minimal extrapolation principle is responsible for this drawback. The estimated enveloping surface fits the data sample like a glove. In practice, the problem of overfitting manifests itself in the fact of observing many evaluated units with an FDH efficiency score of one. Additionally, the prediction or generalization power of FDH is low, reporting bad results when the objective is predicting, for example, the output value of a new non-observed unit. In the traditional framework, the notion of generalization and prediction error is linked to Statistical inference. Of course, if the description of the sample is the main aim, then FDH is a suitable technique. However, researchers and practitioners must be aware of the limitations of techniques, such as FDH, to determine an accurate estimate of the underlying DGP. If we apply the standard FDH, some assessed units in the sample may seem technically efficient, but they are only like this in the sample. In some respects, the usual definition of technical efficiency under FDH is a sample-specific notion or an in-sample estimation of the efficiency score. In contrast, usual ML techniques apply the notion of out-of-sample estimations, focusing on the actual DGP.

The estimates of the underlying production function generated by the FDH technique are step functions. Another well-known non-parametric technique for determining production possibility sets is Data Envelopment Analysis (DEA) by Charnes et al. (1978) and Banker et al. (1984). In this case, the estimates are piece-wise linear functions. In contrast to FDH, DEA also assumes convexity. This means that if two input-output bundles are assumed producible, then any convex combination of them are also feasible. Graphically, FDH could be considered the 'skeleton' of DEA since the convex hull of the frontier estimated by FDH coincides with the DEA frontier (Daraio & Simar, 2005).

Recently, Esteve et al. (2020) have introduced Efficiency Analysis Trees (EAT) as an out-of-sample type FDH, endowing this traditional non-parametric technique in the efficiency measurement area with properties that belong to the modern machine learning field. EAT applies decision trees models for estimating non-overfitted production possibility sets satisfying free disposability in the context of the Free Disposal Hull. The EAT algorithm represents an adaptation of Classification and Regression Trees (CART) by Breiman et al. (1984) for estimating production frontiers through step functions, thereby competing against the standard Free Disposal Hull technique.

In this chapter, we revise the main features of the EAT technique, providing the corresponding code written in Python. Additionally, and as a methodological novelty, we extend the EAT method, which was exclusively defined for implementing the output-oriented radial model in Esteve et al. (2020), to the context of dealing with the (graph) directional distance function (Chambers et al., 1998). Moreover, we will introduce an estimation of production possibility sets, based on the EAT technique, that satisfies the postulate of convexity and, consequently, competes against the standard Data Envelopment Analysis (DEA). All the measures, radial and non-radial, will be illustrated through several empirical examples; one of them with more than 900 units to be assessed. Our contribution has as an objective to evolve FDH and DEA to become data enabled analytics, in the line pointed out by

Zhu (2020). Our out-of-sample versions of FDH and DEA can be seen as data-driven data science tools for performance evaluation.

The chapter is organized as follows. Section 2 is devoted to briefly introducing the EAT technique. In Sect. 3, we extend EAT to deal with the directional distance function, both assuming and dismissing convexity. In Sect. 4, we illustrate how EAT performs through the analysis of a public database in the educational sector, as well as a panel of Taiwanese banks. In Sect. 5, we provide some indications on how to use a code written in Python to execute the algorithms linked to EAT. Finally, Sect. 6 concludes.

2 Efficiency Analysis Trees

Let us consider that we have observed n Decision Making Units (DMUs) that make up the so-called (learning) sample $\aleph = \left\{(\boldsymbol{x}_i, \boldsymbol{y}_i)\right\}_{i=1}^{n}$. In the standard production context, let us assume that DMU$_i$ consumes $\boldsymbol{x}_i = (x_{1i}, \ldots, x_{mi}) \in R_{+}^{m}$ amounts of inputs for the production of $\boldsymbol{y}_i = (y_{1i}, \ldots, y_{si}) \in R_{+}^{s}$ amounts of outputs. Hereinafter, we will use bold for denoting vectors, and non-bold for scalars. The Data Generating Process (DGP) that is behind the generation of these observations is as follows (see Daraio & Simar, 2005). It is assumed that we observe a sample $\aleph = \left\{(\boldsymbol{x}_i, \boldsymbol{y}_i)\right\}_{i=1}^{n}$ of an identically and independently distributed random vector $(\boldsymbol{X}, \boldsymbol{Y})$ with a certain unknown joint distribution with support $\Psi = \left\{(\boldsymbol{x}, \boldsymbol{y}) \in R_{+}^{m+s} : \boldsymbol{x} \text{ can produce } \boldsymbol{y}\right\}$, the so-called (theoretical) technology or production possibility set; where $\boldsymbol{X}$ represents the inputs and $\boldsymbol{Y}$ represents the outputs. Two different approaches have mainly been developed in the literature regarding DGPs: the deterministic frontier model, which assumes that all the observations in $\aleph$ belong to Ψ, and the stochastic frontier model, where random noise allows some observations to be outside of Ψ. The standard FDH, for example, focuses on the deterministic frontier model. For the same reason, the EAT algorithm, which is the data driven out-of-sample extension of the FDH, also works under the same assumption.

Under the standard neoclassical postulates in production theory, the technology Ψ meets a certain list of assumptions, such as monotonicity (free disposability) of inputs and outputs, meaning that if $(\boldsymbol{x},\boldsymbol{y}) \in \Psi$, then $(\boldsymbol{x}',\boldsymbol{y}') \in \Psi$, as long as $\boldsymbol{x}' \geq \boldsymbol{x}$ and $\boldsymbol{y}' \leq \boldsymbol{y}$[1]; or convexity, i.e., if $(\boldsymbol{x},\boldsymbol{y}) \in \Psi$ and if $(\boldsymbol{x}',\boldsymbol{y}') \in \Psi$, then $\lambda(\boldsymbol{x},\boldsymbol{y}) + (1-\lambda)(\boldsymbol{x}',\boldsymbol{y}') \in \Psi, \forall \lambda \in [0, 1]$ (see Fare & Primont, 1995). While standard FDH satisfies only free disposability, Data Envelopment Analysis (DEA) also meets convexity.

[1] Let $\boldsymbol{z} = (z_1, \ldots, z_q)$ and $\boldsymbol{z}' = \left(z_1', \ldots, z_q'\right)$. Hereinafter, $\boldsymbol{z} \leq \boldsymbol{z}'$ means $z_j \leq z_j'$ for all $j = 1, \ldots, q$, and $\boldsymbol{z} < \boldsymbol{z}'$ means $z_j < z_j'$ for all $j = 1, \ldots, q$.

The notion of technical efficiency is related to the measurement of the distance from an interior point of the technology to some special part of the border of Ψ. Specifically, the production frontier or efficient frontier of Ψ is defined as $\partial(\Psi) := \{(x, y) \in \Psi : \hat{x} < x, \hat{y} > y \Rightarrow (\hat{x}, \hat{y}) \notin \Psi\}$. We can find many different types (families) of distances to the efficient frontier in the literature. One of the most well-known technical efficiency measures in the literature is that known as the output-oriented radial model (see Charnes et al., 1978; Banker et al., 1984), which is the inverse of the classical Shephard output distance function (Shephard, 1953). This measure is particularly convenient in settings where inputs are fixed (such as land or family-labor on a farm) and when generating maximum output would be the suitable aim from a technical perspective. In particular, the output-oriented radial model determines the efficiency score for $(x_k, y_k) \in R_+^{m+s}$ by increasing all its outputs equiproportionally while maintaining inputs constant: $\phi(x_k, y_k) = \max\{\phi_k \in R : (x_k, \phi_k y_k) \in \Psi\}$.

Nowadays, there are different nonparametric methods in the literature for estimating Ψ or, equivalently, its efficient frontier. Some of them are those that apply Kernel-based approaches and local regression techniques. See, for example, Du et al. (2013), where the authors propose a kernel smoothing method that can handle multiple shape constraints (e.g., monotonicity) for multivariate functions, generalizing Hall and Huang (2001). Another interesting contribution is Parmeter et al. (2014), who showed how constraint-weighted bootstrapping may be applied to impose smoothness conditions on linear estimates. In particular, these authors estimate an input distance function both parametrically and non-parametrically, resorting in the last case to a local linear generalized kernel regression. See also Henderson and Parmeter (2009). Here, to introduce machine learning techniques to DEA, we start from two of the most noteworthy and usual nonparametric models for calculating production frontiers: FDH and DEA, coupled with an output-oriented efficiency measure.

Free Disposal Hull (Deprins et al., 1984) has attracted the interest of researchers, practitioners and scholars in recent years (see, for example, Tavakoli & Mostafaee, 2019; Barbosa et al., 2019; and Kerstens et al., 2019). Deprins et al. (1984) proposed the following mathematical expression as an estimate of the set Ψ from the (learning) sample $\aleph = \{(x_i, y_i)\}_{i=1}^{n}$:

$$\hat{\Psi}_{FDH} = \left\{(x, y) \in R_+^{m+s} : \exists i = 1, \ldots, n \text{ such that } y \leq y_i, x \geq x_i\right\} \tag{1}$$

Additionally, the efficiency score $\phi(x_k, y_k)$ can be estimated by plugging $\hat{\Psi}_{FDH}$ into $\max\{\phi_k \in R : (x_k, \phi_k y_k) \in \Psi\}$ in place of Ψ. And the optimization problem can be rewritten as a mixed-integer linear optimization program:

$$\phi^{FDH}\left(\boldsymbol{x}_k,\boldsymbol{y}_k\right)=\max\ \phi$$
$$\begin{aligned} & s.t. \\ & \sum_{i=1}^{n}\lambda_i x_{ji}\le x_{jk},\quad j=1,\ldots,m \\ & \sum_{i=1}^{n}\lambda_i y_{ri}\ge \phi y_{rk},\ r=1,\ldots,s \\ & \sum_{i=1}^{n}\lambda_i=1, \\ & \lambda_i\in\{0,1\},\qquad i=1,\ldots,n \end{aligned} \tag{2}$$

The same type of technical measure can be estimated through Data Envelopment Analysis by convexification of the production frontier generated by FDH. Next, we show the Linear Programming model that should be solved in that case:

$$\phi^{DEA}\left(\boldsymbol{x}_k,\boldsymbol{y}_k\right)=\max\ \phi$$
$$\begin{aligned} & s.t. \\ & \sum_{i=1}^{n}\lambda_i x_{ji}\le x_{jk},\quad j=1,\ldots,m \\ & \sum_{i=1}^{n}\lambda_i y_{ri}\ge \phi y_{rk},\ r=1,\ldots,s \\ & \sum_{i=1}^{n}\lambda_i=1, \\ & \lambda_i\ge 0,\qquad i=1,\ldots,n \end{aligned} \tag{3}$$

The difference between models (2) and (3) is the last constraint. DEA only requires non-negativity of the real values of lambda variables, whereas FDH resorts to binary variables.

FDH is a very simple data-driven technique for estimating production frontiers, which is only grounded on the postulate of free disposability. Unfortunately, FDH suffers the problem of overfitting. The efficiency score estimations are overly optimistic (and many observations in the data sample are identified as efficient – i.e., they are located onto the estimated production frontier), and therefore the technique is useful for describing exclusively the observed sample at a frontier level. This is one of the reasons why we talk about 'relative' efficiency when FDH is applied, as a notion that depends on the observed sample. In contrast, a recent approach by Esteve et al. (2020) introduced an extension of the FDH technique, called Efficiency Analysis Trees (EAT), whose aim is not to depend on the sample and, indeed, provides an out-of-sample FDH estimator. Next, we show a graphical example of these two techniques in comparison with the true theoretical production frontier, in the single-output case.

First, through Fig. 1, note how both the FDH and the EAT frontiers (estimates) are step functions. Second, the EAT frontier is located above the standard FDH frontier, but shows segments where both techniques provide the same estimated

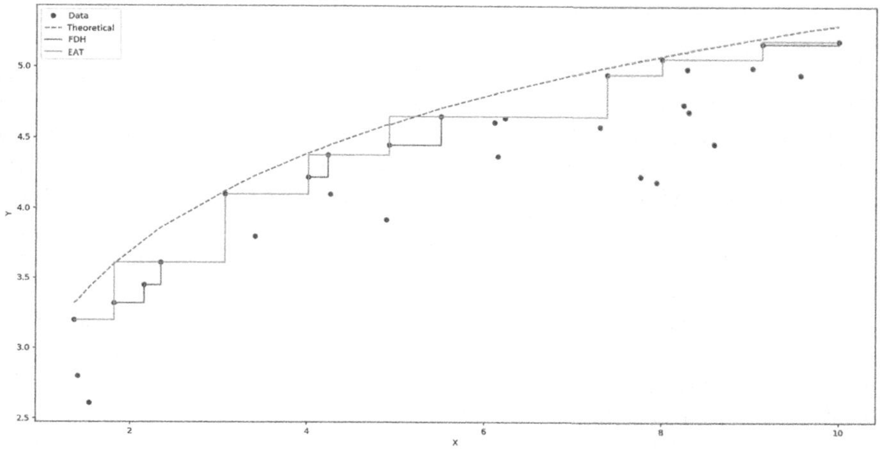

Fig. 1 Graphical illustration of a theoretical production function, the corresponding data sample and the frontier estimates provided by FDH and EAT

value for the output. Third, the EAT frontier really tries to approach the true production function, regardless of the sample size, whereas the FDH approximation to the theoretical frontier is based on convergence. The FDH estimate is very close to the data cloud. So, in general, the more data you collect, the better the approximation to the real production frontier. Mathematically speaking, there is another difference between these two techniques: while the FDH estimate comes from an optimization model, the EAT estimate is generated from an algorithm, as we show next. Finally, note also that the EAT technique does not generate an inner approximation of the theoretical frontier, in contrast to the FDH technique.

Next, we are going to revise the main fundamentals of the EAT technique, when the estimation of a multi-output production frontier is the focus. First of all, we introduce some basic notation and terminology associated with tree structures.

EAT is based on CART (Breiman et al., 1984) for regression, which is a renowned machine learning technique for predicting the value of a numerical response variable from a set of predictors. The success of CART in the specialized literature is, in part, due to the simplicity of the principle that is behind the method: A criterion (usually minimizing the mean squared error) is selected to recursively generate binary partitions of the data until no further meaningful division is possible or a stopping rule holds. Graphically speaking, the final product of this procedure is a tree that begins at the root node, which contains all the observed data, develops through intermediate nodes, and ends at the terminal nodes, also called leaves, which contain disjoint subsets of the observed data. The binary nature of the process is reflected in each parent node, except the leaves, which gives birth to two child nodes. The split of the data at any node is caused by a predictor variable and a threshold for this variable. Given all the possible combinations of each predictor and each threshold, CART builds regression trees by choosing the split that minimizes the sum of the mean squared error (MSE) of the two child nodes. The EAT algorithm works by following the same steps as CART. However, while CART yields prediction on the mean of the response variable conditioned to the values of the predictors, EAT has been designed for providing estimations on the maximum of the response variable (the frontier output in our production context) conditioned to the values of the resources (inputs). Additionally, and this is a key point, the standard algorithm associated with CART randomly selects the next node to be split and the final tree structure defined is independent of the sequence of nodes followed. In contrast, the algorithm related to EAT selects the next node to be split through a heuristic and the final resulting tree depends on the order of nodes followed in the splitting process. Why is the order important for EAT and not for CART? Because the EAT algorithm must guarantee the satisfaction of the postulate of free disposability. For EAT this implies that, sequentially, the prediction provided for the output at a node must be greater than or equal to the prediction provided for certain previous nodes in the tree, as we show later in the text.

2.1 The Efficiency Analysis Trees Algorithm

We start now with the introduction of the main steps of the EAT algorithm. In this respect, let us assume that we have a node t to be split. This node contains a certain subset of the learning sample $\aleph = \{(\boldsymbol{x}_i, \boldsymbol{y}_i)\}_{i=1}^{n}$. The EAT algorithm has to select an input variable j, $j = 1, \ldots, m$, and a threshold $s_j \in S_j$, where S_j is the set of possible thresholds for variable j, such that the sum of the mean squared error (MSE) calculated for the data belonging to the left child node t_L and the MSE corresponding to the data belonging to the right child node t_R is minimized. The data that belong to the left child node t_L are the data in the parent node t such that satisfies the condition $x_j < s_j$, and the data that belong to the right child node t_R are the data in the parent node t that meet $x_j \geq s_j$. Additionally, in the EAT algorithm, the set S_j is defined from the observed values of the input j in the data sample. So, it is empirically defined. Mathematically, the idea consists of selecting the combination (x_j, s_j) which minimizes $R(t_L) + R(t_R)=$ $\frac{1}{n}\sum_{(\boldsymbol{x}_i,\boldsymbol{y}_i)\in t_L}\sum_{r=1}^{s}(y_{ri} - y_r(t_L))^2 + \frac{1}{n}\sum_{(\boldsymbol{x}_i,\boldsymbol{y}_i)\in t_R}\sum_{r=1}^{s}(y_{ri} - y_r(t_R))^2$, where $y_r(t)$ is the way of denoting the estimation of the r-th output at node t. As a vector, it is denoted as $\boldsymbol{y}(t)$. The key is how to define $\boldsymbol{y}(t)$ in each node for satisfying free disposability in the growing process of the tree structure. In this sense, an important notion that was introduced by Esteve et al. (2020) is the Pareto-dominance between nodes, which is a way of transferring the concept of Pareto-dominance between input vectors in the standard efficiency literature to the realm of decision trees.

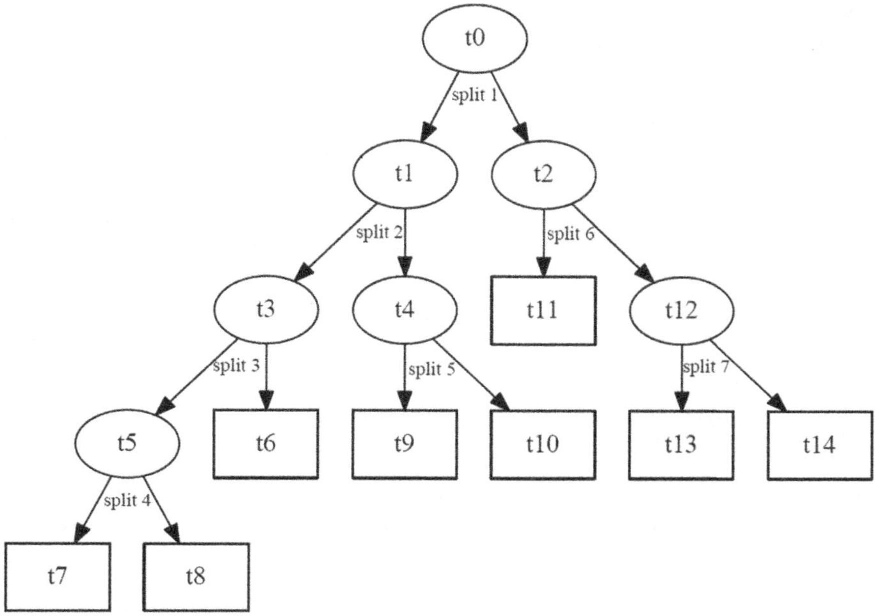

Fig. 2 Example of a tree structure

By the above description of the building process of the tree structure, each node t is defined by the fulfilment of a series of conditions in the input space of the type $\{x_j < s_j\}$ or $\{x_j \geq s_j\}$. In this way, after completing several splits, a certain region in the input space is created in the shape of a segment if $m = 1$ (we deal with only one input), a rectangle if $m = 2$ (we deal with two inputs), a parallelepiped if $m = 3$ (we have three inputs), etc. This region in the input space is the 'support' of node t.

Definition 1 The support of node t is denoted and defined as $\text{supp}(t) = \left\{\boldsymbol{x} \in R_+^m : a_j^t \leq x_j < b_j^t, j = 1, \ldots, m\right\}$, with $a_j^t < b_j^t, \forall j = 1, \ldots, m$.

The parameters a_j^t and b_j^t come from the different thresholds selected during the splitting process. Moreover, given the notion of support of a node, it is possible to define the concept of (input) Pareto-dominance between nodes. Before showing this definition, it is necessary to introduce some previous notation. In this respect, let $k = 1, \ldots, K$ be the number of executed splits in the growing process of a tree. Let T_k be the created tree after the k-th split. Let $\tilde{T}_k$ be the set of terminal nodes in tree T_k. In Fig. 2, T_1 is the (sub)tree composed by $\{t_0, t_1, t_2\}$, with terminal nodes $\tilde{T}_1 = \{t_1, t_2\}$. T_2 could be the tree $\{t_0, t_1, t_2, t_3, t_4\}$ with set of terminal nodes $\tilde{T}_2 = \{t_2, t_3, t_4\}$.

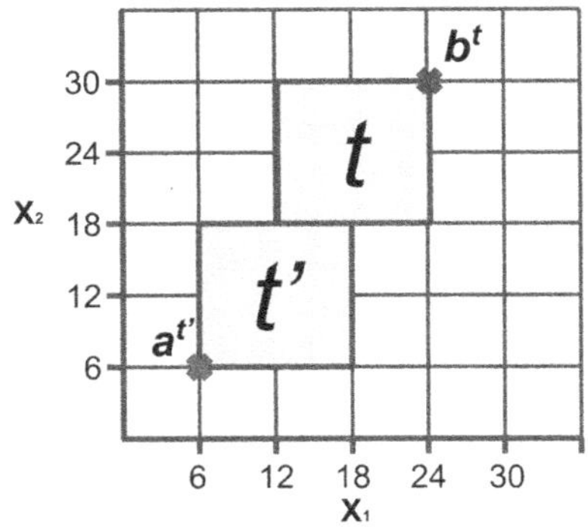

Fig. 3 (Input) Pareto-dominance between nodes

Now, we can introduce the definition of (input) Pareto-dominant nodes of a given node $t \in \tilde{T}_k$.

Definition 2 Let $k = 1, \ldots, K$. Let $t \in \tilde{T}_k$. Then, the set of (input) Pareto-dominant nodes of node t is denoted and defined as $I_{T_k}(t) = \left\{t' \in \tilde{T}_k - t : \exists \boldsymbol{x} \in \text{supp}(t), \exists \boldsymbol{x}' \in \text{supp}\left(t'\right) \text{ such that } \boldsymbol{x}' \leq \boldsymbol{x}\right\}$.

Put in words, an element in $I_{T_k}(t)$ is a node with at least an input vector in its corresponding support, non-necessarily observed, such that it dominates at least an input vector belonging to the support of node t in the Pareto sense. Although implementing this idea may seem difficult, in pratice it is rather easy to do by comparing the components of $\boldsymbol{a}^{t'}$ and $\boldsymbol{b}^t$. In particular, $t' \in I_{T_k}(t)$ if and only if $\boldsymbol{a}^{t'} < \boldsymbol{b}^t$ (see Proposition 2 in Esteve et al., 2020). In Fig. 3, we show a graphical example of dominance in the input space. In the figure, node t' is a Pareto-dominant node of t in the line of Definition 2.

More notation: let $t^* \in \tilde{T}_k$ be a node to be split. Then, the tree associated with this specific split will be denoted as $T(k|\, t^* \to t_L, t_R)$.

Now, we look at how to estimate the outputs in each child node in order to guarantee the satisfaction of free disposability. Before doing that, it is worth mentioning that the estimation of the outputs corresponding to the root node of the tree, t_0, is defined as $y_r(t_0) = \max\{y_{ri} : (\boldsymbol{x}_i, \boldsymbol{y}_i) \in t_0\} = \max\{y_{ri} : i = 1, \ldots, n\}$, $r = 1, \ldots, s$. For any other node $t^* \in \tilde{T}_k$, Esteve et al. (2020) suggest how to estimate the value of the outputs for the left child node as follows:

$$y_r(t_L) = \max\left\{\max\left\{y_{ri} : \left(\boldsymbol{x}_i, \boldsymbol{y}_i\right) \in t_L\right\}, y_r\left(I_{T(k|t^* \to t_L, t_R)}(t_L)\right)\right\}, r = 1, \ldots, s, \tag{4}$$

where $y_r\left(I_{T(k|t^* \to t_L, t_R)}(t_L)\right) = \max\left\{y_r\left(t'\right) : t' \in I_{T(k|t^* \to t_L, t_R)}(t_L)\right\}$ and $y_r(t')$ is the estimation of the output y_r at node $t' \in \tilde{T}(k|t^* \to t_L, t_R)$, $r = 1, \ldots, s$.

Regarding the right child node, the formulation would be very similar to the previous one:

$$y_r(t_R) = \max\left\{\max\left\{y_{ri} : (\boldsymbol{x}_i, \boldsymbol{y}_i) \in t_R\right\}, y_r\left(I_{T(k|t^* \to t_L, t_R)}(t_R)\right)\right\}, r = 1, \ldots, s, \tag{5}$$

where $y_r\left(I_{T(k|t^* \to t_L, t_R)}(t_R)\right) = \max\left\{y_r(t') : t' \in I_{T(k|t^* \to t_L, t_R)}(t_R)\right\}$.

Then, the algorithm selects the best pair (x_{j^*}, s_{j^*}) among all the possible combinations of inputs and its corresponding thresholds in such a way that the sum of the MSE of the left and right child nodes is minimized. Once the split of node t^* is executed, the tree $T(k|t^* \to t_L^*, t_R^*)$ is obtained. The way of updating the parameters a and b corresponding to the supports of the child nodes t_L^* and t_R^* is as follows:

$$\begin{aligned}
a_j^{t_L} &= a_j^{t^*}, \ \forall j = 1, \ldots, m \\
b_j^{t_L} &= b_j^{t^*}, \ \forall j = 1, \ldots, m, \ j \neq j^* \\
b_{j^*}^{t_L} &= s_{j^*}^*, \\
a_j^{t_R} &= a_j^{t^*}, \ \forall j = 1, \ldots, m, \ j \neq j^* \\
a_{j^*}^{t_R} &= s_{j^*}^*, \\
b_j^{t_R} &= b_j^{t^*}, \ \forall j = 1, \ldots, m
\end{aligned} \tag{6}$$

A relevant result that was not proved in Esteve et al. (2020) is that, for the right child node, $y_r(t_R) = y_r(t^*)$. Put in words, the right child node always inherits the output estimation from its corresponding parent node. To prove this result, it is first necessary to establish the following lemma and corollaries.

Lemma 1 If $t' \in I_{T(k|t^* \to t_L, t_R)}(t_L)$, then $t' \in I_{T_k}(t^*)$.

Proof. Note that, by definition, $\tilde{T}(k|t^* \to t_L, t_R) = \left(\tilde{T}_k - t^*\right) \cup \{t_L, t_R\}$. Moreover, $t' \in I_{T(k|t^* \to t_L, t_R)}(t_L)$ if and only if $\boldsymbol{a}^{t'} < \boldsymbol{b}^{t_L}$ (see Proposition 2 in Esteve et al., 2020). By (6), $\boldsymbol{b}^{t_L} \leq \boldsymbol{b}^{t^*}$. This implies that $\boldsymbol{a}^{t'} < \boldsymbol{b}^{t^*}$. Hence, $t' \in I_{T_k}(t^*)$. ■

Corollary 1 $y_r(t_L) \leq y_r\left(I_{T_k}(t^*)\right), r = 1, \ldots, s$.

Proof. By Lemma 1, $I_{T(k|t^* \to t_L, t_R)}(t_L) \subset I_{T_k}(t^*)$. Then, $y_r\left(I_{T(k|t^* \to t_L, t_R)}(t_L)\right) \leq y_r\left(I_{T_k}(t^*)\right), r = 1, \ldots, s$. Additionally, $\max\{y_{ri} : (\boldsymbol{x}_i, \boldsymbol{y}_i) \in t^*\} \geq \max\{y_{ri} : (\boldsymbol{x}_i, \boldsymbol{y}_i) \in t_L\}$. By the EAT algorithm, we know that $y_r(t^*) = \max\left\{\max\left\{y_{ri} : (\boldsymbol{x}_i, \boldsymbol{y}_i) \in t^*\right\}, y_r\left(I_{T_k}(t^*)\right)\right\}, r = 1, \ldots, s$. Finally, by (4), we have that $y_r(t_L) \leq y_r\left(I_{T_k}(t^*)\right), r = 1, \ldots, s$. ■

Corollary 2 $y_r(t_L) \leq y_r(t^*), r = 1, \ldots, s$.

Proof. By the EAT algorithm, $y_r(t^*) = \max\{\max\{y_{ri} : (\boldsymbol{x}_i, \boldsymbol{y}_i) \in t^*\}, y_r(I_{T_k}(t^*))\}$, $r = 1, \ldots, s$. By Corollary 1, $y_r(t_L) \le y_r(I_{T_k}(t^*))$, $r = 1, \ldots, s$. Hence, $y_r(t^*) = \max\{\max\{y_{ri} : (\boldsymbol{x}_i, \boldsymbol{y}_i) \in t^*\}, y_r(I_{T_k}(t^*))\} \ge y_r(t_L)$, $r = 1, \ldots, s$. ■

Proposition 1 $y_r(t_R) = y_r(t^*)$, $r = 1, \ldots, s$.

Proof. $I_{T(k|t^* \to t_L, t_R)}(t_R) = I_{T_k}(t^*) \cup \{t_L\}$ because, by (6), the nodes t_R and t^* share the same parameter $\boldsymbol{b}$ ($b_j^{t_R} = b_j^{t^*}$, $\forall j = 1, \ldots, m$) and $\boldsymbol{a}^{t_L} = \boldsymbol{a}^{t^*} < \boldsymbol{b}^{t^*}$. Then, by (5), $y_r(t_R) = \max\{\max\{y_{ri} : (\boldsymbol{x}_i, \boldsymbol{y}_i) \in t_R\}, y_r(I_{T_k}(t^*)), y_r(t_L)\}$, $r = 1, \ldots, s$. Additionally, by the EAT algorithm, we know that $y_r(t^*) = \max\{\max\{y_{ri} : (\boldsymbol{x}_i, \boldsymbol{y}_i) \in t^*\}, y_r(I_{T_k}(t^*))\}$. We have two scenarios for each component r, $r = 1, \ldots, s$. Scenario A: $y_r(t^*) = y_r(I_{T_k}(t^*))$. In this way, we have that $y_r(I_{T_k}(t^*)) \ge \max\{y_{ri} : (\boldsymbol{x}_i, \boldsymbol{y}_i) \in t^*\}$. If $y_r(I_{T_k}(t^*)) \ge \max\{y_{ri} : (\boldsymbol{x}_i, \boldsymbol{y}_i) \in t^*\}$, then $y_r(I_{T_k}(t^*)) \ge \max\{y_{ri} : (\boldsymbol{x}_i, \boldsymbol{y}_i) \in t_R\}$, because $\max\{y_{ri} : (\boldsymbol{x}_i, \boldsymbol{y}_i) \in t^*\} \ge \max\{y_{ri} : (\boldsymbol{x}_i, \boldsymbol{y}_i) \in t_R\}$. Moreover, by Corollary 1, $y_r(I_{T_k}(t^*)) \ge y_r(t_L)$, $r = 1, \ldots, s$. Hence, $y_r(t_R) = y_r(I_{T_k}(t^*)) = y_r(t^*)$, $r = 1, \ldots, s$, as we wanted to prove. Scenario B: $y_r(t^*) = \max\{y_{ri} : (\boldsymbol{x}_i, \boldsymbol{y}_i) \in t^*\}$. Therefore, $\max\{y_{ri} : (\boldsymbol{x}_i, \boldsymbol{y}_i) \in t^*\} \ge y_r(I_{T_k}(t^*))$. We must consider two additional scenarios: B1 and B2. Scenario B1: $\max\{y_{ri} : (\boldsymbol{x}_i, \boldsymbol{y}_i) \in t^*\} = \max\{y_{ri} : (\boldsymbol{x}_i, \boldsymbol{y}_i) \in t_L\}$. By Lemma 1, $I_{T(k|t^* \to t_L, t_R)}(t_L) \subset I_{T_k}(t^*)$. Then, $y_r(I_{T(k|t^* \to t_L, t_R)}(t_L)) \le y_r(I_{T_k}(t^*))$. Consequently, we have that $y_r(I_{T(k|t^* \to t_L, t_R)}(t_L)) \le \max\{y_{ri} : (\boldsymbol{x}_i, \boldsymbol{y}_i) \in t^*\} = \max\{y_{ri} : (\boldsymbol{x}_i, \boldsymbol{y}_i) \in t_L\}$. Hence, by (4), $y_r(t_L) = y_r(t^*)$. Now, we get that $y_r(t_R) = y_r(t^*)$ since $y_r(t_R) = \max\{\max\{y_{ri} : (\boldsymbol{x}_i, \boldsymbol{y}_i) \in t_R\}, y_r(I_{T_k}(t^*)), y_r(t_L)\}$ and $\max\{y_{ri} : (\boldsymbol{x}_i, \boldsymbol{y}_i) \in t_R\} \le \max\{y_{ri} : (\boldsymbol{x}_i, \boldsymbol{y}_i) \in t^*\}$. Scenario B2: $\max\{y_{ri} : (\boldsymbol{x}_i, \boldsymbol{y}_i) \in t^*\} = \max\{y_{ri} : (\boldsymbol{x}_i, \boldsymbol{y}_i) \in t_R\}$. By Corollary 2, $y_r(t_L) \le y_r(t^*)$. Hence, $y_r(t_R) = \max\{\max\{y_{ri} : (\boldsymbol{x}_i, \boldsymbol{y}_i) \in t_R\}, y_r(I_{T_k}(t^*)), y_r(t_L)\} = y_r(t^*)$. ■

After performing the $k + 1$ split, we have an estimation of output variables for each node belonging to $\tilde{T}_{k+1}$. In particular, we have that $y_{T_{k+1}}(t) = y_{T_k}(t)$ if $t \ne t^*$ and $y_{T_{k+1}}(t_L) = y(t_L)$ and $y_{T_{k+1}}(t_R) = y(t_R)$, where t_L and t_R are the left and right child nodes of node t^*, respectively.

Additionally, in Esteve et al. (2020), the way of selecting the next node to be split during the growing process of the tree was defined as random. However, there are other alternatives that perform better. The order in which the tree is expanded during the application of the algorithm is key because the efficiency results depend on it. A heuristic procedure that works well with EAT (see Esteve et al., 2021) is choosing the node to be split t^* such that the number of its (input) Pareto-dominants nodes is low, i.e., the cardinal of the set $I_{T_k}(t^*)$ is small. Usually, this condition is satisfied for left child nodes. Accordingly, the growing process focuses on splitting left nodes, which are not leaves of the tree, and, once they are all expanded, continues with the right nodes from the top of the tree structure, from left to right. The splitting process continues until additional bipartitions are not possible (if all the data in a node have the same input values) or a predefined stopping rule is satisfied. The most usual stopping rule in decision trees is that proposed by Breiman et al. (1984), which states that if a node has five or less examples (DMUs in our production context), i.e., $n(t) \leq n_{\min} = 5$, then this node must be marked as a leaf node of the tree. And the splitting process must not be applied to that node. Although, by default, the EAT technique follows that stopping rule, $n_{\min}$ could be considered as a hyperparameter of the algorithm to be tuned. Indeed, some references (see, for example, Berk, 2008) talk about resorting to greater $n_{\min}$, taking a value of 10 or, even, more, in the case of quantile regression. In fact, EAT could be reinterpreted as a quantile regression driven to the limit (the maximum), but satisfying, additionally, some properties coming from microeconomic production theory.

After performing the whole process, one gets a tree structure denoted as $T_{\max}$. The estimations for the values of the outputs conditioned to an input profile are excessively optimistic and, therefore, similar to the observed data. Consequently, as FDH, $T_{\max}$ suffers overfitting because the tree is too large. To overcome this problem without previously assuming any probability distribution on the data, Esteve et al. (2020) suggested pruning the tree in a suitable way, as was previously proposed by Breiman et al. (1984) for CART. The pruning process is strongly grounded on the notion of the error-complexity measure $R_\alpha(T)$, which is a combination between a measure of accuracy of the tree, defined as the aggregation of the mean squared error determined in each leaf node, and a measure of the size of the tree (the number of leaf nodes). Additionally, the error-complexity measure depends on a parameter α, which balances the values of the two components of the error: $R_\alpha(T) = R(T) + \alpha \left|\tilde{T}\right|$. The idea behind the pruning process of the tree $T_{\max}$ is based upon minimizing the error-complexity indicator. We refer the reader to Breiman et al. (1984) who provides a comprehensive explanation of the details of the punning procedure, which is essentially grounded on cross-validation. Hereinafter, we will denote the tree resulting from the pruned process as T^*. We suggest the EAT algorithm be always applied with pruning to identify an accurate estimation of the underlying production frontier. Otherwise, resorting to $T_{\max}$ instead of T^*, we would obtain an upper enveloping surface of the learning sample 'describing' exclusively the observed data at a frontier level, as happens with the FDH technique.

2.2 *Efficiency Analysis Trees (EAT) Efficiency Scores: Relating EAT with Free Disposal Hull (FDH) and (Convex) Data Envelopment Analysis (DEA)*

Let us now introduce more notation. Let $\boldsymbol{d}_{T^*}(\boldsymbol{x}) := \boldsymbol{y}_{T^*}(t)$, for $t \in \tilde{T}^*$ such that $\boldsymbol{x} \in \text{supp}(t)$. In words, $\boldsymbol{d}_{T^*}(\boldsymbol{x})$ denotes the multi-dimensional predictor or estimator, depending on our interest for predicting the outputs given a new or non-observed input profile or for estimating the outputs on the underlying production frontier given an observed DMU, defined from the (optimal) tree T^*. From this notion, we can define the production possibility set induced by $\boldsymbol{d}_{T^*}(\boldsymbol{x})$ as:

$$\hat{\Psi}_{T^*} = \left\{(\boldsymbol{x}, \boldsymbol{y}) \in R_+^{m+s} : \boldsymbol{y} \leq \boldsymbol{d}_{T^*}(\boldsymbol{x})\right\}. \tag{7}$$

It can be proved that $\hat{\Psi}_{T^*}$ fulfils the same standard microeconomic postulates as FDH except minimal extrapolation. These results are summarized in Proposition 2.

Proposition 2 $\hat{\Psi}_{T^*}$ satisfies the following properties.

(i) $(\boldsymbol{x}_i, \boldsymbol{y}_i) \in \hat{\Psi}_{T^*}, \forall i = 1, \ldots, n.$
(ii) $\hat{\Psi}_{T^*}$ meets free disposability in inputs and outputs.
(iii) $\hat{\Psi}_{FDH} \subseteq \hat{\Psi}_{T^*}$.

Proof. (i) is true because the EAT algorithm generates an (upper) enveloping surface of the data. (ii) was proved in Esteve et al. (2020). (iii) is true because $\hat{\Psi}_{FDH}$ satisfies (i) and (ii) and, additionally, minimal extrapolation. ■

The third statement in Proposition 2 reinforces the fact that the technology estimated by EAT does not satisfy minimal extrapolation, which ultimately coincides with the application of the parsimony principle of Occam's razor. Instead, EAT applies cross-validation on the observed data for learning what the actual Data Generating Process is like, on the basis of the traditional principle of minimal extrapolation used in FDH and DEA.

The relationship between the mathematical formulation of $\hat{\Psi}_{FDH}$ and the mathematical formulation of $\hat{\Psi}_{T^*}$ is summarized in Proposition 3, which states that the Free Disposal Hull generated from a certain 'virtual' learning sample coincides with $\hat{\Psi}_{T^*}$.

Proposition 3 The FDH technology constructed from the set of points $\hat{\aleph} = \{(\boldsymbol{a}^t, \boldsymbol{d}_{T^*}(\boldsymbol{a}^t))\}_{t \in \tilde{T}^*}$ coincides with $\hat{\Psi}_{T^*}$:

$$\hat{\Psi}_{T^*} = \left\{(\boldsymbol{x}, \boldsymbol{y}) \in R_+^{m+s} : \exists t \in \tilde{T}^* \text{ such that } \boldsymbol{y} \leq \boldsymbol{d}_{T^*}\left(\boldsymbol{a}^t\right), \boldsymbol{x} \geq \boldsymbol{a}^t\right\}. \tag{8}$$

Proof. See Esteve et al. (2020), Proposition 6. ■

The input-output points that make up the set $\hat{\aleph}$ have a special format: $(\boldsymbol{a}^t, \boldsymbol{d}_{T^*}(\boldsymbol{a}^t))$, for all $t \in \tilde{T}^*$. Some of them could coincide with an actual observation, a DMU, but normally it would be a 'virtual' (non-observed) production unit, which consumes the input bundle $\boldsymbol{a}^t$, corresponding to the bottom corner of the support linked to the leaf node t in tree T^*, and produces the output bundle $\boldsymbol{d}_{T^*}(\boldsymbol{a}^t)$, corresponding to the EAT estimation of the set of outputs for $\boldsymbol{a}^t$.

Proposition 3 shows an alternative way of expressing $\hat{\Psi}_{T^*}$ in terms of a FDH-type production possibility set. This formulation makes it easy to define an estimator of the inverse of the Shephard output distance function (Shephard, 1953), i.e., $\phi(\boldsymbol{x}_k, \boldsymbol{y}_k) = \max\{\phi_k \in R : (\boldsymbol{x}_k, \phi_k \boldsymbol{y}_k) \in \Psi\}$, through the EAT technique. To do that, it is enough to substitute Ψ by its EAT estimator $\hat{\Psi}_{T^*}$. And thanks to Proposition 3, we know that $\phi^{EAT}(\boldsymbol{x}_k, \boldsymbol{y}_k) := \max\left\{\phi_k \in R : (\boldsymbol{x}_k, \phi_k \boldsymbol{y}_k) \in \hat{\Psi}_{T^*}\right\}$ can be determined through the following Mixed-Integer Linear Program:

$$
\begin{aligned}
\phi^{EAT}\left(\boldsymbol{x}_k, \boldsymbol{y}_k\right) = \max \ & \phi \\
s.t. \ & \\
& \sum_{t \in \tilde{T}^*} \lambda_t a_j^t \leq x_{jk}, \qquad j = 1, \ldots, m \\
& \sum_{t \in \tilde{T}^*} \lambda_t d_{rT^*}\left(\boldsymbol{a}^t\right) \geq \phi y_{rk}, \ r = 1, \ldots, s \\
& \sum_{t \in \tilde{T}^*} \lambda_t = 1, \\
& \lambda_t \in \{0, 1\}, \qquad t \in \tilde{T}^*
\end{aligned}
\tag{9}
$$

The reformulation of $\hat{\Psi}_{T^*}$ in (7) as a FDH-type technology, based upon a virtual data sample, as has been established in Proposition 3, can be also exploited to determine other kinds of technical efficiency measures. In Sect. 3, we will show how to address the implementation of the directional distance function (Chambers et al., 1998) in the context of Efficiency Analysis Trees.

Finally, in this section, we show how to provide an estimation, based upon the EAT algorithm, of the underlying production possibility set by assuming convexity, as DEA does. By analogy with the existing relationship between FDH and DEA, that is, the convexification of the FDH technology yields DEA technology, it is possible to derive an estimation of Ψ by the convexification of the set $\hat{\Psi}_{T^*}$. Regarding this new result, the convexification of the production possibility set derived from EAT would be as follows:

$$
conv\left(\hat{\Psi}_{T^*}\right) = \left\{ (\boldsymbol{x}, \boldsymbol{y}) \in R_+^{m+s} : \boldsymbol{x} \geq \sum_{t \in \tilde{T}^*} \lambda_t \boldsymbol{a}^t, \boldsymbol{y} \leq \sum_{t \in \tilde{T}^*} \lambda_t \boldsymbol{d}_{T^*}\left(\boldsymbol{a}^t\right), \sum_{t \in \tilde{T}^*} \lambda_t = 1, \lambda \geq 0_{\left|\tilde{T}^*\right|} \right\},
\tag{10}
$$

where $\left|\tilde{T}^*\right|$ denotes the number of elements in the set $\tilde{T}^*$.

And the optimization model that should be solved in order to determine an estimation of the output-oriented radial measure would be as follows:

$$
\begin{aligned}
\phi^{CEAT}\left(\boldsymbol{x}_k, \boldsymbol{y}_k\right) = \max \ & \phi \\
s.t. \ & \\
& \sum_{t \in \tilde{T}^*} \lambda_t a_j^t \leq x_{jk}, \qquad j = 1, \ldots, m \\
& \sum_{t \in \tilde{T}^*} \lambda_t d_{rT^*}\left(\boldsymbol{a}^t\right) \geq \phi y_{rk}, \ r = 1, \ldots, s \\
& \sum_{t \in \tilde{T}^*} \lambda_t = 1, \\
& \lambda_t \geq 0, \qquad t \in \tilde{T}^*
\end{aligned}
\tag{11}
$$

The technique based on the convexification of the EAT production frontier will be called Convexified Efficiency Analysis Trees (CEAT).

3 The Estimation of the Directional Distance Function Through EAT and CEAT

The well-known Data Envelopment Analysis and Free Disposal Hull approaches have witnessed the introduction of many different technical efficiency measures in the last four decades. To name a few, we may cite radial measures (Charnes et al., 1978; Banker et al., 1984); the Russell input and output measures of technical efficiency and their (non-linear) graph extension (see Färe et al., 1985); the hyperbolic measure by Färe et al. (1985); the additive model by Charnes et al. (1985); the weighted additive measures (Lovell & Pastor, 1995; Cooper et al., 1999; Cooper et al., 2011); the directional distance function (Chambers et al., 1998); and the Slacks-Based Measure (Pastor et al., 1999; Tone, 2001). The introduction of many different technical efficiency measures was motivated for two main reasons. First, for any DMU, which under the deterministic model is located inside the estimated production possibility set, there are many possible paths to reach the production frontier, each one of them generating a different technical efficiency measure. Second, in contrast to the parametric-statistical context for estimating production frontiers, DEA lacks a goodness of fit test. In the non-parametric literature linked to DEA and FDH, this drawback has been overcome by asking for the fulfilment of a list of interesting properties from mathematical and economical viewpoints. Regarding the satisfaction of properties in the DEA literature, see Färe and Lovell (1978), Pastor et al. (1999) and, more recently, Russell and Schworm (2018).

Among all the technical efficiency measures, there is one that stands out for its interpretability, flexibility and fulfilment of interesting properties. We are referring to the directional distance function (DDF). Luenberger (1992, 1995) introduced the concept of benefit function in consumption theory and its equivalent notion, the shortage function, in production theory. The shortage function measures the distance, in the direction of a vector $\boldsymbol{g}$, from a production plan to the boundary of the production possibility set, i.e., the shortage function measures the amount by which a DMU is short of reaching the frontier of the technology. In other words, it measures technical inefficiency. Later, Chambers et al. (1998) renamed the shortage function as the directional distance function and endowed it with new functionalities (for example, it is possible to use the DDF for determining an indicator linked to productivity change over time, called the Luenberger productivity indicator – see Chambers et al., 1996, and Chambers & Pope, 1996).

The DDF is easy to interpret in the sense that, given a directional vector $\boldsymbol{g}$, its value represents how many vectors $\boldsymbol{g}$ are necessary to reach the frontier of the technology. A non-negative value indicates that the assessed unit lays inside the reference technology, whereas a negative value indicates that the evaluated unit is located above the frontier. Regarding its flexibility, it is worth noting that the most usual technical efficiency measures, that is, radial measures (i.e., the inverse of the Shephard distance functions), gauge the 'distance' to the production frontier through equiproportional changes either in inputs or in outputs from the unit under evaluation. However, other ways of approaching to the frontier are possible and even desirable, depending on the production context. In this regard, the definition of the directional distance function meant the introduction of a new flexible tool for measuring technical efficiency following any direction to the frontier, as long as inputs were reduced, and outputs were increased. This was a clear evolution and extension of the Shephard distance functions that, additionally, present the limitation of being oriented. Instead, the DDF may be defined as input or output oriented, as the Shephard distance functions, but, additionally, can be non-oriented, also called 'graph'.

Moreover, and regarding the satisfaction of interesting properties, it is worth noting that the DDF has a close relationship with measures of economic inefficiency, which are indicators of interest for researchers when market prices are available. The economic behavior of firms is typically modeled by assuming cost minimization, revenue maximization or profit maximization. Under cost minimization, the Shephard input distance function may be associated with the cost function: the ratio of the minimum cost to the actual cost (cost efficiency) is decomposed into the inverse of the Shephard input distance function and (input) allocative efficiency (Fare & Primont, 1995). Under revenue maximization, a relationship between the Shephard output distance function and the revenue function may be stated: the ratio of actual revenue to the maximum revenue (revenue efficiency) may be decomposed into the Shephard output distance function and (output) allocative efficiency (Fare & Primont, 1995). However, the case of profit maximizing behavior requires that at the profit-maximizing input-output bundle, the chosen input bundle not only minimizes the cost of producing a given output bundle, but also that the chosen output bundle generates the maximum possible revenue from the selected input bundle. In this case, there is no particular reason to select either the input- or output-oriented measure of technical efficiency. Since the profit maximization of a firm generally requires an alteration of both inputs and outputs, it is desirable to use a measure of technical efficiency that requires not only a reduction of inputs but also an increase of outputs. One such measure is the directional distance function that, given a directional vector $\boldsymbol{g} = \left(\boldsymbol{g}^{-}, \boldsymbol{g}^{+}\right) \in R_{+}^{m+s}$, is defined as follows:

$$\beta\left(\boldsymbol{x}_{k}, \boldsymbol{y}_{k}\right) = \max\left\{\beta_{k} \in R : \left(\boldsymbol{x}_{k} - \beta_{k}\boldsymbol{g}^{-}, \boldsymbol{y}_{k} + \beta_{k}\boldsymbol{g}^{+}\right) \in \Psi\right\} \tag{12}$$

If one uses the FDH technique for estimating the value of the directional distance function, then the optimization model that should be solved would be the following:

$$
\begin{aligned}
\beta^{FDH}\left(\boldsymbol{x}_k, \boldsymbol{y}_k\right) = \max\ & \beta_k \\
s.t.\ & \\
& \sum_{i=1}^{n} \lambda_i x_{ji} \le x_{jk} - \beta_k g_j^-, \ j = 1, \ldots, m \\
& \sum_{i=1}^{n} \lambda_i y_{ri} \ge y_{rk} + \beta_k g_r^+, \ r = 1, \ldots, s \\
& \sum_{i=1}^{n} \lambda_i = 1, \\
& \lambda_i \in \{0, 1\}, \qquad i = 1, \ldots, n
\end{aligned}
\tag{13}
$$

This result is obtained by plugging the FDH estimation of the underlying technology, i.e., $\hat{\Psi}_{FDH}$, into the general expression of the directional distance function.

In the same way, it is possible to define the estimation of the DDF based on the EAT algorithm, by substituting Ψ by $\hat{\Psi}_{T^*}$ in (12): $\beta^{EAT}\left(\boldsymbol{x}_k, \boldsymbol{y}_k\right) = \max\left\{\beta_k \in R : \left(\boldsymbol{x}_k - \beta_k \boldsymbol{g}^-, \boldsymbol{y}_k + \beta_k \boldsymbol{g}^+\right) \in \hat{\Psi}_{T^*}\right\}$. Additionally, by Proposition 3, we know that $\beta^{EAT}(\boldsymbol{x}_k,\boldsymbol{y}_k)$ can be calculated through the following Mixed-Integer Linear Program:

$$
\begin{aligned}
\beta^{EAT}\left(\boldsymbol{x}_k, \boldsymbol{y}_k\right) = \max\ & \beta_k \\
s.t.\ & \\
& \sum_{t \in \tilde{T}^*} \lambda_t a_j^t \le x_{jk} - \beta_k g_j^-, \qquad j = 1, \ldots, m \\
& \sum_{t \in \tilde{T}^*} \lambda_t d_{rT^*}\left(\boldsymbol{a}^t\right) \ge y_{rk} + \beta_k g_r^+, \ r = 1, \ldots, s \\
& \sum_{t \in \tilde{T}^*} \lambda_t = 1, \\
& \lambda_t \in \{0, 1\}, \qquad t \in \tilde{T}^*
\end{aligned}
\tag{14}
$$

In contrast to (13), which works with the observed data, model (14) uses the non-observed database $\hat{\aleph} = \left\{\left(\boldsymbol{a}^t, \boldsymbol{d}_{T^*}\left(\boldsymbol{a}^t\right)\right)\right\}_{t \in \tilde{T}^*}$, which comes from the building of the optimal tree structure T^* through the EAT algorithm.

Note that the set of virtual points used in (14) for the DDF is the same as that used in model (9) for determining the radial measure. Other technical efficiency measures could be derived in the same way, by analogy. All of them will share the same set of non-observed points $\hat{\aleph}$. This means that the construction (estimation) of the technology is independent on the technical efficiency measure selected for gauging the performance of a set of units, which seems a natural result.

Finally, let us define the model that should be solved under the application of CEAT. Before doing so, let us recall how the directional distance function is determined through Data Envelopment Analysis (DEA):

$$\begin{aligned}\beta^{DEA}\left(\boldsymbol{x}_k, \boldsymbol{y}_k\right) = \max\ & \beta_k \\ s.t.\ & \\ & \sum_{i=1}^{n} \lambda_i x_{ji} \leq x_{jk} - \beta_k g_j^-,\ j = 1, \ldots, m \\ & \sum_{i=1}^{n} \lambda_i y_{ri} \geq y_{rk} + \beta_k g_r^+,\ r = 1, \ldots, s \\ & \sum_{i=1}^{n} \lambda_i = 1, \\ & \lambda_i \geq 0, \qquad i = 1, \ldots, n\end{aligned} \tag{15}$$

In the case of CEAT, the optimization model is as follows:

$$\begin{aligned}\beta^{CEAT}\left(\boldsymbol{x}_k, \boldsymbol{y}_k\right) = \max\ & \beta_k \\ s.t.\ & \\ & \sum_{t \in \tilde{T}^*} \lambda_t a_j^t \leq x_{jk} - \beta_k g_j^-, \qquad j = 1, \ldots, m \\ & \sum_{t \in \tilde{T}^*} \lambda_t d_{rT^*}\left(\boldsymbol{a}^t\right) \geq y_{rk} + \beta_k g_r^+,\ r = 1, \ldots, s \\ & \sum_{t \in \tilde{T}^*} \lambda_t = 1, \\ & \lambda_t \geq 0, \qquad t \in \tilde{T}^*\end{aligned} \tag{16}$$

Finally, we would like to highlight that our adaptation of well-known machine learning techniques for dealing with efficiency measurement problems is a way of bringing together two fields that so far have been unconnected. This new connection should allow applying the power of machine learning techniques to efficiency problems. Given that standard machine learning techniques are usually utilized for the analysis of big datasets, we understand that our proposal could work in the same line for big data problems in the field of efficiency and productivity.

4 Empirical Studies

We illustrate the results of the bilateral comparison of Efficiency Analysis Tress (EAT) to Free Disposal Hull (FDH), as well as convex EAT (CEAT) to Data Envelopment Analysis (DEA), resorting to two sets of empirical data. Throughout the analysis, we test whether the efficiency distributions obtained through these pairwise models (EAT vs. FDH, and CEAT vs DEA) differ from each other significantly. For this purpose, we perform the test proposed by Simar and Zelenyuk (2006), who adapt the nonparametric test for the equality of two densities, developed by Li (1996). In particular, we rely on algorithm II with 1000 replications, which computes and bootstraps the Li statistic on the efficiency scores, and where either the unitary or zero values of the efficient observations (resulting in the truncation of the efficiency scores) are smoothed by adding a small noise.[2] To compare the distribution, we also resort to box-plot graphs and, following these authors, show the kernel distributions of each pair of efficiency scores. We use Gaussian kernels, employing the reflection method to overcome the issue of a zero-bounded support of the inefficiency scores (Silverman, 1986), and relying on Sheather and Jones's (1991) method to determine the bandwidths.[3]

4.1 PISA Study

4.1.1 Data

Our first application evaluates the performance of 902 Spanish schools based on the results obtained in the 2012 edition of the PISA survey (Programme for International Student Assessment)—see Aparicio et al. (2018). The survey has taken place every 3 years since 2000, so our data corresponds to the fifth wave of the study. In each assessment, mathematics, reading and science is evaluated in a sample of 15 year-old students. These academic achievements represent our output variables: PVMATH (y_1), PVREAD (y_2), and PVSCIENCE (y_3). Characterizing the production function in education from an input perspective is not a simple matter. However, there is an extensive list of potential indicators that can be considered. In this context, most empirical studies attempting to measure the efficiency of schools usually include some measures of human and capital resources (Worthington, 2001; De Witte & Lopez-Torres, 2017). In our empirical study, we identify key inputs from an index representing the quality of school resources (SCMATEDU, x_1), created by PISA analysts from the responses given by school principals regarding several aspects (computers, educational software, calculators, books, audio visual resources or laboratory equipment); the inverse of the student-teacher ratio, i.e., the number of teachers per (hundred) students (PRFAL100, x_2). Moreover, we also consider the average socio-economic status of students in the school (ESCS, x_3) as an additional input, since students are the "raw material" to be transformed through the learning process. ESCS stands for the Economic, Social and Cultural Status, and provides a measure of family background that includes the highest levels of parents' occupation, educational resources and cultural goods at home. Since the original values of SCMATEDU and ESCS presented positive and negative values, these variables have been rescaled to show positive values. Table 1 presents the descriptive statistics for inputs and outputs.

---

[2]Simar and Zelenyuk (2006) develop their algorithm for radial distance functions, in which the efficiency values equal to one are smoothed. We adapt their algorithm to the additive context of the directional distance functions by smoothing the inefficiency scores equal to zero.

[3]To calculate and illustrate the kernel density functions, we rely on the "kdensity" function in STATA 15.1.

Table 1 Descriptive statistics for the PISA data, 902 Spanish schools, 2012

	Average	Median	Max.	Min.	Stand. Dev.
x_1 – SCMATEDU	3.683	3.618	5.576	0.008	0.891
x_2 – PRFAL100	10.145	9.000	90.009	0.719	6.309
x_3 – ESCS	4.396	4.370	5.970	2.190	0.538
y_1 – PVMATH	490.948	494.760	609.620	130.060	46.165
y_2 – PVREAD	490.405	494.450	626.640	297.740	46.657
y_3 – PVSCIE	500.888	505.595	626.510	92.410	44.042

Source: Own elaboration. OECD (2013), OECD (2014)

4.1.2 Tree Growth

We follow the procedure described in Sect. 2, revising the fundamentals of the EAT technique that estimates the maximum of the output vector by splitting the observed sample conditioned to the values of the observed inputs. Figure 4 presents the complete tree that has been obtained by applying the algorithm that follows the heuristic method, in choosing the split of the nodes such that the number of its (input) Pareto-dominants nodes is low, which is normally the case for left-branch (child) nodes. The tree, as well as the scores corresponding to the different EAT, CEAT and standard FDH and DEA efficiency measures have been obtained running the *EATpy* package implemented in Python as described in Sect. 5.

We start the description of tree starting with the root node at the top. We observe that the key variable that minimizes the sum of the squared errors for the left and right branches (or children), $\min\left(R(t_L)+R(t_R)\right)=\frac{1}{n}\sum_{(x_i,y_i)\in t_L}\sum_{r=1}^{s}\left(y_{ri}-y_r\left(t_L\right)\right)^2+\frac{1}{n}\sum_{(x_i,y_i)\in t_R}\sum_{r=1}^{s}\left(y_{ri}-y_r\left(t_R\right)\right)^2$, is the socio-economic status of students in the school (ESCS, x_3), with the left branch containing all schools (636) with ESCS <4.67, which is the optimal threshold $s_3 \in S_3$ splitting the sample ($x_3 < s_3$). Therefore, the right branch, t_2, of the subtree T_1 composed by $\{t_0, t_1, t_2\}$ includes the remaining 266 schools with ESCS ≥4.67 ($x_3 \geq s_3$). We observe that, as proved in Proposition 1, the right node t_2 (Id = 2), inherits the maximum observed values of the output vector of the root node: $y(t_0) = y(t_2) = [609.62, 626.64, 625.51]$. On the contrary, the left branch t_1 (Id = 1) shows the output estimation $y(t_1) = [574.60, 571.24, 595.82]$, defined by eq. (4), and corresponding to the (lower valued) inputs—once the split is performed, which constitute the 'support' of node t_1 (see Definition 1). Consequently, in each node, we find an identification number (Id), the mean squared error (R), the sample size, the split executed for input *j* and threshold, $s_j \in S_j$, and, finally, the estimated output. Note that, from a data visualization perspective, it is possible to illustrate the whole production technology, as opposed to FDH and DEA, which are limited to no more than three dimensions.

Following the algorithm, the tree is subsequently grown (or expanded) by way of consecutive $k + 1$ splits under the Pareto-dominance condition given in

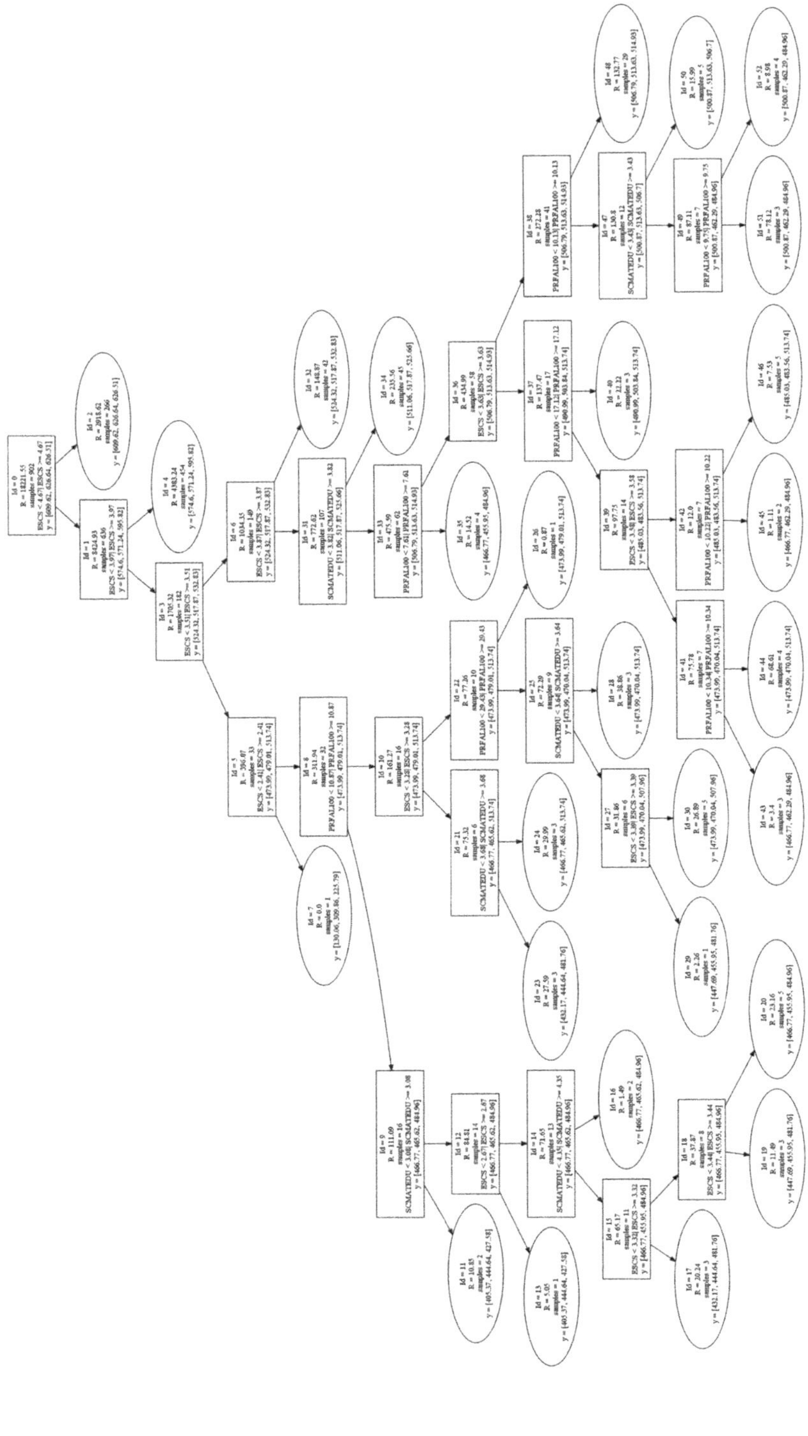

Fig. 4 Pruned EAT for the PISA dataset

Definition 2, and yielding successive nodes $\tilde{T}_{k+1}$. In this empirical application, as many as 52 nodes are calculated (excluding the root node). In Fig. 4, following the shape code presented in Fig. 2, those nodes that are partitioned are represented as rectangles, while end (leaf) nodes are represented as ovals. In the terminal nodes, normally located at the bottom or the sides of the tree, the final estimation of the response variable (the output) is reported. In this way, we can visualize the stepwise frontier by visiting the different branches of the tree from top to bottom. In total there are 25 nodes that are split and 27 leaves. For the case of terminal nodes, the splitting process ends because either the Pareto-dominance criteria cannot be observed, so no more partitions are possible, or the prespecified stopping rule is met ($n(t) \leq n_{\min} = 5$). We remark that the tree represented in Fig. 4 is the result of applying the cross-validation process on the data described in Sect. 2, yielding the 'pruned' (or optimal) version of the tree, T^*. The process prevents the overfitting of the data and allows identifying an accurate version of the production frontier.

4.1.3 Efficiency Results: Output Oriented EAT Vs. FDH and CEAT Vs. DEA

Once we have generated the optimal tree, inducing the production possibility sets for each supporting vector of inputs $\boldsymbol{d}_{T^*}(\boldsymbol{x})$, which jointly characterize the EAT estimator of the technology $\hat{\Psi}_{T^*}$, it is possible to calculate the EAT efficiency scores and compare them to their FDH counterparts. Following common practice in efficiency studies in the educational sector, we adopt an output-oriented approach, by which schools aim at improving their academic results given the available resources. This approach is consistent with the mathematical programs introduced in the first two sections of the chapter.

We start comparing the models for the non-convex estimation of the production technology. Table 2 presents the individual efficiency scores of the ten best and ten worst performing schools, as well as their descriptive statistics for the whole sample at the bottom of the table. Schools are ordered from best performing (smallest scores equal one) to worst performing (largest scores), first according to their EAT scores, and then based on their CEAT values. The EAT output-oriented efficiency scores, $\phi^{EAT}\left(\boldsymbol{x}_k, \boldsymbol{y}_k\right) := \max\left\{\phi_k \in R : \left(\boldsymbol{x}_k, \phi_k \boldsymbol{y}_k\right) \in \hat{\Psi}_{T^*}\right\}$, empirically calculated through program (9), are reported in the second column, while their FDH counterparts, $\phi^{FDH}\left(\boldsymbol{x}_k, \boldsymbol{y}_k\right) := \max\left\{\phi_k \in R : \left(\boldsymbol{x}_k, \phi_k \boldsymbol{y}_k\right) \in \hat{\Psi}_{FDH}\right\}$, calculated though program (2), are presented in the third column. All ten best performing schools are EAT and FDH efficient with unitary efficiency scores, while the scores of the ten worst performing schools illustrate Proposition 2 (iii), by which the FDH technology is enveloped by the EAT technology, and therefore the efficiency scores of the former are either equal or smaller that the latter: $\phi^{FDH}(\boldsymbol{x}_k, \boldsymbol{y}_k) \leq \phi^{EAT}(\boldsymbol{x}_k, \boldsymbol{y}_k)$. This can also be observed in the main statistics. The values of the arithmetic average,

Table 2 Comparing output-oriented efficiency scores: EAT (9) vs. FDH (2), and CEAT (11) vs DEA (3). Selected observations

	Non-Convex		Convex	
Score School	$\phi^{EAT}(\boldsymbol{x}_k, \boldsymbol{y}_k)$ Program (9)	$\phi^{FDH}(\boldsymbol{x}_k, \boldsymbol{y}_k)$ Program (2)	$\phi^{CEAT}(\boldsymbol{x}_k, \boldsymbol{y}_k)$ Program (11)	$\phi^{DEA}(\boldsymbol{x}_k, \boldsymbol{y}_k)$ Program (3)
632	1	1	1	1
635	1	1	1	1
763	1	1	1	1
846	1	1	1.046	1
426	1	1	1.074	1
475	1	1	1.080	1.010
509	1	1	1.087	1.009
625	1	1	1.087	1.012
740	1	1	1.091	1.006
732	1	1	1.116	1.044
250	1.367	1.215	1.376	1.306
548	1.369	1.213	1.369	1.303
840	1.372	1.235	1.413	1.341
306	1.372	1.238	1.508	1.432
877	1.383	1.215	1.523	1.363
510	1.412	1.274	1.412	1.319
854	1.423	1.407	1.589	1.560
480	1.440	1.316	1.440	1.395
620	1.478	1.417	1.682	1.545
343	1.514	1.454	1.550	1.478
Average	1.138	1.064	1.184	1.117
Median	1.133	1.048	1.174	1.106
Max.	1.514	1.454	1.682	1.560
Min.	1.000	1.000	1.000	1.000
Stand. Dev.	0.078	0.068	0.081	0.076

median and maximum of the FDH scores are systematically smaller than their EAT counterparts.

We can compare the rankings and distributions of the efficiency scores *within* the non-convex approach, i.e., EAT vs. FDH. The Spearman rank correlation of both efficiencies, $\rho(\phi^{EAT}, \phi^{FDH}) = 0.725$ ($p = 0.0000$), suggests that their ordering is quite similar. However, looking at the box-plots and kernel density distributions presented on the left-hand side of Fig. 5, it is visually apparent that both distributions are different. The difference in the distributions is also confirmed by the Li test. The obtained statistic, 142.412, allows us to confidently reject the hypothesis of equality of the two density distributions with a p-value of 0.0000.

We now turn our attention to the results obtained for the convexified EAT presented in eq. (10), whose associated CEAT output-oriented efficiency scores $\phi^{CEAT}\left(\boldsymbol{x}_k, \boldsymbol{y}_k\right) := \max\left\{\phi_k \in R : \left(\boldsymbol{x}_k, \phi_k \boldsymbol{y}_k\right) \in conv\left(\hat{\Psi}_{T^*}\right)\right\}$ are calculated

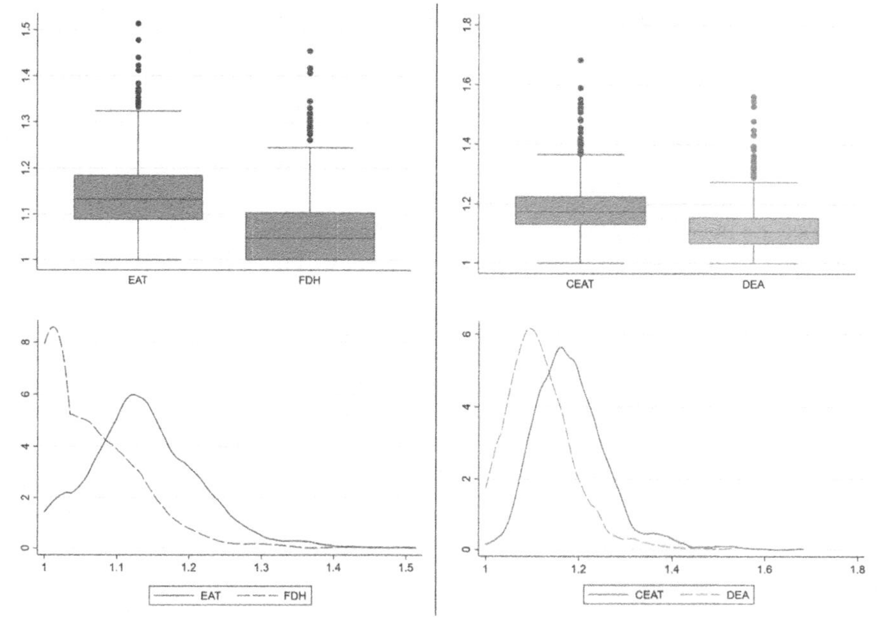

Fig. 5 Box-plots and kernel distributions: EAT (9) vs. FDH (2), and CEAT (11) vs DEA (3)

through program (11), and compare them to their standard DEA counterparts $\phi^{DEA}\left(\boldsymbol{x}_k, \boldsymbol{y}_k\right) := \max\left\{\phi_k \in R : \left(\boldsymbol{x}_k, \phi_k \boldsymbol{y}_k\right) \in \hat{\Psi}_{DEA}\right\}$, solved through program (3). The fourth and fifth rows of Table 2 present the efficiency scores for both series. *Within* convex models, i.e., comparing CEAT vs. DEA, we find that according to Proposition 2 (iii), DEA efficiency scores are either equal or smaller than CEAT scores: $\phi^{DEA}(\boldsymbol{x}_k, \boldsymbol{y}_k) \leq \phi^{CEAT}(\boldsymbol{x}_k, \boldsymbol{y}_k)$. We also check their compatibility in terms of their rank correlation, which is higher than in the non-convex case, with $\rho(\phi^{CEAT}, \phi^{DEA}) = 0.910$ ($p = 0.0000$). But, once again, a visual inspection of the box-plots and kernel distributions on the right hand side of Fig. 5 reveal that both distributions are substantially different. Accordingly, the Li test, valued at 79.135 ($p = 0.0000$), returns that we can reject the hypothesis of equality of densities.

By comparing the distributions of the efficiency scores *between* convex and non-convex models, we highlight that the number of efficient observations under CEAT and DEA is substantially smaller than those corresponding to the non-convex models, EAT and FDH; i.e., convex models allow for a better discrimination among observations. Indeed, while just 4 schools are efficient under CEAT (0.4% of the sample), 26 (2.9%) are efficient under EAT. Looking at the traditional models, convex DEA identifies 34 (3.8%) schools as efficient, while as many as one quarter of the schools, 25.4% (229), are deemed efficient under FDH, showing the futility of using this method to rank efficient benchmarks in a meaningful way. These results show the advantage of using the new methods, particularly the convex efficiency analysis tree, CEAT, to identify production peers and ranking observations. This approach constitutes a real alternative to other competing proposals that rank observations under variable returns to scale, such as the 'Economic Cross-Efficiency' methodology recently introduced by Aparicio and Zofío (2020a, b). In passing, we also remark that the number of efficient observations under non-convex EAT (26) is even smaller than for the convex DEA (34).

Hence, on the one hand we conclude that, within models, the efficiency scores of either the EAT model or its convex version, CEAT, are higher than their corresponding FDH and DEA counterparts; i.e., $\phi^{EAT}(\boldsymbol{x}_k, \boldsymbol{y}_k) \geq \phi^{FDH}(\boldsymbol{x}_k, \boldsymbol{y}_k)$, and $\phi^{CEAT}(\boldsymbol{x}_k, \boldsymbol{y}_k) \geq \phi^{DEA}(\boldsymbol{x}_k, \boldsymbol{y}_k)$. These inequalities extend to their density distributions, which are statistically different, as the Li tests show. On the other hand, between models, results show that the convex EAT yields higher efficiency scores than its non-convex version: $\phi^{CEAT}(\boldsymbol{x}_k, \boldsymbol{y}_k) \geq \phi^{EAT}(\boldsymbol{x}_k, \boldsymbol{y}_k)$, thereby mirroring the well-known result for the DEA and FDH models: $\phi^{DEA}(\boldsymbol{x}_k, \boldsymbol{y}_k) \geq \phi^{FDH}(\boldsymbol{x}_k, \boldsymbol{y}_k)$. These latter results are expected, given that non-convex approximations of the technology envelope the data more tightly that their convex counterparts, resulting in higher efficiency levels (or lower efficiency scores closer to one).

4.2 *Taiwanese Banks*

4.2.1 Data

Our second empirical application uses data of a balanced panel of 31 Taiwanese banks over the period 2006–2010, previously studied by Juo et al. (2015).[4] Following Sealey and Lindley (1977), the production technology corresponds to the so-called intermediation approach by which financial institutions, through labor and capital, collect deposits from savers to produce loans and other earning assets for borrowers. The production technology is represented by three inputs and two outputs. Inputs are FINANCIAL FUNDS (x_1), LABOR (x_2), and physical CAPITAL (x_3). The output vector includes financial INVESTMENTS (y_1) and LOANS (y_2). Table 3 presents descriptive statistics for quantities in 2006 and 2010. All monetary variables are measured in million TWD, while labor is the number of employees. A complete discussion of the statistical sources and variable specifications can be found in Juo et al. (2015).

[4]We are grateful to Juo et al. for sharing the data. We emphasize that we are using them for illustrating the EAT and CEAT methods, and not for revealing any hitherto unknown feature of the Taiwanese banking sector or individual banks. The same dataset has been used to illustrate the decompositions of total factor productivity change using quantities-only and price-based indices by Balk and Zofío (2018), and decomposing the variation in cost by Balk and Zofío (2020).

Table 3 Descriptive statistics for 31 Taiwanese Banks, 2006–2010

	2006					2010				
	Average	Median	Max.	Min.	Stand. Dev.	Average	Median	Max.	Min.	Stand. Dev.
x_1	628,856	304,722	2,133,665	26,162	578,979	795,536	428,995	3,171,493	25,019	768,008
x_2	3781	3388	8463	203	2463	3826	3146	9538	202	2729
x_3	14,623	9686	74,448	494	15,971	13,393	8721	76,576	505	15,185
y_1	111,098	83,272	352,976	2354	108,651	196,808	157,870	904,580	1681	215,063
y_2	506,372	244,739	1,734,526	49,780	468,063	609,489	328,574	2,091,100	66,947	582,854

Source: Own elaboration from Juo et al. (2015)

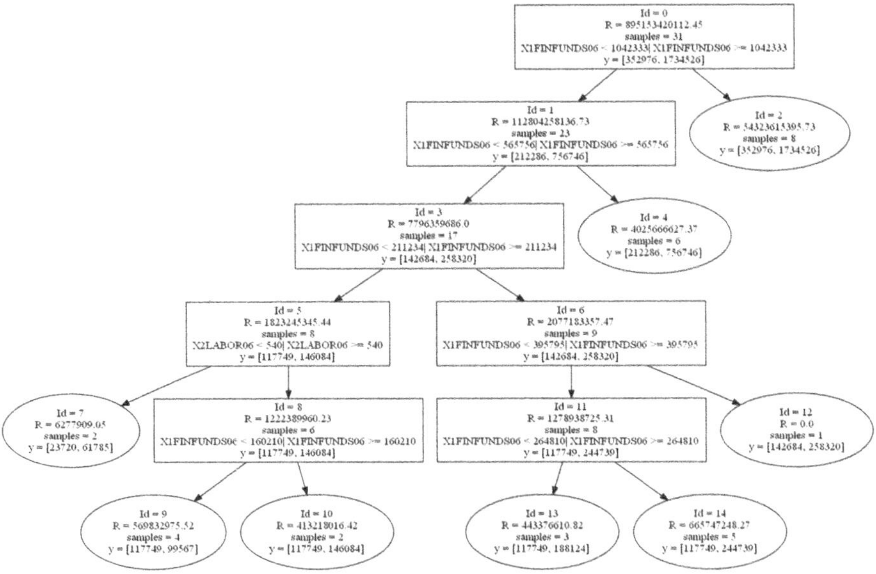

Fig. 6 Pruned EAT for the Taiwanese banks data, 2006

4.2.2 Tree Growth

We follow once again the procedure described in Sect. 2 to generate the tree that allows the calculation of the EAT and CEAT efficiency scores. Figures 6, 7, and 8 present the trees corresponding to the first year of the sample (2006), the middle year (2008) and the last year (2010).[5] On inspecting the three trees, we observe that they share the same structure in terms of the number of splits, 7, until the final leaves are generated. The differences between the trees are small with the number of nodes totaling 14 in 2006, and 12 in 2008 and 2010. Therefore, as compared to the PISA tree in Fig. 4, the trees are less populated. This is the consequence of the smaller number of observations that we have in this empirical application (31), in comparison to the PISA study (902). Consequently, the splitting process is quickly exhausted; in this case because the prespecified minimum number of observations, $n_{\min} = 5$, is reached.

Describing the trees, and regardless of the year, we observe that the first split from the root node is always driven by the first input (FINFUNDS, x_1), minimizing the sum of the squared errors for the left and right branches. As reported at the top of

[5]The trees for 2007 and 2009 are omitted because their structure does not differ substantially from those reported. Nevertheless, they are available upon request, and can be generated by the reader as described in Sect. 5 using the *EATpy* package that allows to replicate the results of this empirical section.

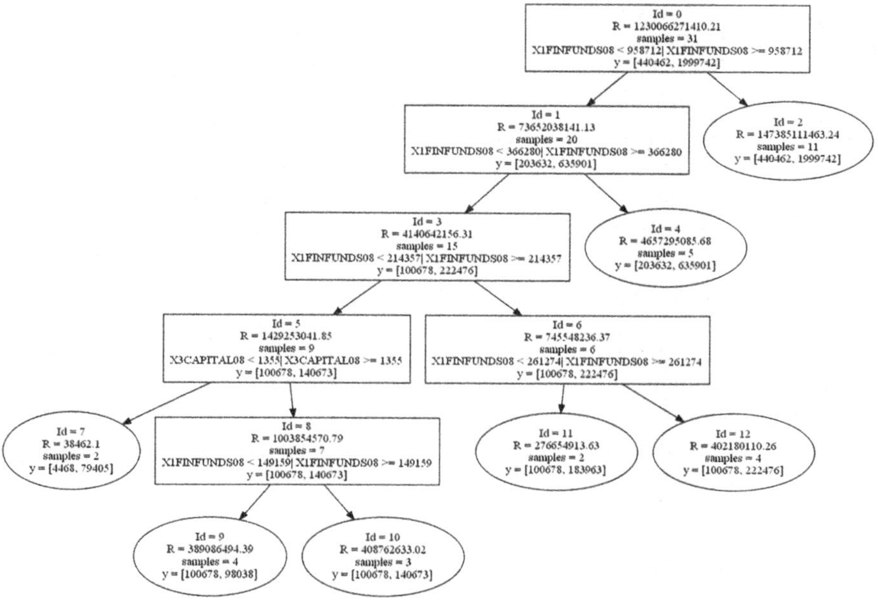

Fig. 7 Pruned EAT for the Taiwanese banks data, 2008

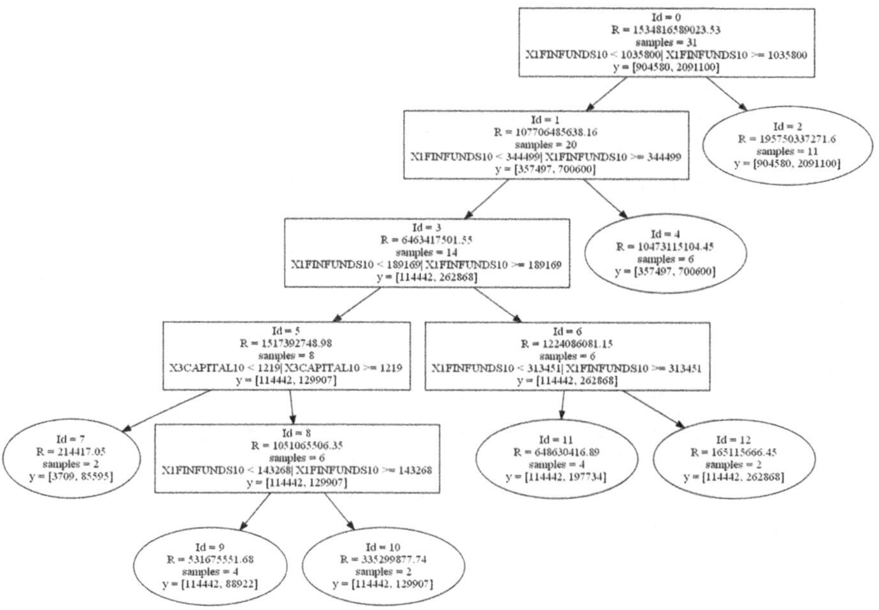

Fig. 8 Pruned EAT for the Taiwanese banks data, 2010

the tree, the thresholds $x_1 < s_1$ for the left branches are FINFUNDS06 < 1,042,333 in 2006, FINFUNDS08 < 958,712 in 2008 and FINFUNDS10 < 1,035,800 in 2010. After this first split, the second one is also driven by the first input in all three years, showing its relevance in the process of growing the EAT tree. It is at the fifth node (Id = 5) where we find the first difference in the trees. While in 2006, LABOR (x_2) is the variable splitting the node for LABOR06 < 540, in 2008 and 2010 it is CAPITAL (x_3), with the following values: CAPITAL08 < 1355 and CAPITAL10 < 1219. In all this process we observe once again that, in accordance with Proposition 1, the right nodes always inherit the maximum observed values of the output vector of the parent node. For example, in 2006, node No. 4, inherits the output values of its parent node No. 1; i.e., $y(t_4) = y(t_1) = [212{,}286; 756{,}746]$. As before, we identify each node numerically, report the mean squared error (R), the sample size, the split executed for the input j with its corresponding threshold $s_j \in S_j$—which, along with the remaining associated input values, constitutes the 'support' of the node (see Definition 1), and, finally, the estimated output. As in the previous application, the final trees constitute the cross-validated or 'pruned' version of the technology, T^*.

4.2.3 Efficiency Results: Directional EAT Vs. FDH and CEAT Vs. DEA

Once the optimal trees T^* have been generated, it is possible to characterize both the non-convex EAT technology $\hat{\Psi}_{T^*}$, as defined in expression (8), and the convex CEAT technology, $conv\left(\hat{\Psi}_{T^*}\right)$, defined in (10). Table 4 reports the descriptive statistics for the inefficiency scores calculated using the directional distance function (DDF) model discussed in Sect. 3.[6] As in previous studies, including Juo et al. (2015), we choose the directional vector whose components are the observed input and output quantities: $\boldsymbol{g} = \left(\boldsymbol{g}^-, \boldsymbol{g}^+\right) = \left(\boldsymbol{x}_k, \boldsymbol{y}_k\right) \in R_+^{m+s}$. This allows the interpretation of the efficiency score as the proportional increase in the observed amount of outputs and the equivalent proportional decrease in the observed amount of inputs, necessary to reach the production frontier. For the non-convex case, we calculate $\beta^{EAT}(\boldsymbol{x}_k, \boldsymbol{y}_k) := \max\left\{\beta_k \in R : \left(\boldsymbol{x}_k - \beta_k \boldsymbol{x}_k, \boldsymbol{y}_k + \beta_k \boldsymbol{y}_k\right) \in \hat{\Psi}_{T^*}\right\}$ and $\beta^{FDH}(\boldsymbol{x}_k, \boldsymbol{y}_k) := \max\left\{\beta_k \in R : \left(\boldsymbol{x}_k - \beta_k \boldsymbol{x}_k, \boldsymbol{y}_k + \beta_k \boldsymbol{y}_k\right) \in \hat{\Psi}_{FDH}\right\}$, solving programs (14) and (13), respectively. The values for both series are shown in the third and fourth columns. What immediately catches our eye is that almost all banks are efficient under the FDH model, with null values for the median and even the arithmetic average. As many as 29 banks are efficient in 2006, with $\beta^{FDH}(\boldsymbol{x}_k, \boldsymbol{y}_k) = 0$, and 30 in 2008 and 2010 (all but one). On the contrary, under the EAT model, the number of efficient banks, $\beta^{EAT}(\boldsymbol{x}_k, \boldsymbol{y}_k) = 0$, are 16 in 2006 and 2008 and 15 in 2010 (about half of the sample).

[6] Results for 2007 and 2009 are available upon request and, once again, can be computed by the interested reader using the *EATpy* package in Python as described in Sect. 5.

Table 4 Comparing directional inefficiency scores: EAT (14) vs. FDH (13), and CEAT (16) vs. DEA(15). Descriptive statistics

		Non-Convex		Convex	
Year	Score Statistic	$\beta^{EAT}(\boldsymbol{x}_k, \boldsymbol{y}_k)$ Program (14)	$\beta^{FDH}(\boldsymbol{x}_k, \boldsymbol{y}_k)$ Program (13)	$\beta^{CEAT}(\boldsymbol{x}_k, \boldsymbol{y}_k)$ Program (16)	$\beta^{DEA}(\boldsymbol{x}_k, \boldsymbol{y}_k)$ Program (15)
	Average	0.065	0.004	0.312	0.111
	Median	0.000	0.000	0.372	0.122
2006	Max.	0.290	0.093	0.552	0.298
	Min.	0.000	0.000	0.000	0.000
	Stand. Dev.	0.093	0.018	0.168	0.096
	Average	0.087	0.000	0.387	0.117
	Median	0.000	0.000	0.476	0.094
2008	Max.	0.322	0.007	0.584	0.364
	Min.	0.000	0.000	0.000	0.000
	Stand. Dev.	0.117	0.001	0.185	0.125
	Average	0.123	0.005	0.406	0.111
	Median	0.034	0.000	0.465	0.056
2010	Max.	0.403	0.168	0.649	0.520
	Min.	0.000	0.000	0.000	0.000
	Stand. Dev.	0.140	0.030	0.165	0.137

Table 4 also presents in the last two columns the inefficiency scores of the convex models $\beta^{CEAT}(\boldsymbol{x}_k, \boldsymbol{y}_k) := \max\left\{\beta_k \in R : \left(\boldsymbol{x}_k - \beta_k \boldsymbol{x}_k, \boldsymbol{y}_k + \beta_k \boldsymbol{y}_k\right) \in conv\left(\hat{\Psi}_{T^*}\right)\right\}$ and $\beta^{DEA}(\boldsymbol{x}_k, \boldsymbol{y}_k) := \max\left\{\beta_k \in R : \left(\boldsymbol{x}_k - \beta_k \boldsymbol{x}_k, \boldsymbol{y}_k + \beta_k \boldsymbol{y}_k\right) \in \hat{\Psi}_{DEA}\right\}$, corresponding to programs (16) and (15), respectively. Now, the number of efficient banks decreases substantially and therefore the descriptive statistics of the inefficiency scores increase. Indeed, the number of efficient banks under CEAT and DEA drops to 3 and 9 in 2006, 3 and 11 in 2008, and 2 and 11 in 2010. Therefore, we conclude once again the high discriminatory power of the CEAT model over its non-convex EAT counterpart and the traditional models. Also, in concordance with Proposition 2 (iii), all EAT and CEAT individual inefficiencies are larger than their FDH and DEA counterparts; i.e., $\beta^{EAT}(\boldsymbol{x}_k, \boldsymbol{y}_k) \geq \beta^{FDH}(\boldsymbol{x}_k, \boldsymbol{y}_k)$ and $\beta^{CEAT}(\boldsymbol{x}_k, \boldsymbol{y}_k) \geq \beta^{DEA}(\boldsymbol{x}_k, \boldsymbol{y}_k)$. Consequently, these inequalities are observed for the descriptive statistics.

We now proceed to compare the rankings and distributions of the directional inefficiency scores *within* the non-convex models, i.e., EAT vs. FDH. Table 5 reports the Spearman rank correlations and Li tests, comparing both pairs of distributions. The low and not significant values of the Spearman rank correlation, $\rho(\beta^{EAT}, \beta^{FDH})$, are unsurprising given the large number of zeros in the FDH distributions, which impedes a meaningful calculation of the rank correlations. Figure 9 presents the box-plots distributions of the inefficiency scores corresponding to the non-convex technologies EAT and FDH for the selected years. Below each box-plot, we also include the kernel distributions of the EAT scores. Those of the FDH scores are

Table 5 Comparing directional inefficiency distributions: Spearman correlations and Li tests

	Non-Convex		Convex	
Statistic Year	Spearman rank$\rho(\beta^{EAT}, \beta^{FDH})$	Li Test	Spearman rank$\rho(\beta^{CEAT}, \beta^{DEA})$	Li Test
2006	0.225 (0.2232)	1.066***(0.0049)	0.8789*** (0.000)	8.628***(0.000)
2008	0.330* (0.0702)	1.652***(0.0030)	0.9161*** (0.0000)	11.170***(0.000)
2010	0.1951 (0.2930)	2.360***(0.0015)	0.8647*** (0.000)	13.077***(0.000)

EAT (14) vs. FDH (13), and CEAT (16) vs. DEA (15)
Notes: p-values in parentheses
Spearman rank correlations: *Denotes statistically significant rank correlation at the 10% level
Li Tests: *** Denotes statistically significant differences between models at the 1% level

omitted because being almost all values equal to zero, their inclusion greatly distorts the scale in the graphs, making it impossible to visually compare the remaining distributions. From the box-plots, it is immediately clear that the EAT and FDH distributions are different for all years, a result that is confirmed by the Li tests reported in Table 5. Their values and significance confirm that the density functions are different.

As opposed to the non-convex case, it is possible to perform a meaningful comparison of the efficiency scores for the convex technologies, CEAT vs. DEA, in terms of their rankings and distributions. Table 5 reports the Spearman rank correlations for the selected years: $\rho(\beta^{CEAT}, \beta^{DEA})$. All coefficients exhibit large and significant values, implying the compatibility of the rankings in both series for all 3 years. Despite the similarity in the rankings, the box-plot representations of the distributions shown in Fig. 9 clearly show that they are different from each other. A result that is further confirmed by the Li tests reported in Table 5, which reject the null hypothesis of the equality of the distributions of the convex models at the 1% level of confidence in all 3 years.

Overall, looking at the descriptive statistics, the box-plots and associated density distributions, it is apparent that the productive efficiency of the Taiwanese banking industry worsens over the 5 years. The average value of the inefficiency scores for the EAT model, i.e., $\beta^{EAT}(\boldsymbol{x}_k, \boldsymbol{y}_k)$, increases from 0.065 in 2006 to 0.123 in 2010, while that for the convex version, i.e., $\beta^{CEAT}(\boldsymbol{x}_k, \boldsymbol{y}_k)$, increases from 0.312 in 2006 to 0.406 in 2010. As expected, this worsening trend cannot be captured by the traditional FDH counterpart, while its convex DEA counterpart does not reflect it as clearly as the new methods based on machine learning.

Therefore, as in the previous empirical application using PISA data, we conclude that, within models, the inefficiency scores obtained under the EAT model or its new convex version, CEAT, are greater than their corresponding FDH and DEA counterparts; i.e., $\beta^{EAT}(\boldsymbol{x}_k, \boldsymbol{y}_k) \geq \beta^{FDH}(\boldsymbol{x}_k, \boldsymbol{y}_k)$, and $\beta^{CEAT}(\boldsymbol{x}_k, \boldsymbol{y}_k) \geq \beta^{DEA}(\boldsymbol{x}_k, \boldsymbol{y}_k)$. Although the ranking of the Taiwanese banks is compatible when comparing the convex models, CEAT vs. DEA (EAT cannot be compared to FDH because this latter

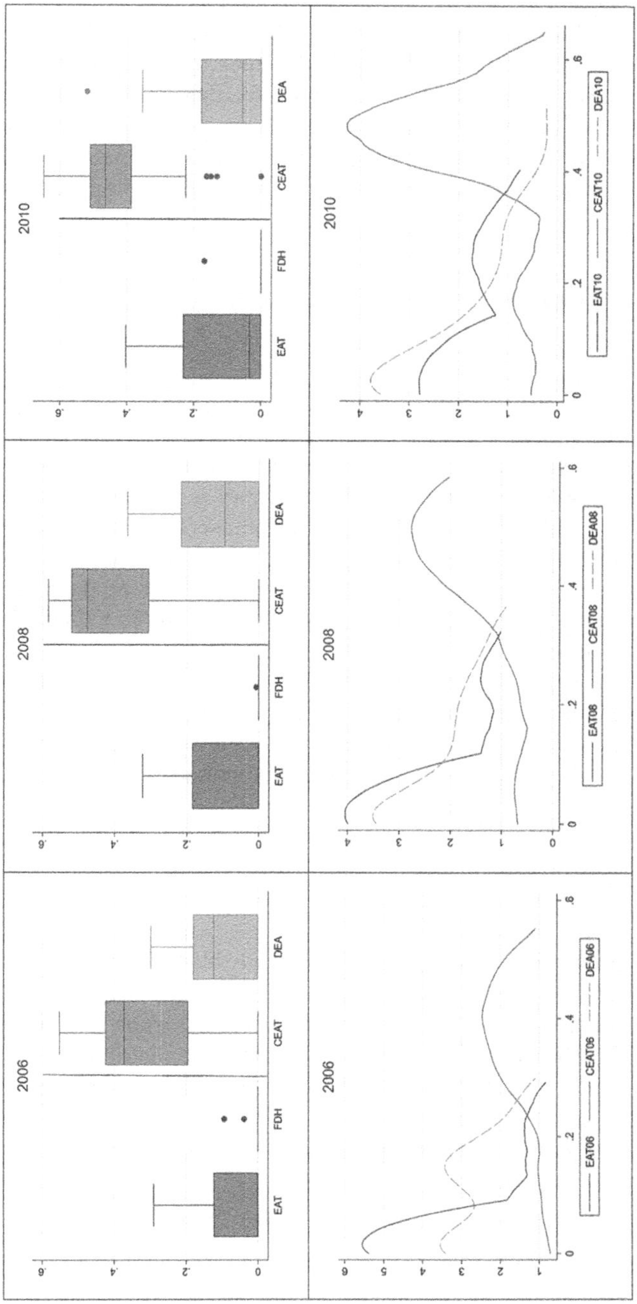

Fig. 9 Box-plots and kernel distributions: EAT (14) vs. FDH (13) and CEAT (16) vs. DEA (15). 2006, 2008 and 2010

model returns almost efficient banks only), their density functions are statistically different, as visually shown by the yearly box-plots and the Li tests. Also, comparing results between the non-convex and the convex models, the former yield inefficiency scores that are systematically higher than the latter: i.e., $\beta^{CEAT}(\boldsymbol{x}_k, \boldsymbol{y}_k) \geq \beta^{EAT}(\boldsymbol{x}_k, \boldsymbol{y}_k)$, just as reported in the literature for their traditional counterparts: $\beta^{DEA}(\boldsymbol{x}_k, \boldsymbol{y}_k) \geq \beta^{FDH}(\boldsymbol{x}_k, \boldsymbol{y}_k)$. Again, these results are expected, given that convex approximations of the technology envelope their non-convex counterparts, resulting in greater inefficiency scores, further away from zero (i.e., lower efficiency levels).

5 The Algorithms EAT and Convex (EAT) in Python

The package implemented in Python, named *EATpy*, is based on the methodology proposed by Breiman et al. (1984) and its adaptation to Efficiency Analysis Trees (Esteve et al., 2020). The different features that it presents are inspired by the decision trees of the Machine Learning library called *scikit-learn* (see https://scikit-learn.org/stable/modules/tree.html). Since this library is widely known among the Python community and our proposal is very similar, users will find it more intuitive in practice. Additionally, other libraries used and necessary for the development of the package are: *pandas*, for the storage and handling of data due to their potential; *numpy*, for vector calculus; *graphviz*, to draw the tree; and *docplex*, for the calculation of the efficiency models.

The *EATpy* package, in turn, contains two libraries. The first one is the implementation of the EAT tree as described in the second section, while the second contains the implementation of the calculations of the efficiency models. Specifically, it includes the BCC output oriented model and the directional distance function model (DDF).

The different steps to follow regarding the installation and execution, as well as its main features are described below.

(a) *Installation*

The *EATpy* package is available on the GitHub platform on the link: https://github.com/MiriamEsteve/EATpy.git

For easy installation on a personal computer is recommended to install *git* (see: https://git-scm.com/downloads) as well as the Anaconda distribution (see: https://www.anaconda.com/products/individual). The steps to be followed are based on these two installations.

Step 1. Open the **Anaconda Prompt** console, go to the desired directory for the installation and enter the command:

```
git clone https://github.com/MiriamEsteve/EATpy.git
```

Step 2. Go to the *EATpy* folder created, using *cd EATpy*, and execute the command:

```
python setup.py install
```

x1	x2	x3	y1	y2
1282	297	14189	65579	28135
2442	564	23923	162310	59819
952	237	12584	67959	65906
1518	457	21927	114289	55152
709	140	9021	43450	27200
994	266	13414	65072	24498
842	191	10633	56507	39123
1353	282	20195	101821	121539
1103	360	17618	83613	32021
422	98	4708	24789	27502
1008	202	10930	51612	31836
507	143	7105	32348	36154
2114	519	24876	133381	50275
3240	812	33058	148291	73983
1981	570	20515	85917	47648

```
x1;x2;x3;y1;y2
1282;297;14189;65579;28135
2442;564;23923;162310;59819
952;237;12584;67959;65906
1518;457;21927;114289;55152
709;140;9021;43450;27200
994;266;13414;65072;24498
842;191;10633;56507;39123
1353;282;20195;101821;121539
1103;360;17618;83613;32021
422;98;4708;24789;27502
1008;202;10930;51612;31836
507;143;7105;32348;36154
2114;519;24876;133381;50275
3240;812;33058;148291;73983
1981;570;20515;85917;47648
```

Fig. 10 Example of a Data Set in Excel and CSV

Following these two steps, we have the *EATpy* package installed on our computer. The examples below are also available in the *EATpy/test* folder that was created during installation.

(b) *Data Set*

The input data set required by all the functionalities (that will be explained later in this section) must meet the condition of being a *DataFrame* data type from the *pandas* library. To do this, the user can start from an Excel document containing as many columns as inputs and as many columns as outputs. They all must include a header with their names, because they will need to be entered in the EATpy package. The user can also start from a CSV file in which the data headers will need to be specified. Figure 10 shows an example of this for each type of file.

Once the chosen file type complies with the indicated structure, it must be converted to a *DataFrame* type from the *pandas* library. The commands are:

– Import the *pandas* library

```
import pandas as pd
```

– Reading and converting Excel document

```
dataset = pd.read_excel("nombre_del_fichero.xlsx", sheet_name
        = "data")
```

– Reading and converting a CSV file

```
dataset = pd.read_csv("nombre_del_fichero.csv", sep = ";")
```

Once these steps are completed, you can start using the *EATpy* library.

(c) *Options*

(i) Importing libraries

All the necessary libraries that we must include in our code are:

```
import eat
import graphviz
```

(ii) Generating simulated data sets

The *EATpy* package includes a simulated data generator module. It is used to exemplify the use of the package. To run it, the user must set the generator seed and dataset size. In the example shown below, the value 1 is set as a seed and 50 DMUs are generated with two inputs and two outputs. As required, the generated data set is a *DataFrame* data type from the *pandas* library:

```
dataset = eat.Data(1, 50).data
```

(iii) Creating the EAT model

Building the EAT model consists of specifying the name of the input and output columns, the stopping rule, and the folder number for the cross-validation process. Once this is done, the model is created and fit to be built.

In this example, the column names of the inputs and outputs of the dataset are listed first. If these do not exist in the dataset, the EAT model returns an error:

```
x = ["x1", "x2"]
y = ["y1", "y2"]
```

Second, the stopping rule and the number of folds (usually 5 or 10) in the cross-validation process are specified:

```
numStop = 5
fold = 5
```

And finally, the creation and adjusting of the EAT model is done. Its input parameters are the dataset, the column names of the inputs, the column names of the outputs, the stopping rule and the number of folds:

```
model = eat.EAT(dataset, x, y, numStop, fold)
model.fit()
```

(iv) Drawing the EAT tree

The drawing of the EAT tree is done using the external *graphviz* library. To do this, the first command generates the *dot_data* file that *graphviz* needs to draw the tree. It also saves it as an image in the working directory:

```
dot_data = model.export_graphviz('EAT')
graph = graphviz.Source(dot_data, filename="tree", format="png")
graph.view()
```

(v) Predictions

The prediction module of the EAT model can be applied on a single data set or on a single record in the data set. To do this, you need the data set or single record from which you want to predict the output variables and the names of the input columns. These names will generally be the same as the initial data set.

In this example, the first 10 records in the initial data set are selected and the name of the inputs is the same as this. Then the EAT model performs the prediction (estimation in this case) and returns the data set with the predicted values. To identify the columns within the data set that belong to the prediction, the prefix "*p_*" is added to their name. For example, the output returned by the code shown below will return a DataFrame with the two inputs ("x1" and "x2") and their two predictions ("p_y1", "p_y2"):

```
x_p = ["x1", "x2"]
data_pred = dataset.loc[:10, x_p]
data_prediction = model.predict(data_pred, x_p)
```

(vi) Calculating efficiency measures

The current *EATpy* version implements two efficiency measures discussed in this study: the output-oriented radial measure and the 'graph' directional distance function with the directional vector being equal the input and output values of the unit under evaluation. These two measures are calculated under EAT, CEAT, and their standard FDH and DEA counterparts. To calculate these measures, the object eat.Scores is built, taking as parameters: the data set, the names of the input columns, the names of the output columns and the EAT tree that has been built in a previous step.

```
mdl_scores = eat.Scores(dataset, x, y, model.tree)
```

Then, the efficiency values associated with the different measures under EAT and CEAT are calculated by executing the following lines:

```
#BCC output oriented under EAT

mdl_scores.BCC_output_EAT()
#BCC output oriented under CEAT

mdl_scores.BCC_output_CEAT()
#Directional Distance Function under EAT

mdl_scores.DDF_EAT()
#Directional Distance Function under CEAT

mdl_scores.DDF_CEAT()
```

The same measures but calculated under FDH are determined through the following code:

```
#BCC output oriented under FDH

mdl_scores.BCC_output_FDH()
#Directional Distance Function under FDH

mdl_scores.DDF_FDH()
```

Finally, the output-oriented radial model and the directional distance function under DEA are calculated by way of:

```
#BCC output oriented under DEA

mdl_scores.BCC_output_DEA()
#Directional Distance Function under DEA

mdl_scores.DDF_DEA()
```

(d) Analysis of the PISA data set

A practical example of the complete operation of the *EATpy* package is detailed below. It corresponds to the PISA results discussed in Sect. 4.1, located in the *EATpy/data set* folder. It can also be downloaded from https://github.com/MiriamEsteve/EATpy/tree/main/data%20set.

```
import pandas

import eat

import graphviz
#Read the data set

dataset = pd.read_excel("PISA.xlsx",
sheet_name = "data")
#Set the names of the input and output columns
x = ["SCMATEDU", "ESCS", "PRFAL100"]

y = ["PVMATH", "PVREAD, "PVSCIE"]
#Set the stopping rule and number of folds for cross-validation

numStop = 5

fold = 5
#Create the EAT model

model = eat.EAT(dataset, x, y, numStop, fold)
model.fit()
#Draw the EAT tree

dot_data = model.export_graphviz('EAT')

graph = graphviz.Source(dot_data, filename="tree", format="png")

graph.view()
#Create object eat.Scores

mdl_scores = eat.Scores(dataset, x, y, model.tree)
#Output-oriented radial scores under FDH

mdl_scores. BCC_output_FDH()
#Output-oriented radial scores under EAT
```

```
mdl_scores.BCC_output_EAT()
#Output-oriented radial scores under DEA

mdl_scores. BCC_output_DEA()
#Output-oriented radial scores under CEAT

mdl_scores.BCC_output_CEAT()
```

6 Conclusions and Future Works

As the general Operations Research community is moving into the data analytics field, Free Disposal Hull (FDH) and Data Envelopment Analysis (DEA) should explore the possibilities that these methods offer, thereby bridging the gap with the field of "Machine Learning". Accordingly, in this chapter, we have revised a very recent approach for estimating production frontiers based on decision trees, the so-called Efficiency Analysis Trees (EAT) by Esteve et al. (2020), and we have extended that methodology for measuring technical efficiency under convex technologies (CEAT) and through the flexible directional distance function. We have also shown how the algorithm associated with EAT is implemented through the *EATpy* package implemented in Python, whose installation, use and interpretation are extensively discussed, thereby making it available to the scientific community. Indeed, we believe that the new package could be useful in practice to analyze both small and big data sets.

We have illustrated how the different methods work by resorting to two real datasets. The first one consists of PISA data observed in 2012 and including 902 Spanish schools that use 3 inputs to achieve 3 academic outputs. The second one studies a dataset of 31 Taiwanese banks observed from 2006 to 2010 and including 3 inputs and 2 outputs. Making use of the dedicated *EATpy* package, we have grown the efficiency analysis trees, upon which we characterize the non-convex EAT and convex CEAT technologies. Here, we follow the algorithm presented in the second section, and describe the structure of the trees, and the splitting process from the root node and subsequent branches, to the final leaves. Relying on these approximations of the production technology, we have calculated output-oriented efficiency scores for the PISA study, and directional inefficiency scores for the Taiwanese banks. The exploratory data analyses on both sets of results confirm the validity of the new approach when estimating productive efficiency, while overcoming the overfitting problem found in standard FDH and DEA methods, and reducing the mean standard error and bias associated with the estimation of the true frontier (Esteve et al., 2020). Moreover, EAT and CEAT methods provide higher discriminatory capacity than their traditional FDH and DEA counterparts, particularly for relatively small samples. Our results show that, between models, EAT and CEAT yield larger inefficiency levels than those obtained under FDH and DEA. This is because the former characterizations of the technology are outer approximations of the latter—i.e., Proposition 2 (iii). As a consequence, although the pairwise EAT-FDH and CEAT-DEA rankings are positively and significantly correlated, their density distributions are statistically different. This result can be easily seen in the bilateral comparison of their box-plots and is confirmed by the Li tests. Within models, the convex version of the efficiency analysis tree, CEAT, as well as the standard DEA method, return larger inefficiencies than their non-convex EAT and FDH counterparts, precisely because, once again, the former envelop the latter.

Finally, we highlight three areas that could be good avenues for further follow-up research. First, techniques based on machine learning are not usually able to explain why they produce a certain estimation of the response variable from a predictor profile. This point contrasts to more traditional statistical models, where a coefficient for each predictor is determined, even in the way of a confidence interval. Instead, machine learning techniques, trying to yield something similar, can determine a ranking of importance of predictor variables. In the context of the Efficiency Analysis Trees, it would imply the identification of the most important inputs for estimating the output variables. Second, another interesting option could be defining how to benchmark units from the information provided by EAT. It is clear how to select input and output targets by applying some specific distance to the production frontier from the evaluated unit. It is related to the input and output coordinates of the projection vector derived from the efficiency measure (e.g., the directional distance function). However, it is more difficult to define the set of peers linked to this projection, something that is rather simple in the case of standard FDH and DEA. In practice, this information is crucial for making inefficient units learn how to become technically efficient. Inefficient units have the opportunity of learning from the managers of actual units that operate in the same sector. However, this is not so clear in the case of the EAT technique, since, for example under convexification, the projection point could not be a convex combination of actual observations (real units). A third research line to be followed could be the estimation of economic efficiency and its drivers (e.g., technical efficiency and allocative or price efficiency). To do that, it is necessary to have information on market prices for inputs and/or outputs. In this respect, the usual optimization programs for determining minimum cost, maximum revenue or maximum profit should be adapted for dealing with the estimation of the production possibility set through EAT or CEAT.

Acknowledgments The authors thank the financial support from the Spanish Ministry of Science and Innovation and the State Research Agency under grant PID2019-105952GB-I00/ AEI / 10.13039/501100011033. Additionally, J.L. Zofío thanks the financial support from the Spanish Ministry of Science and Innovation and the State Research Agency under grant EIN2020-112260/AEI/10.13039/501100011033. This work was also supported by the Spanish Ministry of Science, Innovation and Universities under Grant FPU17/05365. Authors thank Dr. Lidia Ortiz for assisting in the calculation of the Li tests.

References

Aparicio, J., Cordero, J. M., Gonzalez, M., & Lopez-Espin, J. J. (2018). Using non-radial DEA to assess school efficiency in a cross-country perspective: An empirical analysis of OECD countries. *Omega, 79*, 9–20.

Aparicio, J., & Zofío, J. L. (2020a). Economic cross-inefficiency. *Omega, 102374*.

Aparicio, J. & Zofío, J.L. (2020b). New definitions of economic cross-efficiency, in Aparicio, J., Knox Lovell, C.A., Pastor, J.T. & Zhu, J. (edts.) Advances in efficiency and productivity II, Springer Nature's International Series in Operations Research & Management Science: 11-32.

Balk, B. M., & Zofío, J. L. (2018). *The many decompositions of total factor productivity change*. ERIM report series research in management, no. ERS-2018-003-LIS. Erasmus Research Institute of Management, Erasmus University. Available from http://hdl.handle.net/1765/104721

Balk, B. M., & Zofío, J. L. (2020). Symmetric decompositions of cost variation. *European Journal of Operational Research, 185*, 1189–1198.

Banker, R. D., Charnes, A., & Cooper, W. W. (1984). Some models for estimating technical and scale inefficiencies in data envelopment analysis. *Management Science, 30*(9), 1078–1092.

Barbosa, F., Rampazzo, P. C. B., Yamakami, A., & Camanho, A. S. (2019). The use of frontier techniques to identify efficient solutions for the berth allocation problem solved with a hybrid evolutionary algorithm. *Computers & Operations Research, 107*, 43–60.

Berk, R. A. (2008). *Statistical learning from a regression perspective* (Vol. 14). Springer.

Breiman, L., Friedman, J., Stone, C. J., & Olshen, R. A. (1984). *Classification and regression trees*. Taylor & Francis.

Chambers, R. G., Färe, R., & Grosskopf, S. (1996). Productivity growth in APEC countries. *Pacific Economic Review, 1*, 181–190.

Chambers, R. G., & Pope, R. D. (1996). Aggregate productivity measures. *American Journal of Agricultural Economics, 78*(5), 1360–1365.

Chambers, R. G., Chung, Y., & Färe, R. (1998). Profit, directional distance functions, and Nerlovian efficiency. *Journal of Optimization Theory and Applications, 98*, 351–364.

Charnes, A., Cooper, W. W., & Rhodes, E. (1978). Measuring the efficiency of decision making units. *European Journal of Operational Research, 2*(6), 429–444.

Charnes, A., Cooper, W. W., Golany, B., Seiford, L., & Stutz, J. (1985). Foundations of data envelopment analysis for Pareto–Koopmans efficient empirical production functions. *Journal of Econometrics, 30*, 91–107.

Cooper, W. W., Park, K. S., & Pastor, J. T. (1999). RAM: A range adjusted measure of inefficiency for use with additive models, and relations to other models and measures in DEA. *Journal of Productivity Analysis, 11*(1), 5–42.

Cooper, W. W., Pastor, J. T., Borras, F., Aparicio, J., & Pastor, D. (2011). BAM: A bounded adjusted measure of efficiency for use with bounded additive models. *Journal of Productivity Analysis, 35*(2), 85–94.

Daraio, C., & Simar, L. (2005). Introducing environmental variables in nonparametric frontier models: A probabilistic approach. *Journal of Productivity Analysis, 24*(1), 93–121.

Deprins, D., Simar, L., & Tulkens, H. (1984). Measuring labor inefficiency in post offices. In M. Marchand, P. Pestieau, & H. Tulkens (Eds.), *The performance of public enterprises: Concepts and measurements* (pp. 243–267). North-Holland.

De Witte, K., & Lopez-Torres, L. (2017). Efficiency in education. *A review of literature and a way forward, Journal of Operational Research Society, 68*(4), 339–363.

Du, P., Parmeter, C. F., & Racine, J. S. (2013). Nonparametric kernel regression with multiple predictors and multiple shape constraints. *Statistica Sinica*, 1347–1371.

Esteve, M., Aparicio, J., Rabasa, A., & Rodriguez-Sala, J. J. (2020). Efficiency analysis trees: A new methodology for estimating production frontiers through decision trees. *Expert Systems with Applications, 162*, 113783.

Esteve, M., Rodríguez-Sala, J. J., López-Espín, J. J., & Aparicio, J. (2021). Heuristic and backtracking algorithms for improving the performance of efficiency analysis trees. *IEEE Access, 9*, 17421–17428.
Färe, R., & Lovell, C. K. (1978). Measuring the technical efficiency of production. *Journal of Economic Theory, 19*(1), 150–162.
Färe, R., Grosskopf, S., & Lovell, C. A. K. (1985). *The measurement of efficiency of production*. Kluwer Nijhof Publishing.
Fare, R., & Primont, D. (1995). *Multi-output production and duality: Theory and applications*. Kluwer Academic Publishers.
Hall, P., & Huang, L. S. (2001). Nonparametric kernel regression subject to monotonicity constraints. *Annals of Statistics*, 624–647.
Henderson, D. J., & Parmeter, C. F. (2009). Imposing economic constraints in nonparametric regression: Survey, implementation, and extension. *Advances in Econometrics, 25*(2009), 433–469.
Juo, J.-C., Fu, T.-T., Yu, M.-M., & Lin, Y. H. (2015). Profit-oriented productivity change. *Omega, 57*, 176–187.
Kerstens, K., O'Donnell, C., & Van de Woestyne, I. (2019). Metatechnology frontier and convexity: A restatement. *European Journal of Operational Research, 275*(2), 780–792.
Li, Q. (1996). Nonparametric testing of closeness between two unknown distribution functions. *Econometric Reviews, 15*, 261–274.
Lovell, C. K., & Pastor, J. T. (1995). Units invariant and translation invariant DEA models. *Operations Research Letters, 18*(3), 147–151.
Luenberger, D. G. (1992). New optimality principles for economic efficiency and equilibrium. *Journal of Optimization Theory and Applications, 75*(2), 221–264.
Luenberger, D. G. (1995). *Microeconomic theory*. McGraw-Hill.
OECD. (2013). *PISA 2012 results: What students know and can do–student performance in mathematics, reading and science* (Vol. I). PISA, OECD Publishing.
OECD. (2014). *PISA 2012 technical report*. PISA, OECD Publishing.
Parmeter, C. F., Sun, K., Henderson, D. J., & Kumbhakar, S. C. (2014). Regression and inference under smoothness restrictions. *Journal of Productivity Analysis, 41*, 111–129.
Pastor, J. T., Ruiz, J. L., & Sirvent, I. (1999). An enhanced DEA Russell graph efficiency measure. *European Journal of Operational Research, 115*(3), 596–607.
Russell, R. R., & Schworm, W. (2018). Technological inefficiency indexes: A binary taxonomy and a generic theorem. *Journal of Productivity Analysis, 49*(1), 17–23.
Sealey, C. W., Jr., & Lindley, J. T. (1977). Inputs, outputs, and a theory of production and cost at depository financial institutions. *The Journal of Finance, 32*(4), 1251–1266.
Sheather, S., & Jones, M. (1991). A reliable data-based bandwidth selection method for kernel density estimation. *Journal of the Royal Statistical Society, Series B, 53*, 683–690.
Shephard, R. W. (1953). *Cost and production functions*. Princeton University Press.
Silverman, B. W. (1986). *Density estimation for statistics and data analysis*. Chapman and Hall.
Simar, L., & Zelenyuk, V. (2006). On testing equality of distributions of technical efficiency scores. *Econometric Reviews, 25*(4), 497–522.
Tavakoli, I. M., & Mostafaee, A. (2019). Free disposal hull efficiency scores of units with network structures. *European Journal of Operational Research, 277*(3), 1027–1036.
Tone, K. (2001). A slacks-based measure of efficiency in data envelopment analysis. *European Journal of Operational Research, 130*(3), 498–509.
Vapnik, V. (2013). *The nature of statistical learning theory*. Springer.
Worthington, A. C. (2001). An empirical survey of frontier efficiency measurement techniques in education. *Education Economics, 9*(3), 245–268.
Zhu, J. (2020). DEA under big data: Data enabled analytics and network data envelopment analysis. *Annals of Operations Research* (in press).

Hybrid Data Science and Reinforcement Learning in Data Envelopment Analysis

Chia-Yen Lee, Yu-Hsin Hung, and Yen-Wen Chen

1 Introduction

Data envelopment analysis (DEA) is a nonparametric optimization approach and can be used to estimate the production function or evaluate the productive efficiency of each observation (also called decision-making unit, DMU) against best-practice in the production systems (Charnes et al., 1978; Cook et al., 2014). DEA uses a predetermined direction (i.e., orientation) to identify the "position" (i.e., efficiency score) of each DMU in the production possibility set. In the past decade, artificial intelligence (AI), machine learning (ML), and data science (DS) rise, and these emerging technologies contribute to the digital revolution in our society (Lee and Chien, 2020). This study provides a tutorial which introduces hybrid data science and reinforcement learning to enhance the applications in DEA.

The expanding field of data science has been described as "This coupling of scientific discovery and practice involves the collection, management, processing, analysis, visualization, and interpretation of vast amounts of heterogeneous data associated with a diverse array of scientific, translational, and inter-disciplinary applications" from the website of Data Science Initiative funded by University of Michigan in 2015 (Donoho, 2017). Other definitions include "Data science is more than the combination of statistics and computer science—it requires training in how to weave statistical and computational techniques into a larger framework, problem by problem, and to address discipline-specific questions" (Blei & Smyth, 2017) and

C.-Y. Lee (✉) · Y.-H. Hung
Department of Information Management, National Taiwan University, Taipei, Taiwan
e-mail: chiayenlee@ntu.edu.tw

Y.-W. Chen
Institute of Manufacturing Information and Systems, National Cheng Kung University, Tainan City, Taiwan
e-mail: p96081063@mail.ncku.edu.tw

J. Zhu, V. Charles (eds.), *Data-Enabled Analytics*, International Series in Operations Research & Management Science 312, https://doi.org/10.1007/978-3-030-75162-3_4

"the study of learning from data; it studies the methods involved in the analysis and processing of data and proposes technology to improve methods in an evidence-based manner" (Donoho, 2017). In addition, Davenport and Patil (2012) claimed "Data scientists are the people who understand how to fish out answers to important business questions from today's tsunami of unstructured information". Thus, DEA is a methodology of data science to learn from data with multiple inputs and multiple outputs to estimate production functions (Coelli et al., 2005; Fried et al., 2008), capture marginal productivity or shadow prices (Lee, 2017; Lee & Zhou, 2015), identify returns-to-scale (Banker et al., 1996, 2004), or evaluate efficiency scores for performance benchmarking among other uses (Afriat, 1972; Banker et al., 1984; Charnes et al., 1978). Recent studies of the integrated aspects of DEA and ML/DS can be categorized as follows (see Table 1 below).

- *Network DEA*: Used to decompose the efficiency of a production system characterized by a network structure into the efficiency of each sub-component (Kao, 2014; Kao & Hwang, 2008), considering production characteristics such as demand effect (Lee & Johnson, 2011, 2012), lagged production effect (Chen, 2009; Chen & van Dalen, 2010), dynamic system with carry-over (Tone & Tsutsui, 2014), and big data-enabled analytics (Zhu, 2020).
- *Feature selection*: Used to select the subset of features characterizing the concept of target variables for the purpose of model construction and avoiding the curse of dimensionality; the mutually-exclusive and collectively-exhaustive (MECE) selection principle collects and extracts features with non-overlapping and complete information (Lee & Chen, 2018). Feature selection techniques related to DEA include absolute shrinkage and selection operator (LASSO) for variable selection with a small dataset (Lee & Cai, 2020), principal component analysis (PCA) (Adler & Yazhemsky, 2010), dimension reduction (Charles et al., 2019), and mixed integer linear programming (Benítez-Peñaa et al., 2020).
- *Big data*: Used to address the computational burdens when iteratively solving the large-scale linear programming models; in particular, if the dimension and sample's density increase. Generally, the previous studies tried to partition the DMUs into smaller groups and identify all efficient DMUs, and then the efficiency scores of the inefficient DMUs can be estimated. Commonly used methods include hierarchical decomposition procedure (Barr & Durchholz, 1997), build-hull procedure (Dulá, 2011), heuristic procedure (Chen & Cho, 2009; Chen & Lai, 2017), and five-step subsampling framework (Khezrimotlagh et al., 2019)
- *Predictive efficiency analytics*: Used to make efficiency prediction or classification about unknown future events, or estimate the coefficients of the regressors to identify and explain factors which significantly affect the efficiency; the analytics include predictive efficiency (Johnson & Lee, 2017), DEA-discriminant analysis (Sueyoshi, 1999), and two-stage DEA which estimates efficiency in the stage 1 and regresses the efficiency on contextual variable in stage 2 (Johnson & Kuosmanen, 2011, 2012; McDonald, 2009; Simar & Wilson, 2007).

Table 1 Categories of data science and DEA

Category	Topic	Reference
Network DEA	Efficiency decomposition with relational model	Kao and Hwang (2008) and Kao (2014)
	Demand effect	Lee and Johnson (2011) and Lee and Johnson (2012)
	Dynamic systems with lagged production effect or carry-over	Chen (2009), Chen and van Dalen (2010) and Tone and Tsutsui (2014)
	Big data-enabled analytics	Zhu (2020)
Feature selection	Variable selection with small datasets	Lee and Cai (2020)
	Principal component analysis (PCA) and dimension reduction	Adler and Yazhemsky (2010) and Charles et al. (2019)
	Mathematical optimization	Benítez-Peñaa et al. (2020)
Big data	Algorithm	Chen and Cho (2009), Chen and Lai (2017) and Khezrimotlagh et al. (2019)
	Hierarchical decomposition	Barr and Durchholz (1997)
	Build-hull	Dulá (2011)
Predictive efficiency analytics	Predictive efficiency	Johnson and Lee (2017)
	DEA-discriminant analysis	Sueyoshi (1999)
	Two-stage DEA: Efficiency estimation in stage 1; regress the efficiency on contextual variable in stage 2	Simar and Wilson (2007), McDonald (2009), Johnson and Kuosmanen (2011) and (2012)
	Hybrid data science (HDS)	This study
Prescriptive efficiency analytics	Directional distance function with explainable directional vectors (e.g. profit maximization, marginal profit maximization, and Nash profit maximization)	Zofio et al. (2013), Lee (2014), (2016) and Wang et al. (2019)
	Nash equilibrium	Liang et al. (2008), Lee and Johnson (2015a) and Lee (2018)
	Effectiveness and proactive DEA	Yu and Lin (2008), Lee (2015), Lee and Johnson (2015b) and Lee and Johnson (2014)
	Reinforcement learning (RL)	This study

- *Prescriptive efficiency analytics*: Used to make the best decision/action of efficiency optimization in a complex scenario, considering the resource constraints and tradeoffs between objectives; popular methods include directional distance function (Chung et al., 1997; Wang et al., 2019) with explainable directional vectors (e.g., profit-maximized direction (Zofio et al., 2013), marginal-profit-maximized direction (Lee, 2014), Nash-profit-maximized direction (Lee, 2016)), Nash equilibrium (Lee, 2018; Lee & Johnson, 2015a; Liang et al., 2008), effectiveness measure (Lee, 2015; Lee & Johnson, 2015b; Yu & Lin, 2008), and proactive DEA (Lee & Johnson, 2014); our study falls in this category.

This study introduces a hybrid DS framework and reinforcement learning (RL) to enhance the productivity analytics, and contributes to the DEA literature in the era of big data. First, recognizing that it is often difficult to formulate the DEA-estimated frontier as a closed functional form, and that the error (i.e. residual) cannot be interpreted, this study proposes a hybrid DS framework to characterize the production frontier and the error by identifying some basic functions or elements which may inspire and guide the productivity improvement. Second, DEA generally provides an *ex-post* analysis with efficiency evaluation; however, we may suffer if DEA cannot provide the guideline for productivity improvement. RL is a ML/DS technique which can find the optimal policy by taking robust actions over time according to change of state in an uncertain environment for the cumulative reward maximization. This study suggests using RL in DEA to identify the direction/orientation (i.e. action) toward efficiency (i.e. reward) maximization. Third, to our best of knowledge, there are few DEA studies to support productivity analysis by applying hybrid DS and RL.

The remainder of this study is organized as follows. Section "Fundamentals" introduces the fundamentals of DS techniques including symbolic regression (SR), multivariate adaptive regression spline (MARS), tree-based method, and RL. Section "Data Science with DEA" presents our DS in the productivity analytics through data generating process (DGP). Section "Reinforcement Learning with DEA" describes RL in U.S. coal-fired and oil-fired power plants operating in 2004–2019 to drive productivity. Section "Conclusion" concludes.

2 Fundamentals

This section briefly describes several fundamental techniques of DS. SR, MARS, and tree-based method are used to characterize the production frontier and the error, and thus the explainable elements can be extracted for managerial insights. It also describes the use of RL to extract an optimal policy and guide productivity improvement by learning from a dynamic system presenting productivity change over time.

2.1 Symbolic Regression

SR (McKay et al., 1995) is a type of regression analysis that investigates the searching space of symbolical expressions to find the mathematical functional form that best fits a given dataset by randomly recombining the variables, functions, arithmetic operators, and constants in a syntax tree structure through genetic programming (Koza, 1994). Genetic programming is an evolutionary algorithm, extended from the genetic algorithm inspired by Darwin's natural selection. It conducts an iterative procedure including population initialization (i.e. a number of syntax trees are randomly generated), crossover and mutation, fitness evaluation, selection, and termination condition. Genetic programming builds a syntax tree to represent the mathematical expression, and each tree built represents a solution. Figure 1 illustrates a syntax tree with a mathematical expression $x_1x_2 + \exp\left(x_1^3\right)$, where the variables and constants are shown as leave nodes and the arithmetic operators and functions are shown as internal nodes. Genetic programming shows a random search between exploration and exploitation in an iterative procedure to avoid the local optimum and approach the approximate global optimum. Each syntax tree represents a solution. GP adopts crossover and mutation to generate the offspring (i.e. new solutions) from the parents, and the selection mechanism evaluates the fitness value of each solution to identify the ones which will be retained to form the new population in the next evolutionary iteration.

2.2 Tree-Based Method

Tree-based method (also called decision tree) partitions the feature space into a set of rectangles or cubes to fit each leaf node with a classification or regression model in the tree structure. The tree structure uses an If-Then rule extraction or

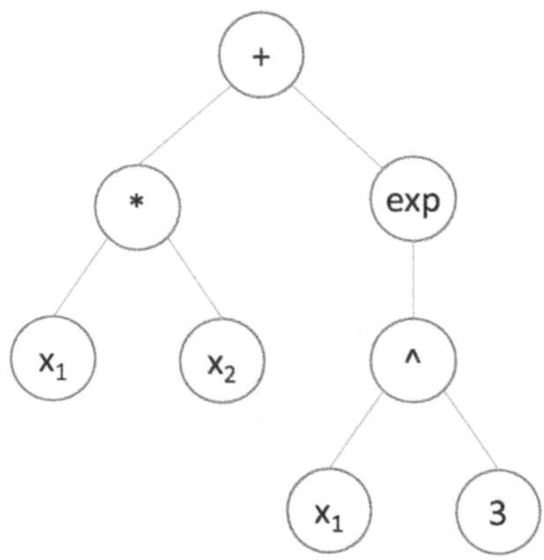

Fig. 1 Syntax tree of genetic programming

tree visualization. Breiman et al. (1984) proposed a decision tree algorithm called classification and regression trees (CART), which can be used for both categorical and continuous labels (i.e. response variable). CART constructs a hierarchy based on the values of the feature variables and partitions the measurement space sequentially into the disjoint subgroups that are the maximum possible internal homogeneous outcomes associated with the observations in this subgroup. CART builds each level of the tree by selecting the split associated with the feature variable that gives the best separation of the classes by measuring the impurity of the subgroups with the Gini index. CART also provides the importance ranking of variables that are suitable for the variable selection via a tree model with interaction effect.

Figure 2 illustrates an example of a manufacturer who has 25 lots with 15 good products (i.e. Pass) and 10 defective products (i.e. NG). CART uses two variables, Temperature and Pressure, from the production dataset to construct the tree structure, extract the rules, and visualize the partitions of thefeature spaces. The tree partially considers the interaction effect between variables (i.e., If Temperature > 580 °C and Pressure ≤ 1 kPa, then Pass). CART also ranks the variables in importance (i.e. Temperature close to the root node is more important than Pressure in Fig. 2). The depth of the tree is post-pruned via cost complexity pruning.

Random Forest is an ensemble method that aggregates multiple weak classifiers (e.g., CART) for prediction using the subsets sampling (i.e. bootstrapping) (Breiman, 2001; Liaw & Wiener, 2002). RF builds a decision tree for each bootstrapping subset and aggregates the trees by unweighted majority vote to predict the response variable (i.e. label). It can handle thousands of variables (even with missing values) and it is commonly used for feature selection.

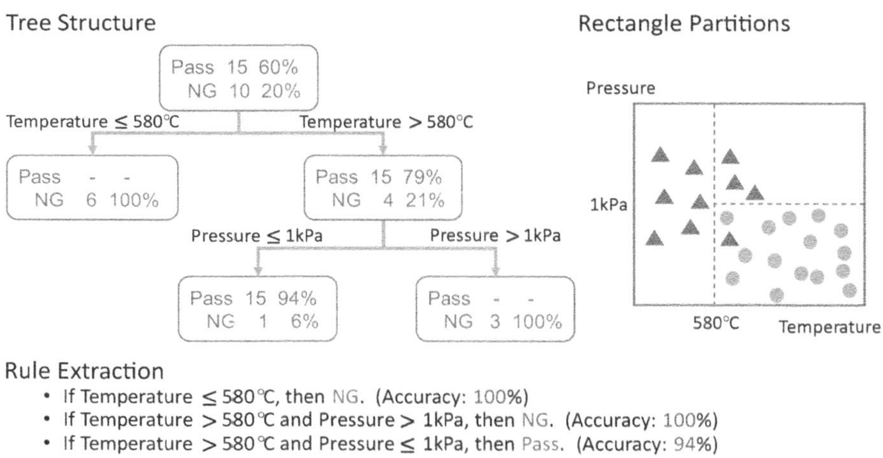

Fig. 2 Tree-based methcd, structure, rule, and partition visualization

2.3 Multivariate Adaptive Regression Spline (MARS)

Multivariate Adaptive Regression Spline (MARS) introduced by Friedman (1991) extends linear model to solve non-linear relationship for modeling high dimensional data. MARS formulates the contribution of each predictor variable to the target variable using a sequence of piece-wise linear regression splines. The general form is:

$$\hat{f}(x) = \sum_{i=1}^{k} a_i B_i(\mathbf{x}), \tag{1}$$

where $\mathbf{x}$ denotes the vector of predictor data, a_i are the estimated coefficients of the ith basis functions B_i, and k is the number of basis functions in the model. Fundamentally, a general MARS model is a weighted sum of the basis functions that are defined as:

$$B_i(\mathbf{x}) = \begin{cases} 1, & i = 1 \\ \prod_{j=1}^{J_i} \left[s_{ji} \bullet \left(x_{v(ji)} - t_{ji} \right) \right]_+, & i = 2, 3, \ldots \end{cases}, \tag{2}$$

where $(\bullet)_+ := \max(0, \bullet)$, J_i is the degree of the interaction of basis functions B_i, $(x_{v(ji)} - t_{ji})_+$ or $(t_{ji} - x_{v(ji)})_+$ is the form of the hinge function, sign indicator s_{ji} equals ± 1, and $v(ji)$ gives the index of the predictor variable which is being split on the knots t_{ji}. Figure 3 illustrates a schematic diagram of a basis function when the index i is not equal to 1 and the knot t is predefined.

The fitting process of the MARS can be separated into a forward stepwise procedure and a backward stepwise procedure. MARS starts with a model which consists of a constant basis function, and then adds the basis functions in pairs that produce the maximum reduction in some lack-of-fit criterion, e.g., residual sum-of-squares (RSS). This forward stepwise procedure stops when the number of the basis functions reaches the predefined fitness condition. The backward stepwise procedure eliminates the basis function that contributes the least to the model via generalized cross validation, until the model reaches the most appropriate fitting

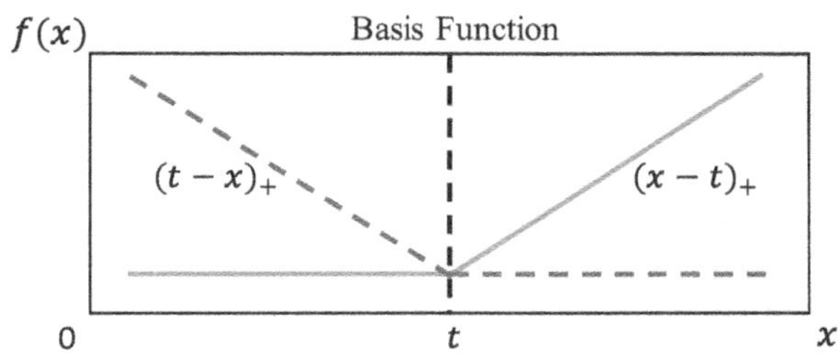

Fig. 3 Schematic diagram of a basis function

state (i.e. an optimal balance between the RSS and the number of hinge-function knots used). Simply stated, MARS provides the trade-off between goodness-of-fit and model complexity. Although MARS models are more flexible than linear regression models and perform variable selection automatically, their computation times can be higher. Furthermore, the sequential dependence problem occurs because MARS depends on the sequence of basis functions added/removed.

2.4 Reinforcement Learning

Reinforcement Learning (RL) enables an agent to learn in an interactive and uncertain environment using trial-and-error and feedback from its own actions. RL obtains the optimal policy which maps the agent's perceived state of the environment to action for expected reward maximization (Chunming et al., 2015; Puterman, 2005). For each time epoch $t \in T$, the agent makes a decision (i.e. an action a) in an observed state s after taking action a, the agent receives a reward $R(s^{'} \mid s, a)$, and the environment transits from current state s to a next state $s^{'}$ with the transition probability $P(s^{'} \mid s, a)$. A Markov decision process (MDP) is defined if all the elements in RL mentioned above are well-defined and provided. Specifically, the state transitions of the MDP satisfy the Markov memoryless property; the new state $s^{'}$ only depends on current state s and is conditionally independent of all previous states and actions. The MDP provides a mathematical framework to describe an agent interacting with an uncertain environment under stochastic decisions (e.g., 80% takes one action, but 20% randomly takes other actions).

The goal of an MDP is to find an optimal policy $\pi*$ to maximize the expected total reward from state-value function $V_{\pi}(s)$, which is defined as value of state s under a policy π:

$$V_{\pi}(s) = \mathbb{E}_{\pi}\left[\sum_{t=0}^{T} \gamma^{t} R\left(s_t^{\prime} | s_t\right)\right], \tag{3}$$

where $\mathbb{E}_{\pi}[\bullet]$ denotes the expected value of a random variable, and γ is the discounting factor. Similarly, an action-value function $Q_{\pi}(s, a)$ can be defined as:

$$Q_{\pi}(s, a) = \mathbb{E}_{\pi}\left[\sum_{t=0}^{T} \gamma^{t} R\left(s_t^{\prime} | s_t\right)\right]. \tag{4}$$

If optimal policy $\pi*$ is to maximize $V_{\pi}(s)$, which is denoted as $V*(s)$, then $V*(s)$ at each state can be found recursively by solving Bellman optimal equations (i.e. dynamic programming formulations):

$$
\begin{aligned}
V^*(s) &= \max_a \sum_{s'} Q^*(s,a) = \max_a \mathbb{E}_{\pi^*}\left[\sum_{t=0}^{T} \gamma^t R\left(s_t^{'}|s_t\right)\right] \\
&= \max_a \mathbb{E}_{\pi^*}\left[R_t + \gamma \sum_{t=0}^{T} \gamma^t R\left(s_{t+1}^{'}|s_{t+1}\right)\right] = \max_a \mathbb{E}\left[R_t + \gamma V_{t+1}^*(s)\right] \\
&= \max_a \sum_{s'} P\left(s'|s,a\right)\left[R\left(s'|s,a\right) + \gamma V^*\left(s'\right)\right].
\end{aligned} \tag{5}
$$

Similarly, the Bellman optimal equation for $Q^*(s,a)$ is defined as:

$$
\begin{aligned}
Q^*(s,a) &= \mathbb{E}\left[R_t + \gamma \max_a Q_{t+1}^*(s,a)\right] \\
&= \sum_{s'} P\left(s'|s,a\right)\left[R\left(s'|s,a\right) + \gamma \max_{a'} Q^*\left(s',a'\right)\right] \\
&= \sum_{s'} P\left(s'|s,a\right)\left[R\left(s'|s,a\right) + \gamma V^*\left(s'\right)\right].
\end{aligned} \tag{6}
$$

The equations above show backwards induction usually used in dynamic programming (i.e. the optimal action and the optimal value function can first be determined at the end of an epoch, and the optimal path can then be derived recursively from the last to the first epoch using a Bellman equation).

For any finite MDP, Q-learning is a model-free RL that can be used to learn the optimal policy maximizing the expected reward without modelling the environment (i.e. without knowing $P(s' \mid s,a)$ or $R(s' \mid s,a)$). The Q-table generated by Q-learning, which lists the expected rewards for each state and action, makes it easy to choose the best action. Q-learning tries to add the reward obtained from the current state to the maximum reward achievable from the future states. The potential future reward is calculated by a weighted sum of the expected rewards of all future steps from now on that will influence the current action. Typically, the value iteration is used to update the expected reward using a Bellman equation. Specifically, the expected reward is updated iteratively by calculating a weighted average of the old expected-reward value $Q(s,a)$ and the new sample $R\left(s'|s,a\right) + \gamma \max_{a'} Q\left(s',a'\right)$, when obtaining the new information (i.e. observing the new sample):

$$
Q(s,a) \leftarrow (1-\alpha)\,Q(s,a) + \alpha\left[R\left(s'|s,a\right) + \gamma \max_{a'} Q\left(s',a'\right)\right], \tag{7}
$$

where α is the learning rate. This method is also called sample-based Q-value iteration. If all relations among state, action, transition probability, and reward are well-defined, then the MDP is suggested; otherwise Q-learning is suggested for partial information or for a complicated environment.

3 Data Science with DEA

This section introduces the data generating process (DGP) (i.e. Monte Carlo simulations) to validate the prediction performance and interpretability of symbolic regression (SR), regression tree (RT), and MARS.[1] First, we develop a hybrid DS framework to estimate the production frontier. Second, we compare the DS methods using DGP for three pre-determined scenarios of the production functions with an inefficiency term. Third, we estimate the frontier by considering a composite error term consisting of noise and inefficiency. Our purpose is to identify a mathematical form or some elements of frontier to enhance the interpretability.

3.1 Hybrid Data Science Framework

Our proposed hybrid DS framework includes three methods which potentially hybridize model-based and model-free nature. A model-based approach considers a pre-determined mathematical functional form and estimates the coefficients (or parameters) of the function (e.g. regression-based approach), and a model-free approach that obtains a successful model-fitting without requiring a functional form prior (e.g. k-nearest neighbor). The model-based approach, however, is subject to model misspecification, while the black box that is generally built into the model-free approach makes the model intractable and untransparent. Specifically, we use symbolic regression to estimate the production frontier, and then conduct the residual analysis by using regression tree and MARS. Below, Fig. 4 illustrates the flowchart of the hybrid DS framework we develop to improve interpretation.

DGP with inefficiency term and composite error term are used for Monte Carlo simulation, respectively. The sample size n is 100 and 1000. The dataset is randomly split into a training dataset with 80% samples and a testing dataset with 20% samples. We assume three scenarios of the production functions $y = f(x)$, and all inputs are randomly sampled from the uniform distribution $Uni(1, 10)$. Noise v is a draw from a normal distribution with $Normal$(0,0.1) or $Normal$(0,0.3). Inefficiency is a draw from a half-normal distribution with $Normal^{+}$(0,0.1) or $Normal^{+}$(0,0.3). We use DEA to identify the efficient DMUs of the training dataset, apply symbolic regression to fit them, and calculate their residuals. To perform the residual diagnostics, we use three hypothesis tests typically used in regression analysis (Walpole et al., 2012): normality, independency, and homogeneity of variance. If the testing does not reject null hypothesis (i.e. residual is random noise), we use the symbolic regression to fit the frontier for functional-form identification and prediction; otherwise, we use the regression tree and MARS to investigate the pattern of the residual. We run 100 trials (i.e. replications M) for each scenario of

[1] We only discuss the continuous independent variable and response variable; thus, the regression tree is used in CART.

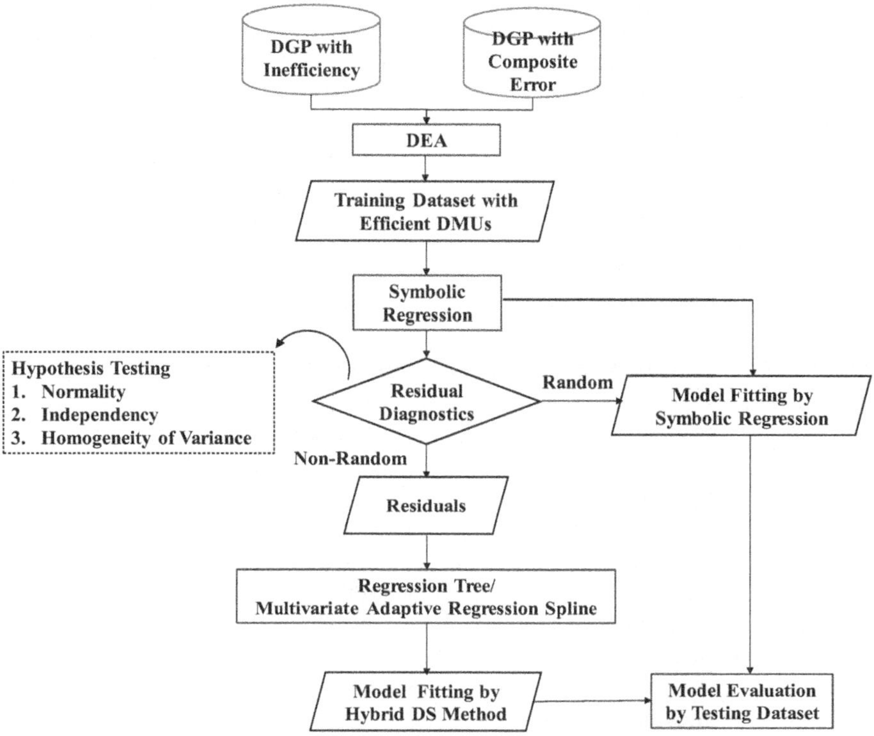

Fig. 4 Hybrid DS framework for production frontier estimation in DEA

the production frontiers and evaluate the model performance of the testing dataset with the metrics root-mean-square error (RMSE), mean absolute error (MAE), and symmetric mean absolute percentage error (sMAPE):

$$\mathrm{RMSE} = \frac{1}{M}\sum_{l=1}^{M}\sqrt{\frac{1}{n}\sum_{k=1}^{n}\left(y_{lk} - \hat{y}_{lk}\right)^2} \tag{8}$$

$$\mathrm{MAE} = \frac{1}{nM}\sum_{l=1}^{M}\sum_{k=1}^{n}\left|y_{lk} - \hat{y}_{lk}\right| \tag{9}$$

$$\mathrm{sMAPE} = \frac{1}{nM}\sum_{l=1}^{M}\sum_{k=1}^{n}\frac{\left|y_{lk} - \hat{y}_{lk}\right|}{\left(\left|y_{lk}\right| + \left|\hat{y}_{lk}\right|\right)/2} \times 100\%, \tag{10}$$

where y_{lk} is the DEA-efficient DMUs generated from GDP in the testing dataset, and $\hat{y}_{lk}$ is predicted in the testing dataset by the proposed hybrid DS which fits the DEA-efficient points in the training dataset.

We compare the five methods: SR, RT, MARS, hybrid method with SR and RT (Hybrid w/ SR + RT), and hybrid method with SR and MARS (Hybrid w/SR + MARS). For SR we specify the following grammar settings for production frontier estimation: operator $\{+, -, \times, \div, \wedge\}$, function {log, square root (sqrt)}, and constant $\{0.1, 0.2, \ldots, 0.9\} \bigoplus \{1, 2, \ldots, 10\}$.We set the number of iterations at 10,000 for SR.

3.2 Frontier Estimation with an Inefficiency Term

This section introduces the DGP with an additive inefficiency term, identifies the efficient DMU by DEA, and uses the proposed hybrid DS framework to estimate the DEA frontier. We assume three scenarios (A, B, and C) and select three functional forms from Kuosmanen and Johnson (2010) as shown in Table 2. For illustration, Fig. 5 only shows that the DGP with an inefficiency term u~$Normal^{+}(0, 0.3)$ generates 1000 samples, which is divided into 800 random samples for the training dataset and 200 samples for the testing dataset, while Fig. 6 shows the efficient DMUs (i.e. efficiency score equals 1) on the DEA frontier identified using an output-oriented variable-returns-to-scale (VRS) model (Banker et al., 1984).

The five DS methods are applied to fit the efficient DMUs for DEA frontier estimation. Figures 7 through 11 illustrate the frontier estimation with an inefficiency term u~$Normal^{+}(0, 0.3)$ using the five DS methods (RT, MARS, SR, Hybrid with

Table 2 Production function with an additive inefficient term

Scenario	Inputs	Functional form
A	x	$y = \ln(x) + 3 - u$
B	x	$y = \ln(x) + 3 + (x)^{1/2} - u$
C	x_1, x_2	$y = 0.1x_1 + 0.1x_2 + 0.3(x_1x_2)^{1/2} - u$

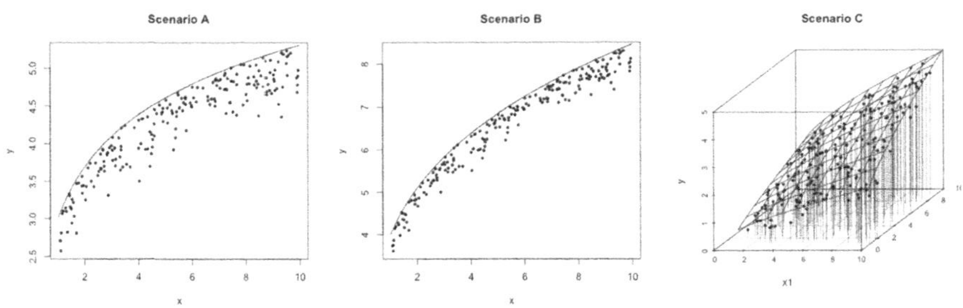

Fig. 5 DGP with an inefficiency term u~$Normal^{+}(0,0.3)$ in the testing dataset

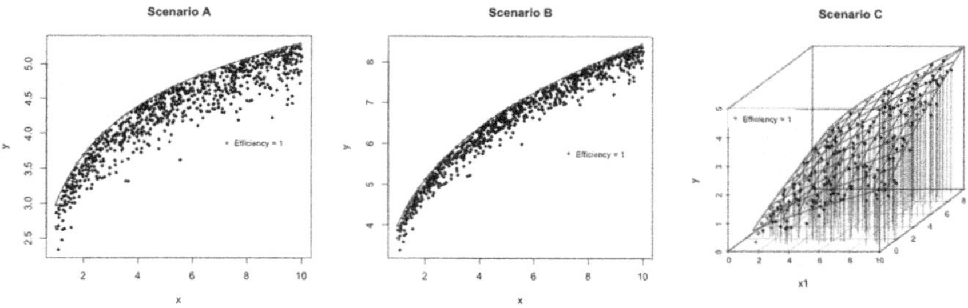

Fig. 6 Efficient DMUs identified on the DEA-VRS frontier

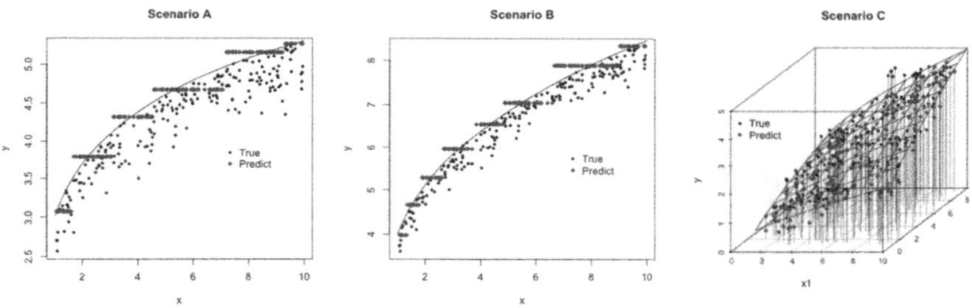

Fig. 7 Regression tree for the DEA-VRS frontier

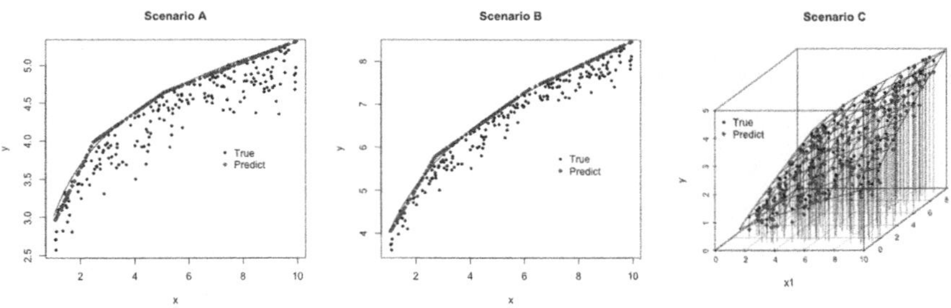

Fig. 8 MARS for the DEA-VRS frontier

SR + RT, and Hybrid with SR + RT), respectively. Tables 3 through 5 list the performance metrics of RMSE, MAE, and sMAPE of the five methods considering the combination of the sample size {100, 1000} and the variance of inefficiency with {0.1, 0.3}. Scenario A does not trigger the hybrid methods via residual analysis, and MARS effectively formulates a DEA-VRS frontier.[2] In Scenario B, Hybrid

[2]Scenario A is not included in Figures 10 and 11 with hybrid methods because the hypothesis testing of the residual analysis shows random noise.

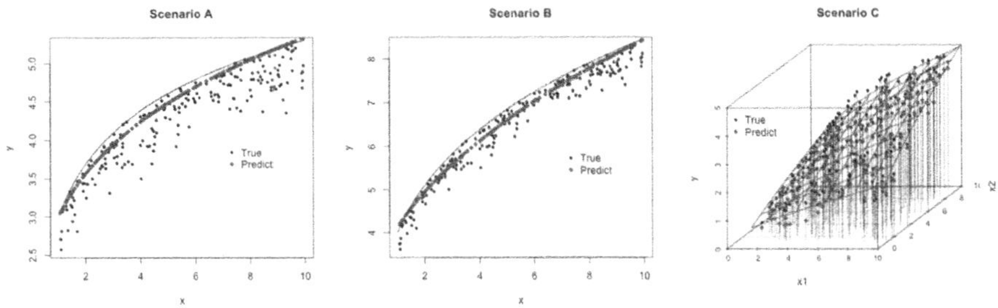

Fig. 9 Symbolic regression for the DEA-VRS frontier

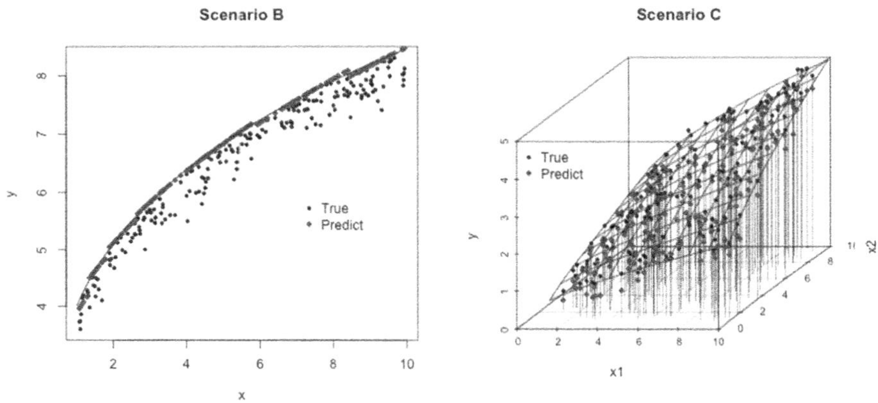

Fig. 10 Hybrid with SR − RT for the DEA-VRS frontier

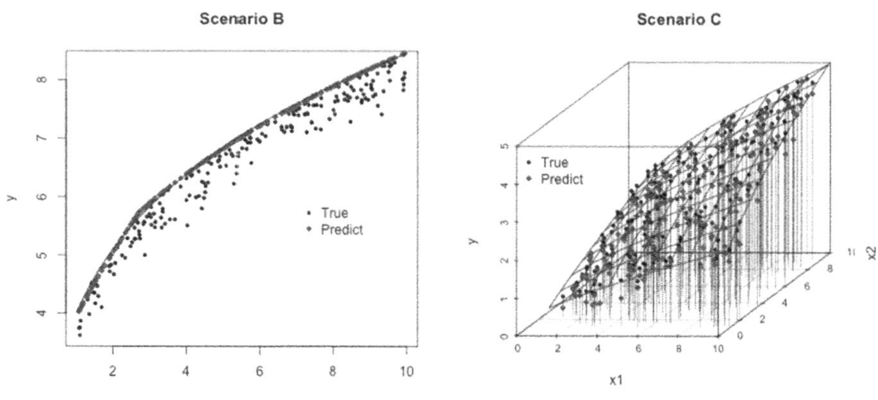

Fig. 11 Hybrid with SR + MARS for the DEA-VRS frontier

Table 3 RMSE of the five DS methods

			Mean (Standard Deviation)				
Scenario	Sample size	Var. of Ineff.	RT	MARS	SR	Hybrid w/ SR + RT	Hybrid w/ SR + MARS
A	100	0.1	0.234 (0.069)	0.077 (0.026)	**0.076** (0.041)	–	–
A	1000	0.1	0.115 (0.024)	**0.054** (0.015)	0.066 (0.048)	–	–
A	100	0.3	0.324 (0.095)	**0.129** (0.063)	0.137 (0.065)	–	–
A	1000	0.3	0.171 (0.043)	**0.091** (0.039)	0.093 (0.049)	–	–
B	100	0.1	0.386 (0.096)	0.101 (0.031)	0.211 (0.138)	0.094 (0.098)	**0.074** (0.086)
B	1000	0.1	0.18 (0.031)	0.072 (0.018)	0.215 (0.101)	0.041 (0.021)	**0.032** (0.017)
B	100	0.3	0.536 (0.148)	0.143 (0.058)	0.223 (0.101)	0.138 (0.062)	**0.137** (0.078)
B	1000	0.3	0.267 (0.056)	0.107 (0.038)	0.236 (0.119)	0.074 (0.037)	**0.069** (0.038)
C	100	0.1	0.471 (0.104)	**0.074** (0.022)	0.213 (0.081)	0.149 (0.052)	0.094 (0.055)
C	1000	0.1	0.299 (0.048)	0.056 (0.01)	0.213 (0.059)	0.085 (0.03)	**0.046** (0.015)
C	100	0.3	0.532 (0.131)	**0.175** (0.055)	0.256 (0.092)	0.216 (0.063)	0.194 (0.07)
C	1000	0.3	0.365 (0.059)	0.119 (0.031)	0.231 (0.071)	0.146 (0.036)	**0.114** (0.033)

w/ SR + MARS is the best performer and in Scenario C, MARS and Hybrid w/ SR + MARS are the best performers. The results imply that MARS and Hybrid w/SR + MARS support a complicated model-fitting of the production frontier.

For estimating the functional form, SR provides some useful managerial insights in Table 6. With the 100 trials, we count the grammar frequency and find that SR effectively identifies the grammar which is the component in the true production function in Table 6. In Scenario A, SR identifies Log with the frequency 74%–81%; in Scenario B, SR identifies Log and Square Root with higher frequencies; and in Scenario C, it identifies Square Root with the frequency 71%–77%. Table 6 is also interesting because SR shows the frequency 100% to identify the interaction effect (e.g., $x_1 \times x_2$) from DEA frontier in Scenario C. The results validate the use of SR as a data mining tool to extract useful insights from the dataset characterizing the production frontier.

Table 4 MAE of the five DS methods

Scenario	Sample size	Var. of Ineff.	Mean (Standard Deviation)				
			RT	MARS	SR	Hybrid w/ SR + RT	Hybrid w/ SR + MARS
A	100	0.1	0.189 (0.057)	**0.061** (0.019)	0.062 (0.033)	–	–
A	1000	0.1	0.092 (0.02)	**0.04** (0.012)	0.052 (0.039)	–	–
A	100	0.3	0.268 (0.084)	**0.102** (0.051)	0.103 (0.043)	–	–
A	1000	0.3	0.138 (0.033)	0.067 (0.023)	**0.061** (0.034)	–	–
B	100	0.1	0.32 (0.085)	0.081 (0.023)	0.169 (0.091)	0.07 (0.055)	**0.053** (0.046)
B	1000	0.1	0.149 (0.027)	0.055 (0.014)	0.175 (0.077)	0.029 (0.016)	**0.022** (0.012)
B	100	0.3	0.448 (0.131)	0.113 (0.046)	0.181 (0.08)	**0.103** (0.044)	0.105 (0.064)
B	1000	0.3	0.219 (0.047)	0.081 (0.025)	0.19 (0.092)	0.049 (0.021)	**0.044** (0.02)
C	100	0.1	0.382 (0.092)	**0.059** (0.017)	0.171 (0.069)	0.116 (0.04)	0.072 (0.035)
C	1000	0.1	0.237 (0.039)	0.043 (0.007)	0.165 (0.051)	0.063 (0.021)	**0.033** (0.01)
C	100	0.3	0.43 (0.111)	**0.134** (0.043)	0.206 (0.08)	0.167 (0.051)	0.147 (0.053)
C	1000	0.3	0.289 (0.046)	0.079 (0.017)	0.179 (0.059)	0.103 (0.027)	**0.076** (0.02)

3.3 Frontier Estimation with a Composite Error Term

This section describes the DGP with a composite error term. Again, we assume three scenarios of the production functions with an additive composite error term and the three functional forms from Kuosmanen and Johnson (2010). For illustration, Table 7 only shows that the DGP with a noise term $v \sim Normal(0, 0.1)$ and an inefficiency term $u \sim Normal^{+}(0, 0.3)$ generates 1000 samples, and 200 samples for the testing dataset is shown as Fig. 12. The efficient DMUs (i.e. efficiency score equals 1) on the DEA frontier, then, are identified via output-oriented VRS model. However, if the number of efficient DMUs identified by DEA is insufficient, then the model-fitting by SR is poor with larger variations. Therefore, we suggest using the output-oriented non-increasing-returns-to-scale (NIRS) model (Seiford & Zhu, 1999) to identify the efficient DMUs, or generating the data randomly via a convex combination between two adjacent efficient DMUs (e.g. 10 efficient observations simulated). Figure 13 illustrates the results of combining them.

Table 5 sMAPE of the five DS methods

Scenario	Sample size	Var. of Ineff.	Mean (Standard Deviation)				
			RT	MARS	SR	Hybrid w/ SR + RT	Hybrid w/ SR + MARS
A	100	0.1	0.047 (0.015)	**0.015** (0.006)	0.015 (0.009)	–	–
A	1000	0.1	0.023 (0.005)	**0.011** (0.003)	0.013 (0.01)	–	–
A	100	0.3	0.068 (0.023)	**0.026** (0.014)	0.027 (0.012)	–	–
A	1000	0.3	0.036 (0.01)	0.018 (0.007)	**0.017** (0.009)	–	–
B	100	0.1	0.054 (0.016)	0.014 (0.005)	0.028 (0.019)	0.012 (0.013)	**0.01** (0.011)
B	1000	0.1	0.025 (0.004)	0.01 (0.003)	0.029 (0.014)	0.005 (0.003)	**0.004** (0.002)
B	100	0.3	0.076 (0.024)	0.02 (0.009)	0.032 (0.015)	**0.018** (0.008)	0.019 (0.012)
B	1000	0.3	0.038 (0.009)	0.015 (0.005)	0.033 (0.016)	**0.009** (0.004)	0.009 (0.005)
C	100	0.1	0.168 (0.042)	**0.03** (0.011)	0.084 (0.042)	0.063 (0.031)	0.039 (0.024)
C	1000	0.1	0.115 (0.019)	0.024 (0.005)	0.09 (0.036)	0.038 (0.017)	**0.02** (0.007)
C	100	0.3	0.218 (0.058)	**0.083** (0.036)	0.117 (0.053)	0.101 (0.047)	0.091 (0.044)
C	1000	0.3	0.165 (0.031)	0.063 (0.02)	0.12 (0.045)	0.079 (0.026)	**0.062** (0.022)

Table 6 Results of functional form estimation by SR

Scenario	Sample size	Var. of Ineff	Grammar Frequency			Best formulation of SR
			Log	Square Root	Power	
A	100	0.1	**81%**	53%	5%	$y = \ln(x) + 3$
A	1000	0.1	**74%**	63%	6%	$y = \ln(x) + \ln(1/(\ln(0.9)\ln(0.5)))$
A	100	0.3	**80%**	49%	4%	$y = \ln(x) + (8)^{1/2}$
A	1000	0.3	**76%**	44%	5%	$y = \ln(x) + 3$
B	100	0.1	76%	**81%**	26%	$y = (10/3)\ln(x + 2 + \ln(\ln(5)))$
B	1000	0.1	**81%**	80%	25%	$y = \ln(x) + \ln(3)(7)^{1/2} + (x)^{1/2}$
B	100	0.3	**81%**	78%	24%	$y = \ln(x) + 1 + \ln(6) + (x)^{1/2}$
B	1000	0.3	74%	**84%**	25%	$y = \ln(x) + 3 + (x)^{1/2}$
C	100	0.1	67%	**73%**	20%	$y = 0.5(x_1x_2)^{1/2}$
C	1000	0.1	54%	**77%**	17%	$y = (x_1x_2/2)^{1/2}$
C	100	0.3	69%	**75%**	18%	$y = ((\ln((\ln(3))^{1/2}))^{1/2}x_1x_2)^{1/2}$
C	1000	0.3	54%	**71%**	9%	$y = ((x_1 + 0.7)^{1/2}x_2/5)^{1/2}$

Table 7 Production function with an additive composite error term

Scenario	Inputs	Functional form
A	x	$y = \ln(x) + 3 + \varepsilon - u$
B	x	$y = \ln(x) + 3 + (x)^{1/2} + \varepsilon - u$
C	x_1, x_2	$y = 0.1x_1 + 0.1x_2 + 0.3(x_1x_2)^{1/2} + \varepsilon - u$

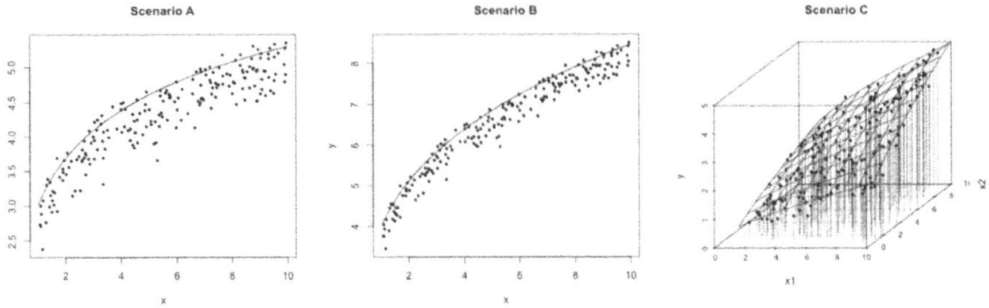

Fig. 12 DGP with a composite error term in the testing dataset

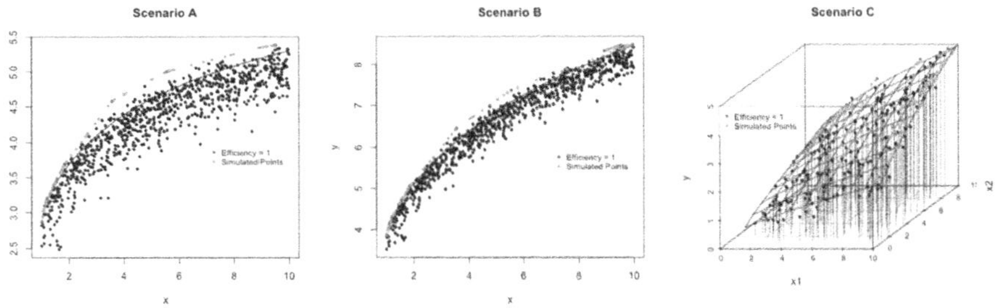

Fig. 13 Efficient DMUs identified on the DEA-NIRS frontier

The five DS methods are applied to fit the efficient DMUs for DEA frontier estimation. Figures 14 through 18 illustrate the frontier estimation with a noise term *v~Normal*(0, 0.1) and an inefficiency term *u~Normal*$^+$(0, 0.3) using the five DS methods (RT, MARS, SR, Hybrid with SR + RT, and Hybrid with SR + RT), respectively. Tables 8 through 10 list the performance metrics of RMSE, MAE, and sMAPE of the five methods considering the combination of the sample size {100, 1000}, and (variance of noise, variance of inefficiency) equals (0.1, 0.3) and (0.3, 0.1).

Again, MARS and the hybrid methods perform well if the noise is small (equal to 0.1). If the noise becomes larger, then RT is better for Scenario A, Hybrid w/SR + RT is better for Scenario B, and SR is better for Scenario C, with sample size equal to 1000. The variability of the performance results when a larger variance of noise generates some outliers which DEA is sensitive to. Note that in Scenario

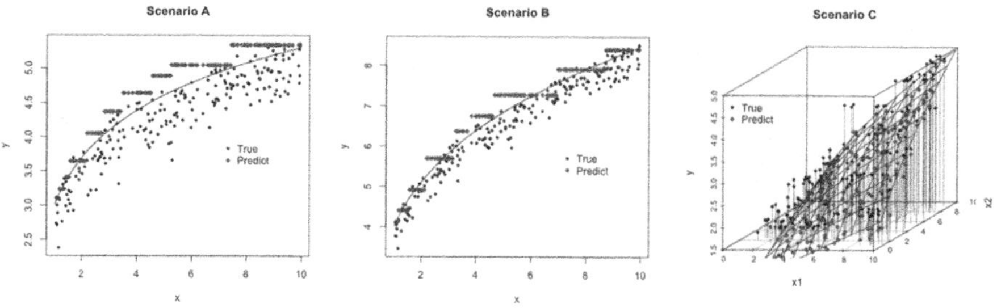

Fig. 14 Regression tree for the DEA-NIRS frontier

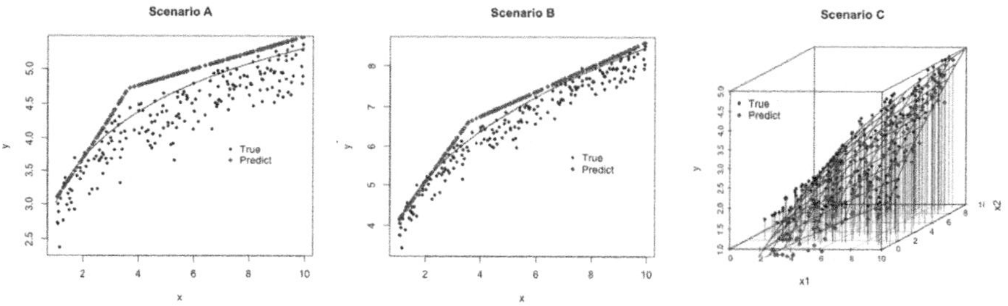

Fig. 15 MARS for DEA-VRS frontier

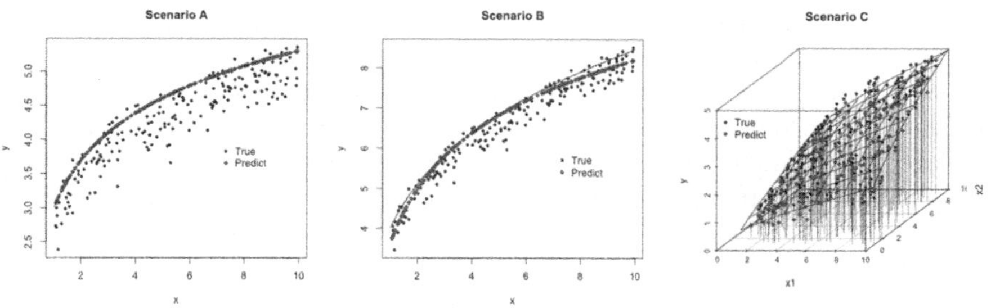

Fig. 16 Symbolic regression for DEA-NIRS frontier

C with 1000 sample size and a larger variance of noise, the mean of the RMSE of the hybrid methods (i.e., 0.488 and 0.501) could be worse than only using SR (i.e., 0.393). The reason could be that the hybrid methods intend to fit the data randomly generated via convex combination between two adjacent efficient DMUs. In other words, the hybrid methods may partially suffer from overfitting in this case with high variance of noise. Nevertheless, the standard deviations of the RMSE of the hybrid methods perform well (i.e., 0.047 and 0.044).

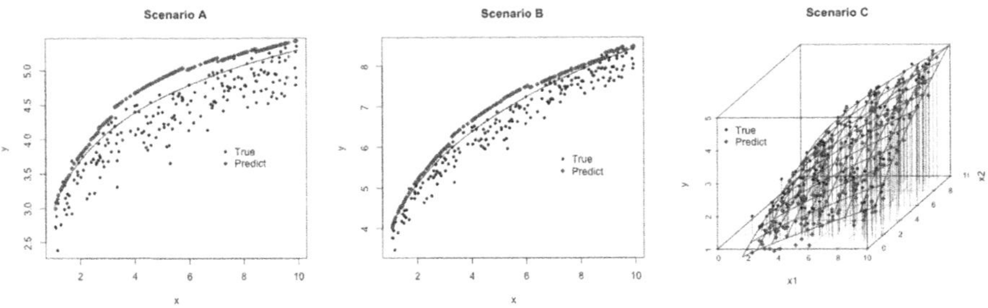

Fig. 17 Hybrid with SR + RT for the DEA-NIRS frontier

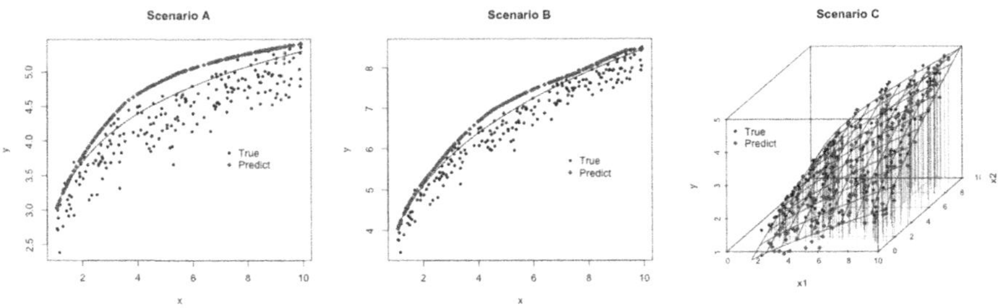

Fig. 18 Hybrid with SR + MARS for the DEA-NIRS frontier

For estimating the functional form, SR provides some useful managerial insights in Table 11. With the 100 trials, we count the grammar frequency and the results show that SR can still identify the grammar which is the component in the true production function in Table 7. However, the noise in the composite error term undermines the performance. For Scenario A, if the noise is 0.3, the Log cannot be identified well; particularly, Square Root dominates others in sample size equal to 1000. For Scenario B, a small noise (i.e. 0.1) shows more balance between Log and Square Root, whereas a larger noise (i.e. 0.3) shows the domination of Square Root. For Scenario C, a larger noise shows a misspecification of the production frontier (i.e. higher frequency for Log term), and SR also completely identifies the interaction effect (e.g., $x_1 \times x_2$) from the DEA frontier in Scenario C.

In conclusion, the five DS methods support the estimation of the production frontier and its functional form. For illustration, we can use them to identify the grammar term and then use ordinary least squares (OLS) to estimate the coefficient of each term in a regression equation. Doing so can partially address the model misspecification issue in a regression analysis, particularly for a complicated functional form of the production function. Nevertheless, the noise term will affect the methods' performance because the DEA frontier is sensitive to the outlier.

Table 8 RMSE of the five DS methods

Scenario	Sample size	Var. of noise	Var. of Ineff.	Mean (Standard Deviation)				
				RT	MARS	SR	Hybrid w/ SR + RT	Hybrid w/ SR + MARS
A	100	0.1	0.3	0.306 (0.035)	**0.265** (0.005)	0.29 (0.043)	0.299 (0.004)	0.314 (0.006)
A	1000	0.1	0.3	0.165 (0.025)	0.154 (0.006)	0.235 (0.028)	0.146 (0.008)	**0.146** (0.007)
A	100	0.3	0.1	0.594 (0.04)	**0.554** (0.017)	0.667 (0.103)	0.626 (0.017)	0.598 (0.032)
A	1000	0.3	0.1	**0.267** (0.023)	0.382 (0.032)	0.463 (0.069)	0.289 (0.023)	0.308 (0.039)
B	100	0.1	0.3	0.292 (0.072)	**0.26** (0.016)	0.385 (0.193)	0.291 (0.02)	0.309 (0.044)
B	1000	0.1	0.3	0.198 (0.039)	0.15 (0.006)	0.348 (0.17)	**0.14** (0.013)	0.149 (0.063)
B	100	0.3	0.1	0.55 (0.069)	**0.506** (0.011)	0.706 (0.134)	0.567 (0.031)	0.554 (0.036)
B	1000	0.3	0.1	0.424 (0.038)	0.295 (0.024)	0.453 (0.125)	**0.261** (0.023)	0.282 (0.036)
C	100	0.1	0.3	0.371 (0.046)	**0.274** (0.016)	0.359 (0.123)	0.276 (0.047)	0.313 (0.052)
C	1000	0.1	0.3	0.388 (0.037)	**0.203** (0.01)	0.273 (0.056)	0.208 (0.051)	0.204 (0.023)
C	100	0.3	0.1	0.59 (0.073)	0.542 (0.167)	0.485 (0.058)	**0.453** (0.062)	0.583 (0.2)
C	1000	0.3	0.1	0.673 (0.082)	0.503 (0.05)	**0.393** (0.08)	0.488 (0.047)	0.501 (0.044)

4 Reinforcement Learning with DEA

This section describes using RL to identify the optimal policy for the DMUs which are moving toward better productive efficiency by deciding to take a capacity expansion (EXP) action, a do nothing (D/N) action, a contraction (CON) action, or an other (O) action. Our empirical study uses monthly data (EIA, 2020a) collected from 498 coal-fired and oil-fired power plants operating from 2004 through 2019 in the US. We assume a relatively controllable combination of inputs' consumption and a desire to minimize the use of inputs to produce a given level of output. We use the input-oriented VRS DEA estimate the efficiency score in each individual month.

We start by defining the following elements in RL. Let State (S) be a composite state with capacity state (CS) and efficiency state (ES) (i.e. a vector $s \in S = (CS, ES)$). $CS = \{1, 2, 3, 4\}$ represents the small, medium, large, and very large sizes, respectively, of nameplate capacity according to the quartile rankings of the nameplate capacity level among all DMUs in the same month. To categorize the scale sizes, we consider a plant's nameplate capacity as an exogenously fixed

Table 9 MAE of the five DS methods

Scenario	Sample size	Var. of noise	Var. of Ineff.	Mean (Standard Deviation)				
				RT	MARS	SR	Hybrid w/ SR + RT	Hybrid w/ SR + MARS
A	100	0.1	0.3	0.265 (0.031)	**0.229** (0.004)	0.238 (0.053)	0.253 (0.005)	0.264 (0.005)
A	1000	0.1	0.3	0.128 (0.02)	0.123 (0.007)	0.174 (0.025)	**0.107** (0.007)	0.115 (0.008)
A	100	0.3	0.1	0.541 (0.031)	**0.518** (0.013)	0.592 (0.084)	0.576 (0.014)	0.552 (0.023)
A	1000	0.3	0.1	**0.189** (0.019)	0.339 (0.029)	0.423 (0.059)	0.262 (0.022)	0.278 (0.031)
B	100	0.1	0.3	0.243 (0.056)	**0.235** (0.019)	0.329 (0.162)	0.247 (0.017)	0.266 (0.034)
B	1000	0.1	0.3	0.159 (0.03)	0.116 (0.008)	0.284 (0.134)	**0.101** (0.011)	0.114 (0.051)
B	100	0.3	0.1	**0.467** (0.054)	0.474 (0.008)	0.616 (0.122)	0.512 (0.024)	0.507 (0.025)
B	1000	0.3	0.1	0.331 (0.041)	0.26 (0.019)	0.399 (0.113)	**0.233** (0.021)	0.252 (0.031)
C	100	0.1	0.3	0.323 (0.035)	0.234 (0.013)	0.297 (0.096)	**0.231** (0.039)	0.269 (0.038)
C	1000	0.1	0.3	0.316 (0.035)	0.167 (0.008)	0.22 (0.041)	0.163 (0.028)	**0.161** (0.016)
C	100	0.3	0.1	0.507 (0.052)	0.444 (0.1)	0.406 (0.049)	**0.363** (0.049)	0.461 (0.103)
C	1000	0.3	0.1	0.566 (0.071)	0.434 (0.038)	**0.332** (0.062)	0.418 (0.05)	0.443 (0.035)

input factor, and do not consider it as an input or output for efficiency estimation by DEA. We define $ES = \{1, 2, 3, 4, 5\}$ to represent the different levels of efficiency scores, where $\{1, 2, 3, 4\}$ indicates low, medium, high, and very high, respectively, according to the quartile rankings of efficiency scores among all the DMUs in the same month, and $\{5\}$ indicates the efficient DMUs (i.e. efficiency score equals 1). We consider total coal consumption (short tons) and total oil consumption (MMBtu) as the two inputs, and electricity net generation from coal (MWh) and electricity net generation from oil (MWh) as the two outputs (EIA, 2020b).[3] The data is collected from EIA (2020b).

We define four actions for efficiency improvement by RL. Action $A = \{\text{E}, \text{N}, \text{C}, \text{O}\}$ indicates {Expansion (E), Do Nothing (N), Contraction (C), Other (O)} regarding the amount input and output adjustments. Expansion refers to increases in all inputs

[3]Coal consists of anthracite coal, bituminous coal, lignite coal, refined coal, coal-based synfuel, subbituminous coal, and waste/other coal. We sum all types and ignore coal quality. We do the same for oil.

Table 10 sMAPE of the five DS methods

Scenario	Sample size	Var. of noise	Var. of Ineff.	Mean (Standard Deviation)				
				RT	MARS	SR	Hybrid w/ SR + RT	Hybrid w/ SR + MARS
A	100	0.1	0.3	0.069 (0.008)	**0.061** (0.001)	0.064 (0.012)	0.067 (0.001)	0.07 (0.001)
A	1000	0.1	0.3	0.036 (0.005)	0.036 (0.002)	0.049 (0.008)	**0.031** (0.002)	0.033 (0.002)
A	100	0.3	0.1	0.128 (0.007)	**0.123** (0.003)	0.14 (0.018)	0.136 (0.003)	0.131 (0.005)
A	1000	0.3	0.1	**0.044** (0.005)	0.072 (0.006)	0.098 (0.014)	0.056 (0.005)	0.06 (0.007)
B	100	0.1	0.3	0.043 (0.01)	**0.042** (0.003)	0.058 (0.029)	0.046 (0.003)	0.049 (0.006)
B	1000	0.1	0.3	0.031 (0.006)	0.024 (0.001)	0.055 (0.032)	**0.021** (0.002)	0.024 (0.011)
B	100	0.3	0.1	0.078 (0.009)	**0.077** (0.001)	0.1 (0.019)	0.084 (0.004)	0.083 (0.004)
B	1000	0.3	0.1	0.05 (0.007)	0.041 (0.003)	0.068 (0.021)	**0.037** (0.003)	0.04 (0.006)
C	100	0.1	0.3	0.179 (0.02)	**0.128** (0.007)	0.168 (0.048)	0.139 (0.028)	0.156 (0.026)
C	1000	0.1	0.3	0.233 (0.014)	0.139 (0.005)	0.181 (0.066)	0.134 (0.032)	**0.122** (0.013)
C	100	0.3	0.1	0.229 (0.024)	0.21 (0.054)	0.197 (0.02)	**0.164** (0.032)	0.218 (0.064)
C	1000	0.3	0.1	0.311 (0.043)	0.302 (0.043)	**0.19** (0.044)	0.306 (0.037)	0.317 (0.036)

and outputs between two consecutive months. Do Nothing refers to minor changes in all inputs and outputs, Contraction refers to decreases in all inputs and outputs, and Others refers to actions that have different directions on several dimensions (e.g. decrease in oil input but increase in power generated by oil) between two consecutive months. Thus, we can observe a transition from current state s to next state $s^{'}$ by taking an action. A transition forms a sample, and we can define and calculate their rewards based on historical data. For the Reward Function $R(s^{'} \mid s, a)$, we define the reward value by observing the transition of efficiency state (ES) and subtract current state ES from next state $ES^{'}$ to obtain the improvement or deterioration upon efficiency. Thus, we can evaluate the reward of an action through a transition from current state to next state by taking some action.

Having defined the state, action, and reward, we apply the Q-learning algorithm and use sample-based Q-value iterations to derive the policy. The policy provides a mapping from state to action for maximizing the expected reward in the long run.

Table 11 Results of functional form estimation by SR

Scenario	Sample size	Var. of noise	Var. of Ineff.	Grammar Frequency Log	Square Root	Power	Best formulation of SR
A	100	0.1	0.3	**84%**	42%	5%	$y = \ln(x) + 3$
A	1000	0.1	0.3	**71%**	67%	5%	$y = \ln(x) + 3$
A	100	0.3	0.1	**76%**	75%	26%	$y = (20(\ln(x))^{1/2})^{1/2}$
A	1000	0.3	0.1	20%	**93%**	7%	$y = (x)^{1/2} + 3$
B	100	0.1	0.3	72%	**85%**	27%	$y = \ln(x) - \ln(0.1) + (x/0.7)^{1/2}$
B	1000	0.1	0.3	68%	**80%**	24%	$y = \ln(x) + 3 + (x)^{1/2}$
B	100	0.3	0.1	42%	**88%**	14%	$y = 4 + (2.2x)^{1/2}$
B	1000	0.3	0.1	47%	**86%**	18%	$y = \ln(20) + (x)^{1/2}/(0.3)^{1/2}$
C	100	0.1	0.3	77%	**81%**	33%	$y = 0.5(x_1x_2)^{1/2}$
C	1000	0.1	0.3	73%	**85%**	29%	$y = 0.5(x_1x_2)^{1/2}$
C	100	0.3	0.1	**94%**	71%	32%	$y = \ln((x_1 - 0.7)x_2)$
C	1000	0.3	0.1	**86%**	55%	24%	$y = (x_1x_2)^{0.7/2}$

Table 12 Q-table generated by RL

CS	*ES*	A = {E}	A = {N}	A = {C}	A = {O}
1	1	0.207925	**0.259827**	0.247454	0.247455
1	2	0.053062		−0.59801	**0.065946**
1	3	−0.64039		−0.00081	**−0.0003**
1	4				
1	5				**−3.56602**
2	1	0.353099		0.90479	**0.950029**
2	2	0.07586		0.082142	**0.093918**
2	3	−0.14942	**0.034326**	0.027462	0.027461
2	4	**1.38E-09**		1.17E-09	
2	5				
3	1	**1.505918**		1.244807	1.285925
3	2	**0.694352**		0.289623	0.651908
3	3	0.346487	0.3631	**0.433109**	0.346487
3	4	−0.02083		−0.11027	**0.000994**
3	5				
4	1	**1.980073**		1.557297	1.590364
4	2	1.016086		0.669024	**1.261347**
4	3	**1.244244**		0.537502	0.486681
4	4	**0.331699**		−0.45473	0.327232
4	5	**−0.01117**		−0.16335	−0.66439

Given the learning rate $\alpha = 0.8$ and the discounted factor $\gamma = 0.8$, Table 12 illustrates the resulting Q-table, and Fig. 19 illustrates the optimal policy derived.[4]

[4] The blanks in Table 12 and the three blank states in Fig. 19 indicate missing data.

Figure 19 provides the following managerial insights:

- It is preferable to take an Expansion or Others action for better efficiency improvement.
- It is preferable to take an Expansion action for larger capacity states (e.g. CS = 4).
- It is preferable to take an Expansion action if the DMU with a relatively larger capacity (CS = 3) is poor in efficiency (e.g. ES = 1 or 2).
- It is preferable to take several actions {E,N,C,O} if the DMUs are in the middle level including (CS = 3, ES = 3), (CS = 3, ES = 4), (CS = 2, ES = 3), and (CS = 2, ES = 4).
- It is preferable to take a Do Nothing action for DMUs with small nameplate capacity and poor efficiency (CS = 1, ES = 1), although action {C, O} can also be suggested with slightly worse rewards in Table 12.
- It is not preferable to take a Contraction. We suppose that the power plants should drive productivity to fulfill the demand along with the economic growth of industry development.
- Only the DMUs with (CS = 4, ES = 5) can show the appropriate action extracted successfully by RL ((CS = 2, ES = 5) and (CS = 3, ES = 5) are blank or (CS = 1, ES = 5) with negative reward in Table 12), which implies that it is difficult to investigate the actions when a DMU is on the frontier. Figure 19 shows that DMUs with small capacity are not likely to be efficient, and Table 12 shows that states (1, 3), (2, 4), (3, 4), and (4, 5) have expected rewards close to zero.

Figure 20 illustrates the convergence diagrams of RL. To illustrate, there is a drop in reward for state (4, 1) after the Baxter Wilson plant's nameplate capacity decreased in Mississippi in 2017, even though the inputs, outputs, and efficiency are similar to previous years. Figure 21 illustrates the convergence diagrams of RL after we eliminate the Baxter Wilson plant and retrain the RL process. Without Baxter Wilson, the curve of state (4, 1) decreases in reward. Figure 22 shows that Baxter Wilson actually contributed a positive effect on reward, by decreasing capacity but still maintaining efficiency after 2017. We infer that the drop in the convergence diagram in Fig. 20 was probably caused by the plant's decrease in nameplate capacity in state (4, 1).

Fig. 19 Optimal policy in DEA

CS=4	A=E	A=O	A=E	A=E	A=E
CS=3	A=E	A=E	A=C	A=O	
CS=2	A=O	A=O	A=N	A=E	
CS=1	A=N	A=O	A=O		A=O
	ES=1	ES=2	ES=3	ES=4	ES=5

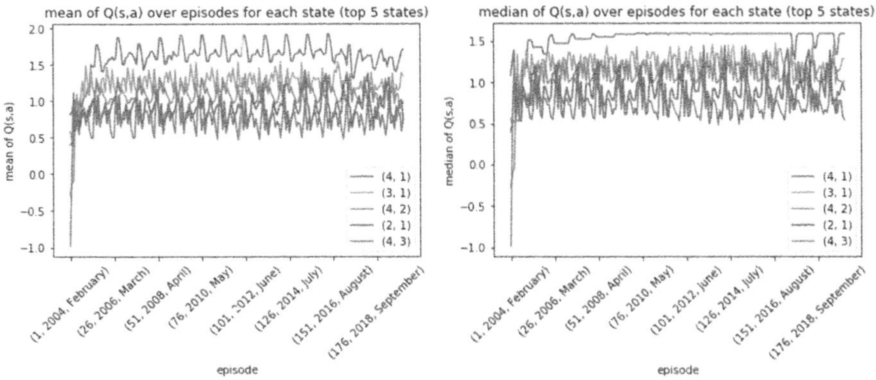

Fig. 20 Convergence plot with mean and median of Q-values for top 5 states with higher Q-values

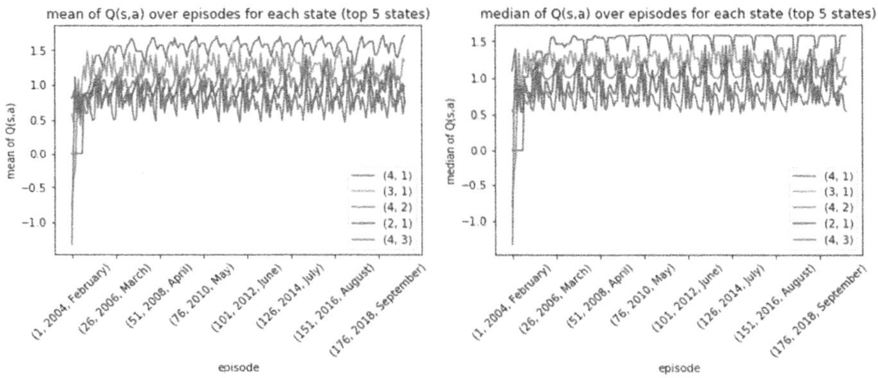

Fig. 21 Convergence plot with mean and median of Q-values for top 5 states with higher Q-values excluding Baxter Wilson

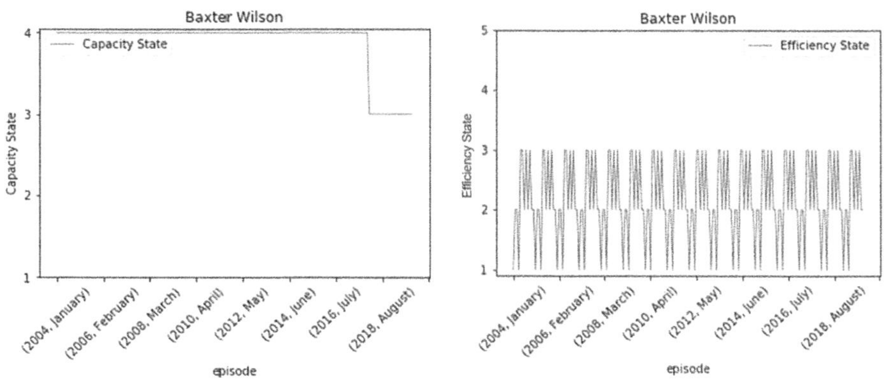

Fig. 22 Capacity state and efficiency state of Baxter Wilson power plant

5 Conclusion

This study proposed a hybrid DS framework and RL used with DEA to support the functional form identification of the production frontier while RL generated the optimal policy by taking resource allocation actions. In fact, both DS and RL techniques complement efficiency analysis. They enhance the interpretation of the production frontier, and guide the efficiency improvement rather than just estimating the "score" of efficiency. That is, this study emphasizes on planning over evaluation.

The proposed hybrid DS framework effectively fits the DEA production frontier through simulations of DGP. We use symbolic regression to estimate the production frontier, and then conduct the residual analysis by using regression tree and MARS. Particularly, SR also effectively identities the grammar which is the component in the true production function. The interaction effect between two inputs is successfully identified, and the results validate the use of SR as a data mining tool to extract useful managerial insights. In addition, RL is used to derive the optimal policy for the DMUs moving toward better productive efficiency. An empirical study of the US coal-fired and oil-fired power plants operating from 2004 through 2019 is conducted to validate the RL with DEA for efficiency estimation. Through investigating the capacity expansion, do nothing, contraction and other regarding the amount input and output adjustments, the results show that it is preferable to take an Expansion or Others action for better efficiency improvement. We conclude that the proposed hybrid DS and RL are useful in DEA and that it can guide productivity improvement strategy.

Acknowledgments This research was funded by Ministry of Science and Technology (MOST108-2221-E-006 -223 -MY3), Taiwan.

References

Adler, N., & Yazhemsky, E. (2010). Improving discrimination in data envelopment analysis: PCA-DEA or variable reduction. *European Journal of Operational Research, 202*, 273–284.

Afriat, S. N. (1972). Efficiency estimation of production functions. *International Economic Review, 13*(3), 568–598.

Banker, R. D., Chang, H., & Cooper, W. W. (1996). Equivalence and implementation of alternative methods for determining returns to scale in data envelopment analysis. *European Journal of Operational Research, 89*, 473–481.

Banker, R. D., Charnes, A., & Cooper, W. W. (1984). Some models for estimating technical and scale inefficiencies in data envelopment analysis. *Management Science, 30*(9), 1078–1092.

Banker, R. D., Cooper, W. W., Seiford, L. M., Thrall, R. M., & Zhu, J. (2004). Returns to scale in different DEA models. *European Journal of Operational Research, 154*, 345–362.

Barr, R. S., & Durchholz, M. L. (1997). Parallel and hierarchical decomposition approaches for solving large-scale data envelopment analysis models. *Annals of Operations Research, 73*, 339–372.

Benítez-Peñaa, S., Bogetoft, P., & Morales, D. R. (2020). Feature selection in data envelopment analysis: A mathematical optimization approach. *Omega, 96*, 102068.

Blei, D. H., & Smyth, P. (2017). Science and data science. *PNAS, 114*(33), 8689–8692.
Breiman, L. (2001). Random forests. *Machine Learning, 45*, 5–32.
Breiman, L., Friedman, J., Stone, C. J., & Olshen, R. A. (1984). *Classification and regression trees*. CRC press.
Charles, V., Aparicio, J., & Zhu, J. (2019). The curse of dimensionality of decision-making units: A simple approach to increase the discriminatory power of data envelopment analysis. *European Journal of Operational Research, 279*, 929–940.
Charnes, A., Cooper, W. W., & Rhodes, E. (1978). Measuring the efficiency of decision making units. *European Journal of Operational Research, 2*(6), 429–444.
Chen, C.-M. (2009). A network-DEA model with new efficiency measures to incorporate the dynamic effect in production networks. *European Journal of Operational Research, 194*, 687–699.
Chen, C.-M., & van Dalen, J. (2010). Measuring dynamic efficiency: Theories and an integrated methodology. *European Journal of Operational Research, 203*, 749–760.
Chen, W.-C., & Cho, W.-J. (2009). A procedure for large-scale DEA computations. *Computers & Operations Research, 36*(6), 1813–1824.
Chen, W.-C., & Lai, S.-Y. (2017). Determining radial efficiency with a large data set by solving small-size linear programs. *Annals of Operations Research, 250*(1), 147–166.
Chung, Y. H., Färe, R., & Grosskopf, S. (1997). Productivity and undesirable outputs: A directional distance function approach. *Journal of Environmental Management, 51*(3), 229–240.
Chunming, L., Xin, X., & Dewen, H. (2015). Multiobjective reinforcement learning: A comprehensive overview. *IEEE Transactions on Systems, Man, and Cybernetics: Systems, 45*(3), 385–398.
Coelli, T. J., Rao, D. S. P., O'Donnell, C. J., & Battese, G. (2005). *An introduction to efficiency and productivity analysis* (2nd ed.). Springer.
Cook, W. D., Tone, K., & Zhu, J. (2014). Data envelopment analysis: Prior to choosing a model. *Omega, 44*, 1–4.
Davenport, T. H., & Patil, D. J. (2012). Data scientist: The sexiest job of the 21st century. *Harvard Business Review, 90*(10), 70–76.
Donoho, D. (2017). 50 years of data science. *Journal of Computational and Graphical Statistics, 26*(4), 745–766.
Dulá, J. H. (2011). A method for data envelopment analysis. *INFORMS Journal on Computing, 23*(2), 284–296.
EIA (U.S. Energy Information Administration). (2020a). Form EIA-860 detailed data with previous form data (EIA-860A/860B). Available at https://www.eia.gov/electricity/data/eia860/ (Accessed 26 July 2020).
EIA (U.S. Energy Information Administration). (2020b). Form EIA-923 detailed data with previous form data (EIA-906/920). Available at https://www.eia.gov/electricity/data/eia923/ (Accessed 26 July 2020).
Fried, H. O., Lovell, C. A. K., & Schmidt, S. S. (2008). *The measurement of productive efficiency and productivity growth*. Oxford University Press.
Friedman, J. H. (1991). Multivariate adaptive regression splines. *The Annals of Statistics, 19*(1), 1–67.
Johnson, A. L., & Kuosmanen, T. (2011). One-stage estimation of the effects of operational conditions and practices on productive performance: Asymptotically normal and efficient, root-N consistent StoNEZD method. *Journal of Productivity Analysis, 36*(2), 219–230.
Johnson, A. L., & Kuosmanen, T. (2012). One-stage and two-stage DEA estimation of the effects of contextual variables. *European Journal of Operational Research, 220*(2), 559–570.
Johnson, A. L., & Lee, C.-Y., (2017). Predictive efficiency analysis. Book chapter edited in: Kaoru Tone, (Editor), *Advances in DEA theory and applications: with extensions to forecasting models*, (pp. 404–418), John Wiley & Sons Ltd.
Kao, C. (2014). Network data envelopment analysis: A review. *European Journal of Operational Research, 239*(1), 1–16.

Kao, C., & Hwang, S. (2008). Efficiency decomposition in two-stage data envelopment analysis: An application to non-life insurance companies in Taiwan. *European Journal of Operational Research, 185*(1), 418–429.

Khezrimotlagh, D., Zhu, J., Cook, W. D., & Toloo, M. (2019). Data envelopment analysis and big data. *European Journal of Operational Research, 274*, 1047–1054.

Koza, J. R. (1994). Genetic programming as a means for programming computers by natural selection. *Statistics and Computing, 4*, 87–112.

Kuosmanen, T., & Johnson, A. L. (2010). Data envelopment analysis as nonparametric least-squares regression. *Operations Research, 58*(1), 149–160.

Lee, C.-Y. (2014). Meta-data envelopment analysis: Finding a direction towards marginal profit maximization. *European Journal of Operational Research, 237*(1), 207–216.

Lee, C.-Y. (2015). Distinguishing operational performance in power production: A new measure of effectiveness by DEA. *IEEE Transactions on Power Systems, 30*(6), 3160–3167.

Lee, C.-Y. (2016). Nash-profit efficiency: A measure of changes in market structures. *European Journal of Operational Research, 255*(2), 659–663.

Lee, C.-Y. (2017). Directional marginal productivity: A foundation of meta-data envelopment analysis. *Journal of the Operational Research Society, 68*(5), 544–555.

Lee, C.-Y. (2018). Mixed-strategy Nash equilibrium in data envelopment analysis. *European Journal of Operational Research, 266*(3), 1013–1024.

Lee, C.-Y., & Cai, J.-Y. (2020). LASSO variable selection in data envelopment analysis with small datasets. *Omega: The International Journal of Management Science, 91*, 102019.

Lee, C.-Y., & Chen, B.-S. (2018). Mutually-exclusive-and-collectively-exhaustive feature selection scheme. *Applied Soft Computing, 68*, 961–971.

Lee, C.-Y., & Chien, C.-F. (2020). Pitfalls and protocols of data science in manufacturing practice. *Journal of Intelligent Manufacturing*. https://doi.org/10.1007/s10845-020-01711-w

Lee, C.-Y., & Johnson, A. L. (2011). A decomposition of productivity change in the semiconductor manufacturing industry. *International Journal of Production Research, 49*(16), 4761–4785.

Lee, C.-Y., & Johnson, A. L. (2012). Two-dimensional efficiency decomposition to measure the demand effect in productivity analysis. *European Journal of Operational Research, 216*(3), 584–593.

Lee, C.-Y., & Johnson, A. L. (2014). Proactive data envelopment analysis: Effective production and capacity expansion in stochastic environments. *European Journal of Operational Research, 232*(3), 537–548.

Lee, C.-Y., & Johnson, A. L. (2015a). Measuring efficiency in imperfectly competitive markets: An example of rational inefficiency. *Journal of Optimization Theory and Applications, 164*(2), 702–722.

Lee, C.-Y., & Johnson, A. L. (2015b). Effective production: Measuring of the sales effect using data envelopment analysis. *Annals of Operations Research, 235*(1), 453–486.

Lee, C.-Y., & Zhou, P. (2015). Directional shadow price estimation of CO2, SO2 and NOx in the United States coal power industry 1990-2010. *Energy Economics, 51*, 493–502.

Liang, L., Wu, J., Cook, W. D., & Zhu, J. (2008). The DEA game cross-efficiency model and its Nash equilibrium. *Operations Research, 56*(5), 1278–1288.

Liaw, A., & Wiener, M. (2002). Classification and regression by randomforest. *R News, 2*(3), 18–22.

McDonald, J. (2009). Using least squares and tobit in second stage DEA efficiency analyses. *European Journal of Operational Research, 197*(2), 792–798.

McKay, B., Willis, M. J., & Barton, G. W., (1995). *Using a tree structured genetic algorithm to perform symbolic regression*. First international conference on genetic algorithms in engineering systems: Innovations and applications. Sheffield, UK.

Puterman, M. L. (2005). *Markov decision processes: Discrete stochastic dynamic programming* (2nd ed.). John Wiley & Sons.

Seiford, L. M., & Zhu, J. (1999). An investigation of returns to scale in data envelopment analysis. *Omega, 27*, 1–11.

Simar, L., & Wilson, P. W. (2007). Estimation and inference in two-stage, semi-parametric models of production processes. *Journal of Econometrics, 136*(1), 31–64.

Sueyoshi, T. (1999). DEA-discriminant analysis in the view of goal programming. *European Journal of Operational Research, 115*, 564–582.

Tone, K., & Tsutsui, M. (2014). Dynamic DEA with network structure: A slacks-based measure approach. *Omega, 42*, 124–131.

Walpole, R. E., Myers, R. H., Myers, S. L., & Ye, K. E., (2012). Probability and Statistics for Engineers and Scientists, 9 ed.: Pearson.

Wang, K., Xian, Y., Lee, C.-Y., Wei, Y.-M., & Huang, Z. (2019). On selecting directions for directional distance functions in a non-parametric framework: A review. *Annals of Operations Research, 278*(1–2), 43–76.

Yu, M. M., & Lin, E. T. J. (2008). Efficiency and effectiveness in railway performance using a multi-activity network DEA model. *Omega, 36*(6), 1005–1017.

Zhu, J. (2020). DEA under big data: Data enabled analytics and network data envelopment analysis. *Annals of Operations Research*. https://doi.org/10.1007/s10479-020-03668-8

Zofio, J. L., Paster, J. T., & Aparicio, J. (2013). The directional profit efficiency measure: On why profit efficiency is either technical or allocative. *Journal of Productivity Analysis, 40*(3), 257–266.

Aggregation of Outputs and Inputs for DEA Analysis of Hospital Efficiency: Economics, Operations Research and Data Science Perspectives

Bao Hoang Nguyen and Valentin Zelenyuk

1 Introduction

Envelopment estimators in general and data envelopment analysis (DEA) in particular, have been widely recognised as a powerful tool for performance analysis.[1] Over the four decades since the seminal work of Charnes et al. (1978), DEA has been applied to study the efficiency and productivity of decision making units (DMUs) in various economic sectors. Among these, the healthcare industry, especially hospitals, has been one of the most active sectors of research, where many state-of-the-art developments of DEA have been utilised to provide empirical insights into a large number of papers (e.g., see reviews in Hollingsworth 2003, 2008; O'Neill et al. 2008; Kohl et al. 2019).

Despite its popularity, the application of DEA in empirical works has been more challenging, especially in the modern era of analytics, where big data becomes the norm. The challenge comes from the well-known fact that DEA, as with virtually any nonparametric estimator, suffers from the 'curse of dimensionality'. That is

[1]Stochastic frontier analysis (SFA) (Meeusen and van Den Broeck 1977; Aigner et al. 1977) is another popular method for productivity and efficiency analysis. Recently, more advanced methodologies have been developed in the literature such as stochastic DEA (Simar and Zelenyuk 2011), order-m frontiers (Cazals et al. 2002), order-α quantile frontiers (Aragon et al. 2005; Daouia and Simar 2007), to mention a few.

B. H. Nguyen
School of Economics, University of Queensland, Brisbane, QLD, Australia

V. Zelenyuk (✉)
School of Economics and Centre for Efficiency and Productivity Analysis, University of Queensland, Brisbane, QLD, Australia
e-mail: v.zelenyuk@uq.edu.au

J. Zhu, V. Charles (eds.), *Data-Enabled Analytics*, International Series in Operations Research & Management Science 312, https://doi.org/10.1007/978-3-030-75162-3_5

the accuracy and the discrimination power of DEA decrease when the dimension of input-output space increases. The issue is even worse in the context of big wide data, where the number of inputs and outputs is more than the number of observations: DEA becomes practically infeasible!

Interestingly, the challenge of high dimensionality has been there in the analysis of hospital efficiency long before the wave of modern big data. Since its first application in the U.S. in the early 1980s, hospital product in many countries has been defined using the diagnosis-related group (DRG) classification scheme (see more detailed discussions in Fetter 1991). Under the classification system, hospital product is classified into hundreds of DRGs, grouping together patients with similar diagnoses who require similar hospital services. Each of these hundreds of DRGs represents a separate type of output of hospitals and counts into the dimension of the production space, which usually turns out to be larger than the typical sample sizes in many applied works in the field.

To make DEA models feasible, a natural and common approach in the literature is to use the so-called casemix weighted episodes (or discharges or patient days). A casemix weighted episode is a weighted sum of the number of episodes in each DRG with the weight being the DRG's casemix cost weight, which can be viewed as the relative price of each DRG in the funding systems. This approach is actually an example of the price-based aggregation that is one of the most powerful techniques for dimension reduction to deal with the 'curse of dimensionality' in the context of DEA with big wide data.[2] The other powerful technique is the principle component analysis (PCA)-based aggregation,[3] which is based on the eigendecomposition of the moment (or correlation) matrix of inputs or outputs or their subsets.[4] The performance of these two techniques in the context of DEA with big data have recently been investigated in the literature, with intensive Monte Carlo simulations therein showing that the gain of dimension reduction is usually more than the information lost (e.g., see Wilson 2018; Zelenyuk 2020).

Back to the case of output aggregation in hospital efficiency analysis, some questions are of interest, but are usually overlooked by DEA practitioners, and now may be a good time to revisit these given the increasing attention to DEA as "data enabled analytics" under big data (Zhu 2020). In particular, some of these questions include: (1) For price-based aggregation, what weights should be used, e.g., a constant weight system or a weight system varying across years? (This question is typically relevant to the case of short panel data when researchers usually pool data across years to estimate a grand frontier) (2) What about PCA-based aggregation,

---

[2]E.g., see Zelenyuk (2020) for evidence about the performance of price-based aggregation for dimension reduction in the context of DEA with big data.

[3]E.g., see Wilson (2018) for evidence about the performance of PCA-based aggregation for dimension reduction in the context of DEA with big data.

[4]Another promising technique that has been recently applied into the DEA context for dimension reduction is Least Absolute Shrinkage and Selection Operator (LASSO) (see more discussions in Lee and Cai 2020; Chen et al. 2021, and references therein).

i.e., are the estimated efficiency scores robust with respect to the different dimension reduction methods?

In this study, we examine these empirical questions using a fairly disaggregated dataset of public hospitals in Queensland, Australia, which, among other things, consists of 153,472 data points for 704 different types of inpatient procedures across 109 hospitals in 2 years. We find that with a small variation in prices across years, the choice of price (weight) systems does not significantly affect the estimated efficiency scores. Moreover, the estimated efficiency scores from the DEA models are also robust with respect to the two different aggregation approaches. The robustness of the results suggests that PCA-based aggregation can be viewed as a viable alternative for DEA practitioners who are unable to/or unwilling to use the price-based approach, e.g., due to unavailable or unreliable price information.

The rest of the paper is structured as follows. Section 2 discusses methodologies including envelopment estimators and aggregation methods. Section 3 briefly reviews the statistical methods to analyse the efficiency scores. Section 4 describes data and variables used in the study. Section 5 compares the estimated efficiency scores among different aggregation approaches. Section 6 provides concluding remarks.

2 Methodology

2.1 Envelopment Estimators

Let $\mathscr{X}^n = \{(x^k, y^k) \mid k = 1, \cdots, n\}$ be a set of input-output allocations of a sample of n DMUs, where $x^k = (x_1^k, \cdots, x_N^k) \in \Re_+^N$ is an $N-$dimensional vector of inputs and $y^k = (y_1^k, \cdots, y_M^k) \in \Re_+^M$ is an $M-$dimensional vector of outputs.[5]

Under the assumption of Constant Returns to Scale (CRS), the DEA-estimator of the Debreu-Farrell output oriented technical efficiency score of a DMU with input-output allocation of (x^j, y^j) is given by[6]

[5]It is worth mentioning here that in this study we pooled data over years to estimate a grand frontier, so n here is total number of observations across years.

[6]There are also other measures of technical efficiency such as the Debreu-Farrell input-oriented measure, the Russell measure, the directional distance function, the slack-based measure, etc. Due to the limit of space we focus on the Debreu-Farrell output oriented technical efficiency here, and the interested readers can find detailed discussions about the other measures of technical efficiency in Sickles and Zelenyuk (2019, Ch. 3 & Ch. 8).

$$\widehat{OTE}_{DEA-CRS}\left(x^j, y^j | \mathscr{X}^n\right) \equiv \max_{\theta, z^1, \cdots, z^n} \theta$$
$$\text{s.t.}$$
$$\sum_{k=1}^{n} z^k y_m^k \geq \theta y_m^j, \; m = 1, \cdots, M, \tag{1}$$
$$\sum_{k=1}^{n} z^k x_l^k \leq x_l^j, \; l = 1, \cdots, N,$$
$$\theta \geq 0, \; z^k \geq 0, \; k = 1, \cdots, n.$$

One can add an additional constraint, $\sum_{k=1}^{n} z^k = 1$, into Eq. (1) to obtain the DEA-estimator under the assumption of Variable Returns to Scales (VRS), specifically

$$\widehat{OTE}_{DEA-VRS}\left(x^j, y^j | \mathscr{X}^n\right) \equiv \max_{\theta, z^1, \cdots, z^n} \theta$$
$$\text{s.t.}$$
$$\sum_{k=1}^{n} z^k y_m^k \geq \theta y_m^j, \; m = 1, \cdots, M,$$
$$\sum_{k=1}^{n} z^k x_l^k \leq x_l^j, \; l = 1, \cdots, N, \tag{2}$$
$$\sum_{k=1}^{n} z^k = 1,$$
$$\theta \geq 0, \; z^k \geq 0, \; k = 1, \cdots, n.$$

Another popular envelopment estimator is the so-called Free Disposal Hull (FDH) estimator recently developed by Deprins et al. (1984). The FDH estimator of the Debreu-Farrell output oriented technical efficiency score can be obtained by solving the following integer program problem

$$
\begin{aligned}
\widehat{OTE}_{FDH}\left(x^j, y^j \mid \mathcal{X}^n\right) \equiv & \max_{\theta, z^1, \cdots, z^n} \theta \\
& \text{s.t.} \\
& \sum_{k=1}^{n} z^k y_m^k \geq \theta y_m^j, \; m = 1, \cdots, M, \\
& \sum_{k=1}^{n} z^k x_l^k \leq x_l^j, \; l = 1, \cdots, N, \\
& \sum_{k=1}^{n} z^k = 1, \\
& \theta \geq 0, \; z^k \in \{0, 1\}, \; k = 1, \cdots, n.
\end{aligned} \tag{3}
$$

The formulas discussed above are the envelopment estimators of true Debreu-Farrell output oriented technical efficiency scores, defined by

$$OTE\,(x, y|\Psi) \equiv \max_{\theta} \{\theta \geq 0|\,(x, \theta y) \in \Psi\}, \tag{4}$$

where Ψ is the set characterising the technology, defined as

$$\Psi \equiv \left\{(x, y) \in \Re_+^N \times \Re_+^M \,|y \text{ can be produced by } x\right\}. \tag{5}$$

The technology set is assumed to satisfy the standard regularity conditions of the production theory (Shephard 1953, 1970; Färe and Primont 1995).

Due to the limit of space we focus here on these most popular DEA models, while there are many other variations of DEA, which also allow modelling the 'undesirable outputs' and 'congesting inputs', incorporating network structure, or imposing various weight restrictions, etc.[7]

2.2 'Curse of Dimensionality'

In this section, we will discuss the methodological issues relating to a large dimension of input-output space in the context of envelopment analysis.

As we can see from Eqs. (1) to (3), when the dimension of input-output space $(N + M)$ is larger than sample size (n), the models become practically infeasible. Another issue can also arise when $N + M$ is substantially lower than n, yet still large enough to significantly reduce the discrimination power of the models. In such

[7]See Sickles and Zelenyuk (2019, Ch. 8) for more details and related discussion.

a scenario, the large dimension possibly makes the estimated results not practically useful (i.e., all DMUs may attain an efficiency score of or nearly 100%).[8] More importantly, the accuracy of envelopment estimation depends on the dimension of the input-output space in such a way that the rate of convergence to the truth decreases exponentially when the dimension increases. It is well-known in the literature (e.g., see Kneip et al. 1998) that under appropriate assumptions, the DEA/FDH estimators are consistent with the convergence rate depending on the dimension of input-output space as follows

$$\widehat{OTE}(x, y) - OTE(x, y) = O_p\left(n^{-\kappa}\right), \tag{6}$$

where $O_p\left(n^{-\kappa}\right)$ denotes the order of magnitude and κ depends on the estimator. Specifically, we have $\kappa = 2/(N + M)$ for DEA-CRS estimator, $\kappa = 2/(N + M + 1)$ for DEA-VRS estimator and $\kappa = 1/(N + M)$ for FDH estimator.

2.3 Aggregation of Inputs and Outputs Before DEA

Aggregation of all or some inputs and/or outputs, *inter alia*, can be viewed as a useful method (in some cases, the only method) to deal with the 'curse of dimensionality' in the context of envelopment analysis.[9] For example, in hospital efficiency analysis, labour input is composed of many professionals (e.g., medical officers, nurses, diagnostic and health professionals, etc.). Although it is desirable (and might be feasible in many cases) to include all these labour categories as separate inputs in DEA models to account for the difference in labour composition across hospitals, the estimated efficiency scores from such models might not be reliable due to the issue of the 'curse of dimensionality' as discussed in Sect. 2.2. On the output side, we also face a similar issue, but more severe. Under the DRG classification system, hospital product is classified into hundreds of DRGs. Each of these hundreds of DRGs represents a separate type of output of hospitals and counts into the dimension of the production space, which usually turns out to be larger than the typical sample sizes in many applied works in the field, making DEA models practically infeasible (e.g., in the working sample of the current study, there are 704 DRGs but only 218 observations).

In this section we will briefly discuss the two popular approaches to aggregate inputs/outputs which are: (1) PCA-based aggregation and (2) Price-based aggrega-

[8] Also see a related discussion in Charles et al. (2019) who also proposed a novel approach to increase the discriminatory power of DEA in the presence of the 'curse of dimensionality'.

[9] One can also consider the order-α quantile frontier and order-m frontier estimators as alternative approaches for the frontier estimation since these estimators do not suffer from the 'curse of dimensionality'. E.g., see an application of the order-α quantile frontier estimators to hospital data in Nguyen and Zelenyuk (2021b).

tion.[10] We will illustrate for the context of the aggregation of some $\hat{M} \leq M$ outputs, and without loss of generality, assume that these outputs are the 1st-output to the $\hat{M}$th-output.[11]

2.3.1 PCA-Based Aggregation

The idea of PCA-based aggregation in the DEA context is to find linear combinations of inputs/outputs that best represent their variation based on the eigendecomposition of their moment or correlation matrix. The application of PCA in the context of efficiency analysis goes back at least to Zhu (1998), Adler and Golany (2001, 2007), Mouchart and Simar (2002), and Daraio and Simar (2007). In particular, Zhu (1998) appears to be the first who utilised PCA as a method for evaluating relative performance of DMUs and compared it to DEA. Meanwhile, Adler and Golany (2001, 2007) employed PCA as a dimension reduction method based on the eigendecomposition of the correlation matrix of inputs or outputs. Mouchart and Simar (2002) and Daraio and Simar (2007) also utilised PCA for dimension reduction in the context of DEA, but their approach based on the eigendecomposition of the moment matrix of inputs and outputs. It is worth noting here that PCA-based aggregation using a correlation matrix has an issue in that the first eigenvector (corresponding to the largest eigenvalue) of a correlation matrix may contain both positive and negative values, thus it is not practically useful for the context of envelopment analysis (see more discussions in Wilson 2018). To avoid this issue, as in Wilson (2018), we follow Mouchart and Simar (2002) and Daraio and Simar (2007) to utilise the PCA-based aggregation using the eigendecomposition of the moment matrix.

Let $\hat{\mathbf{Y}}$ be an $n \times \hat{M}$ matrix stacking the 1st-output to the $\hat{M}$th-output across all n DMUs.[12] $\hat{\mathbf{Y}}$ can be viewed as a cloud of n data points in an $\hat{M}$−dimensional space. The aggregation question here is to find the best subspace of dimension $\tilde{M}$ ($\tilde{M} < \hat{M}$) through the origin to project the cloud onto. Now let us define $\mathscr{B} = \{u_1, \ldots, u_{\tilde{M}}\}$ as an orthonormal basis for the subspace, i.e., $u_{\tilde{m}}$ ($\tilde{m} = 1, \ldots, \tilde{M}$) are all unit vectors and are orthogonal to each other. A natural approach to determine the best subspace is to find the optimal orthonormal basis to solve the following "least square" problem

$$\min_{u_1,\ldots,u_{\tilde{M}}} \sum_{k=1}^{n} \left(\left\| \hat{y}^k \right\|^2 - \| p_{\hat{y}^k} \|^2 \right),$$

where $\hat{y}^k$ is the kth row of $\hat{\mathbf{Y}}$ and $p_{\hat{y}^k}$ is the projection of $\hat{y}^k$ onto the subspace.

[10] More detailed discussions can be found in Wilson (2018) and Zelenyuk (2020).

[11] Similar description can be given for the case of aggregating many inputs.

[12] Since the efficiency measure is invariant to the units of measurement, it is advised to standardise each column of $\hat{\mathbf{Y}}$ by its standard deviation before the aggregation.

It is well known in mathematics (statistics, optimisation theory, etc.) that the optimal solution for the above problem is the set of orthogonal eigenvectors, $\Lambda_1, \Lambda_2, \dots \Lambda_{\tilde{M}}$, of $\hat{\mathbf{Y}}^\top \hat{\mathbf{Y}}$, which corresponds to the first $\tilde{M}$ largest eigenvalues, $\lambda_1 \geq \lambda_2 \geq \cdots \geq \lambda_{\tilde{M}}$ (e.g., see Härdle and Simar 2020). As in Härdle and Simar (2020), $\Lambda_{\tilde{m}}, \tilde{m} = 1, \dots, \tilde{M}$, is called the $\tilde{m}-$th factorial axis, and the coordinate of each DMU on the $\tilde{m}-$th factorial axis

$$z_{\tilde{m}}^k = \left(\hat{y}^k\right)^\top \Lambda_{\tilde{m}}, \; k = 1, \cdots, n, \tag{7}$$

is called the $\tilde{m}-$th factorial variable. The point $z^k = \left(z_1^k, \dots, z_{\tilde{M}}^k\right)$ is then the representation of $\hat{y}^k$ in the subspace of dimension $\tilde{M}$. Moreover, the quality of the factorial representations of the original data in the subspace of dimension $\tilde{M}$ can be measured by

$$\delta_{\tilde{M}} := \frac{\sum_{m=1}^{\tilde{M}} \lambda_m}{\sum_{m=1}^{\hat{M}} \lambda_m}. \tag{8}$$

The parameter $\delta_{\tilde{M}}$ represents the percentage of the inertia in the original data that can be explained by the $\tilde{M}$ factorial variables. And as discussed in Daraio and Simar (2007) and Wilson (2018), if the correlation among the $\hat{M}$ outputs is high, then the majority of variation in the data can be explained by the first few factorial variables (i.e., high value of $\delta_{\tilde{M}}$ with a small value of $\tilde{M}$). In this study, we will consider a special case of PCA-based approach, where all $\hat{M}$ outputs are aggregated into an aggregated measure of dimension 1—the first factorial variable.

2.3.2 Price-Based Aggregation

Let us assume that all n DMUs face the same output prices for the $\hat{M}$ outputs, denoted as $\left(p_1, \cdots, p_{\hat{M}}\right) \in \Re_{++}^{\hat{M}}$. We then can use the output prices as weights to linearly aggregate the $\hat{M}$ outputs. Specifically, in Eqs. (1)–(3), for each of the DMUs, we replace the first $\hat{M}$ outputs by their sub-revenue

$$r^k = \sum_{m=1}^{\hat{M}} p_m y_m^k, \; k = 1, \cdots, n. \tag{9}$$

Besides being very simple, the price-based aggregation approach takes into account the economic valuation of the outputs or inputs (reflected through the prices), which, in principle, might be very different from the weights implied through statistical approaches, such as the PCA-based approach. On the other hand, a limitation of the price-based aggregation is that in practice, some information about the prices might be unavailable.

Under certain conditions, the estimated efficiency scores obtained using the aggregated output are upward biased (compared to the case of using disaggregated outputs), but the bias is bounded by the allocative efficiency. Moreover, it is worth noting here that the assumption of the same output price is for establishing the exact theoretical properties.[13] In practice, when the prices vary to some extent, this price-aggregation approach may still perform well, as confirmed by the Monte Carlo simulations in Zelenyuk (2020). In this study, we also examine the robustness of results with different price systems: same prices vs. varying prices.

3 Statistical Methods to Analyse the Efficiency Scores

After obtaining DEA/FDH estimates of efficiency scores of individual DMUs, researchers usually perform the statistical inference to explore more about the efficiency level of the population or its some groups of interest, for example, comparing the efficiency of different groups in the population by analysing their means, or their densities, or studying the determinants of the efficiency by using regression analysis. Some standard exploratory tools from statistics/econometrics such as the nonparametric statistical tests (e.g., Wilcoxon rank-sum test, Kolmogorov-Smirnov test, etc.), the ordinary least squares regression, or the censored (tobit) regression, were employed to further analyse the efficiency scores from DEA/FDH models.[14] The statistical inference, however, needs to be carefully adapted to the DEA/FDH context given the well-known fact that the DEA/FDH estimates are biased and are serially correlated in a complicated unknown way (Sickles and Zelenyuk 2019; Simar and Wilson 2007, 2015).

The adapted methods for statistical inference based on the estimated DEA/FDH efficiency scores have been recently developed and become more popular in the literature thanks to the important works of Simar and Wilson (1998, 2000), Kneip et al. (2008) (for the bootstrap-based statistical inference for individual efficiency scores), Simar and Zelenyuk (2007) (for the bootstrap–based statistical inference for the weighted mean of efficiency scores), Simar and Zelenyuk (2006) (for the kernel density estimator–based tests), Simar and Wilson (2007) (for the bootstrap truncated regression), Kneip et al. (2015, 2016), Simar and Wilson (2020) (for the central limit theorems for the simple mean of efficiency scores and related tests), Simar and Zelenyuk (2018) (for the central limit theorems for the weighted mean

[13]E.g., see Färe and Grosskopf (1985), Färe et al. (2004) and Zelenyuk (2020) for detailed discussions.

[14]See more discussion about the application as well as the caveats of these methods in the context of DEA/FDH in Grosskopf (1996) and Simar and Wilson (2007).

of efficiency scores), and Simar and Zelenyuk (2020) (for the improvement in finite sample approximation by the central limit theorems), among others.

In this section, we will focus our discussion on the kernel density estimation and related statistical tests as well as the analysis of simple means and weighted means of estimated efficiency scores, which we will apply in the next section.

3.1 Kernel Density Estimation and Related Tests

To discuss the kernel density estimation, let us denote $\{u_i\}_{i=1}^{n}$ as a set of the realisations of a random variable U and denote f_U as the density function of the variable. Following Rosenblatt (1956), the estimate of f_U at a point u can be obtained by

$$\hat{f}_h(u) = \frac{1}{nh}\sum_{i=1}^{n} K\left(\frac{u_i - u}{h}\right), \tag{10}$$

where $K(\cdot)$ and h are suitable kernel and bandwidth, respectively.

The application of the kernel density estimation for estimating the densities of efficiency scores in the context of DEA/FDH, however, faces two issues. The first issue is the discontinuity problem, that is the probability of having an observation which is fully efficient (i.e., its estimated efficiency score equals one) is always different from zero by construction. The second issue is the problem of bounded support that is the efficiency score is bounded with most of the mass of its distribution being around the boundary, leading to the inconsistency of the kernel density estimator at the boundary.[15] To overcome these issues, one can remove from the sample those observations whose estimated efficiency scores equal to one and then apply the Silverman's (1986) reflection method to estimate the density. Specifically, the kernel density estimate of DEA/FDH efficiency scores at a point u can be obtained by

$$\hat{f}_h^R(u) = \begin{cases} \frac{1}{\tilde{n}h_R}\sum_{i=1}^{\tilde{n}}\left[K\left(\frac{\hat{u}_i - u}{h_R}\right) + K\left(\frac{(2-\hat{u}_i)-u}{h_R}\right)\right], & u \geq 1 \\ 0, \text{ otherwise} \end{cases}, \tag{11}$$

where $\{\hat{u}_i\}_{i=1}^{\tilde{n}}$ is a set of estimated output-oriented efficiency scores whose values are greater than unity ($\tilde{n} < n$), h_R is the bandwidth selected for the reflected sample (i.e., $\{\hat{u}_1, ..., \hat{u}_{\tilde{n}}, 2-\hat{u}_i, ..., 2-\hat{u}_{\tilde{n}}\}$).[16] As discussed in Sickles and Zelenyuk (2019), the choice of the kernel $K(\cdot)$ is not very important and one can obtain a good fit

[15]See more discussions in Sickles and Zelenyuk (2019).

[16]The formula can be easily adapted to the case of input-oriented efficiency scores.

when using popular kernels such as Gaussian or Epanechnikov. On the other hand, the selection of the bandwidth is much more important, thus it is advised to use more advanced procedures (e.g., the method of Sheather and Jones (1991), the cross-validation approach, etc.) to select the bandwidth although the Silverman (1986) robust rule-of-thumb bandwidth can give a fairly good fit.

In addition to estimating and visualising the density of the estimated efficiency scores, DEA/FDH practitioners are also interested in comparing the densities of efficiency scores of different groups of DMUs in the population using the statistical tests. Here, we will focus our discussion on the Li (1996) test—the test based on the kernel density estimator. To simplify the notations, let us consider only two groups of DMUs, say, group A and Z, with their efficiency realisations and efficiency density function being $\{u_i^A\}_{i=1}^{n_A}$ and $f_A(\cdot)$, $\{u_i^Z\}_{i=1}^{n_Z}$ and $f_Z(\cdot)$, respectively. Formally, the null and alternative hypotheses of the test can be stated as follows

$$\begin{aligned} H_0 &: f_A(u) = f_Z(u), \forall u \text{ in the support of } U^A \text{ and } U^Z \\ H_1 &: f_A(u) \neq f_Z(u), \text{ on a set of positive measures.} \end{aligned}$$

The Li (1996) test statistics can be obtained by

$$\hat{L}_{n_A,n_Z,h} = n_A h^{\frac{1}{2}} \hat{J}_{n_A,n_Z,h} / \hat{\sigma}_{\gamma,h} \xrightarrow[\text{under } H_0]{d} N(0,1), \tag{12}$$

where

$$\begin{aligned} \hat{J}_{n_A,n_Z,h} =& \frac{1}{n_A(n_A-1)h} \sum_{i=1}^{n_A} \sum_{k=1,k\neq i}^{n_A} K\left(\frac{u_{A,i}-u_{A,k}}{h}\right) \\ &+ \frac{1}{n_Z(n_Z-1)h} \sum_{i=1}^{n_Z} \sum_{k=1,k\neq i}^{n_Z} K\left(\frac{u_{Z,i}-u_{Z,k}}{h}\right) \\ &- \frac{1}{n_Z(n_A-1)h} \sum_{i=1}^{n_A} \sum_{k=1,k\neq i}^{n_Z} K\left(\frac{u_{A,i}-u_{Z,k}}{h}\right) \\ &- \frac{1}{n_A(n_Z-1)h} \sum_{i=1}^{n_Z} \sum_{k=1,k\neq i}^{n_A} K\left(\frac{u_{Z,i}-u_{A,k}}{h}\right) \end{aligned} \tag{13}$$

$$
\begin{aligned}
\hat{\sigma}^2_{\gamma,h} =2 \Bigg\{ & \frac{1}{n_A^2 h} \sum_{i=1}^{n_A} \sum_{k=1}^{n_A} K\left(\frac{u_{A,i} - u_{A,k}}{h}\right) \\
& + \frac{\gamma_n^2}{n_Z^2 h} \sum_{i=1}^{n_Z} \sum_{k=1}^{n_Z} K\left(\frac{u_{Z,i} - u_{Z,k}}{h}\right) \\
& + \frac{\gamma_n}{n_A n_Z h} \sum_{i=1}^{n_A} \sum_{k=1}^{n_Z} K\left(\frac{u_{A,i} - u_{Z,k}}{h}\right) \\
& + \frac{\gamma_n}{n_Z n_A h} \sum_{i=1}^{n_Z} \sum_{k=1}^{n_A} K\left(\frac{u_{Z,i} - u_{A,k}}{h}\right) \Bigg\} \int K^2(u)du
\end{aligned}
\tag{14}
$$

where: $K(\cdot)$ and h are suitable kernel and suitably selected bandwidth, respectively. $\gamma_n = n_A/n_Z$ and $\lim_{n_A \to \infty} \gamma_n = \gamma$, $\gamma \in (0, \infty)$.

In addition to the problem of discontinuity discussed above, the main issue of applying Li (1996) test (or any tests based on the kernel density estimator or Kolmogorov-Smirnov test) in the context of DEA/FDH is that we are not using the true efficiency scores in the test, but their DEA/FDH estimates, and these estimates are biased and are serially correlated in a complicated unknown way. Simar and Zelenyuk (2006) propose two approaches to remedy the issues. The two approaches are based on bootstrapping the Li (1996) test statistics from: either (1) the sample of estimated efficiency scores with those efficiency scores whose value equal unity being removed (the first approach), or (2) the sample of the 'smoothed' estimated efficiency scores, i.e., 'smoothing' away from the boundary those efficiency scores whose value equal unity by adding a small uniform noise to them (the second approach) (see the bootstrap algorithms in Simar and Zelenyuk 2006).

3.2 *Analysis of Simple and Weighted Means*

To analyse the overall tendency of efficiency in the population, researchers often estimate and perform the statistical inference based on the simple mean of DEA/FDH efficiency scores of a random sample, $\mathcal{S}^n = \{(X^i, Y^i) \mid i = 1, \cdots, n\}$, i.e.,[17]

$$
\bar{\theta}_n = \frac{1}{n} \sum_{i=1}^{n} \widehat{OTE}\left(X^i, Y^i | \mathcal{S}^n\right). \tag{15}
$$

[17]We change the notations to the capital letters to highlight that the statistical inference for the population mean needs to be based on the statistical properties of the DEA/FDH estimators of efficiency scores at the random points.

This simple estimator for the population mean has a drawback that it does not take the size and thus the economic influence of individual DMUs in the population into account. One way to account for the relative economic importance of individual DMUs in the population is to utilise a weighted mean of DEA/FDH efficiency scores, e.g.,

$$\hat{\tau}_n = \sum_{i=1}^{n} \widehat{OTE}\left(X^i, Y^i | \mathscr{S}^n\right) \times W^i, \quad W^i = \frac{pY^i}{\sum_{i=1}^{n} pY^i}, \tag{16}$$

where W^i is the weight, and p is a row vector of output prices. The weighted mean formulated above is nothing but an envelopment estimator of the output oriented aggregate technical efficiency proposed by Färe and Zelenyuk (2003),

$$\tau_n = \sum_{i=1}^{n} OTE\left(X^i, Y^i\right) \times W^i, \quad W^i = \frac{pY^i}{\sum_{i=1}^{n} pY^i}. \tag{17}$$

As discussed in Färe and Zelenyuk (2003), the main advantage of this aggregate measure is that the weights are economically meaningful since they are derived from the economic optimisation principle.[18]

It is well-known in the literature that the conventional central limit theorems (CLT) fail to apply to $\bar{\theta}_n$ and $\hat{\tau}_n$ since these estimators inherit the statistical properties of DEA/FDH estimators, which are biased and the bias is of a higher order than the variance when the number of inputs and outputs increases. Kneip et al. (2015) have recently developed the new CLTs for the simple mean of DEA/FDH efficiency scores by correcting the bias and controlling the convergence rates for both bias and variance. Simar and Zelenyuk (2018) extended the important work of Kneip et al. (2015) to the case of the weighted mean of DEA/FDH efficiency scores and developed the new CLTs for this estimator. Simar and Zelenyuk (2020) further improved the finite sample approximation by these new CLTs by proposing to use the bias-corrected estimators of variances. These important works have laid theoretical foundation for many useful statistical inference methods involving DEA/FDH estimators. In this section, we will briefly summarise the important results in these studies focusing on the CLTs and confidence interval estimation.

3.2.1 Central Limit Theorems and Confidence Interval for Simple Means

The first important result developed in Kneip et al. (2015) is the statistical properties of DEA/FDH estimators at a random point, which can be summarised as follows

[18]Here we focus our discussion on the analysis of simple and weighted means of DEA/FDH efficiency scores in the whole population. One can find the similar discussion for the case of multiple groups in the population in Nguyen and Zelenyuk (2021a).

Table 1 Rate of convergence of envelopment estimators

Estimators	κ	$R_{n,\kappa}$
DEA-CRS	$2/(N+M)$	$O\left(n^{-3\kappa/2}(\log n)^{\alpha_1}\right)$
DEA-VRS	$2/(N+M+1)$	$O\left(n^{-3\kappa/2}(\log n)^{\alpha_2}\right)$
FDH	$1/(N+M)$	$O\left(n^{-2\kappa}(\log n)^{\alpha_3}\right)$

The values of $\alpha_j,\ j=1,2,3$, are greater than one and can be found in Kneip et al. (2015)

$$E\left[\widehat{OTE}\left(X^i,Y^i|\mathscr{S}^n\right)-OTE\left(X^i,Y^i\right)\right]=Cn^{-\kappa}+R_{n,\kappa}, \tag{18}$$

$$E\left[\left(\widehat{OTE}\left(X^i,Y^i|\mathscr{S}^n\right)-OTE\left(X^i,Y^i\right)\right)^2\right]=o\left(n^{-\kappa}\right), \tag{19}$$

$$\begin{aligned}&\Big|COV\Big[\widehat{OTE}\left(X^i,Y^i|\mathscr{S}^n\right)-OTE\left(X^i,Y^i\right),\\&\qquad\widehat{OTE}\left(X^j,Y^j|\mathscr{S}^n\right)-OTE\left(X^j,Y^j\right)\Big]\Big|=o\left(n^{-1}\right),\end{aligned} \tag{20}$$

when $n\to\infty$ and under regularity conditions specified in Kneip et al. (2015). The values of the constant term C, the convergence rate κ and the remainder term $R_{n,\kappa}$ are different for different estimators and depend on the dimension of the production space (see Table 1).

With this important result, Kneip et al. (2015) establish the new CLTs for the simple mean of DEA/FDH efficiency scores. Specifically, under regularity conditions specified in Kneip et al. (2015), and for $N+M\le 5$ if DEA-CRS is used, for $N+M\le 4$ if DEA-VRS is used, and for $N+M\le 3$ if FDH is used, the CLTs for the simple mean of DEA/FDH efficiency scores are given by

$$\frac{\sqrt{n}}{\hat{\sigma}_\theta}\left(\bar{\theta}_n-\hat{B}_{\bar{\theta}_n,\kappa}-\mu_\theta+R_{n,\kappa}\right)\xrightarrow{d}N(0,1), \tag{21}$$

where $\mu_\theta=E\left[OTE\left(X^i,Y^i\right)\right]$ is the population mean, $\hat{\sigma}_\theta$ is the empirical version of the population standard deviation, $\sigma_\theta=\sqrt{VAR\left[OTE\left(X^i,Y^i\right)\right]}$, and $\hat{B}_{\bar{\theta}_n,\kappa}$ is the generalized jackknife estimator of the bias of $\bar{\theta}_n$, which will be discussed later.

Alternatively, for $\kappa<1/2$, the new CLTs are given by

$$\frac{\sqrt{n_\kappa}}{\hat{\sigma}_\theta}\left(\bar{\theta}_{n,\kappa}-\hat{B}_{\bar{\theta}_n,\kappa}-\mu_\theta+R_{n,\kappa}\right)\xrightarrow{d}N(0,1) \tag{22}$$

where $\bar{\theta}_{n,\kappa}$ is a subsample version of $\bar{\theta}_n$, in the sense that the averages are taken over a random subsample $\mathscr{S}_{n_\kappa}\subset\mathscr{S}^n$ of size $n_\kappa=\lfloor n^{2\kappa}\rfloor$. Formally

$$\bar{\theta}_{n,\kappa}=n_\kappa^{-1}\sum_{\left\{i\,\middle|\,(X^i,Y^i)\in\mathscr{S}_{n_\kappa}\right\}}\widehat{OTE}\left(X^i,Y^i|\mathscr{S}^n\right). \tag{23}$$

The procedure to obtain the generalized jackknife estimator of the bias of $\bar{\theta}_n$ in Kneip et al. (2015) can be summarised as follows. First, let us consider a random split, say random split ℓ, of the sample into two parts with their sizes being $m_1 = \lfloor n/2 \rfloor$ and $m_2 = n - m_1$, and let us denote the random subset of size m_1 of $\mathscr{S}^n$ as $\mathscr{S}^{(1)}_{m_1,\ell}$, and the set of remainders in $\mathscr{S}^n$ as $\mathscr{S}^{(2)}_{m_2,\ell}$.

For $j \in \{1, 2\}$, let

$$\bar{\theta}^{(j)}_{m_j,\ell} = (m_j)^{-1} \sum_{(X^i,Y^i)\in \mathscr{S}^{(j)}_{m_j,\ell}} \widehat{OTE}\left(X^i, Y^i | \mathscr{S}^{(j)}_{m_j,\ell}\right). \tag{24}$$

Now, let us define

$$\bar{\theta}^*_{n,\ell} = \frac{1}{2}\left(\bar{\theta}^{(1)}_{m_1,\ell} + \bar{\theta}^{(2)}_{m_2,\ell}\right), \tag{25}$$

then

$$\left(2^{\kappa} - 1\right)^{-1}\left(\bar{\theta}^*_{n,\ell} - \bar{\theta}_n\right) \tag{26}$$

provides an estimator of the bias term of $\bar{\theta}_n$. To reduce the variance of the estimator, one can repeat the above operations for L times and average (26) to obtain $\hat{B}_{\bar{\theta}_n,\kappa}$

$$\hat{B}_{\bar{\theta}_n,\kappa} = L^{-1}\sum_{\ell=1}^{L}\left(2^{\kappa} - 1\right)^{-1}\left(\bar{\theta}^*_{n,\ell} - \bar{\theta}_n\right). \tag{27}$$

Using the new CLTs, one can construct the $(1-\alpha)$ 100% confidence interval for the simple mean in the population as

$$\left[\bar{\theta}_n - \hat{B}_{\bar{\theta}_n,\kappa} \pm z_{1-\alpha/2}\frac{\hat{\sigma}_\theta}{\sqrt{n}}\right], \tag{28}$$

when (21) can be applied (here $z_{1-\alpha/2}$ is the $(1-\alpha/2)$ quantile of the standard normal distribution), and

$$\left[\bar{\theta}_{n,\kappa} - \hat{B}_{\bar{\theta}_n,\kappa} \pm z_{1-\alpha/2}\frac{\hat{\sigma}_\theta}{\sqrt{n_\kappa}}\right], \tag{29}$$

when (22) can be applied.

Moreover, it is worth mentioning here that the bias of envelopment estimators is also a source of bias in estimating variance of individual efficiencies. Specifically, Simar and Zelenyuk (2020) show that the empirical version of variance, $\hat{\sigma}^2_\theta$, underestimates the population variance, σ^2_θ. To improve the accuracy of statistical inference in finite samples, Simar and Zelenyuk (2020) propose to use the following bias-corrected version of the estimator of variance

$$\tilde{\sigma}_{\theta}^{2} = \hat{\sigma}_{\theta}^{2} + \hat{B}_{\bar{\theta}_{n,\kappa}}^{2}. \tag{30}$$

As discussed in Simar and Zelenyuk (2020), the new CLTs also apply if $\tilde{\sigma}_{\theta}$ is used in place of $\hat{\sigma}_{\theta}$ in (21) and (22). Therefore, the $(1-\alpha)$ 100% confidence interval for the simple mean in the population can also be obtained by

$$\left[\bar{\theta}_{n} - \hat{B}_{\bar{\theta}_{n,\kappa}} \pm z_{1-\alpha/2}\frac{\tilde{\sigma}_{\theta}}{\sqrt{n}}\right], \tag{31}$$

when (21) can be applied, and

$$\left[\bar{\theta}_{n,\kappa} - \hat{B}_{\bar{\theta}_{n,\kappa}} \pm z_{1-\alpha/2}\frac{\tilde{\sigma}_{\theta}}{\sqrt{n_{\kappa}}}\right], \tag{32}$$

when (22) can be applied.

3.2.2 Central Limit Theorems and Confidence Intervals for Weighted Means

To develop the CLTs for the weighted mean, $\hat{\tau}_n$, Simar and Zelenyuk (2018) first propose an alternative representation of the aggregate efficiency, τ_n, as

$$\tau_n = \frac{(1/n)\sum_{i=1}^{n} pY^{i\partial}}{(1/n)\sum_{i=1}^{n} pY^{i}} = \frac{(1/n)\sum_{i=1}^{n} Z^{i\partial}}{(1/n)\sum_{i=1}^{n} Z^{i}}, \tag{33}$$

where, $Z^i \equiv pY^i$ is the revenue of DMU i, and $Z^{i\partial} \equiv pY^{i\partial} = OTE\left(X^i, Y^i\right) pY^i$ can be viewed as the monetary value of the projection of Y^i on the production frontier. τ_n then can be viewed as an estimator of the population parameter τ, which is a ratio of two population means, i.e.,

$$\tau = \frac{\mu_1}{\mu_2}, \tag{34}$$

where $\mu_1 = E\left[Z^{i\partial}\right]$ and $\mu_2 = E\left[Z^i\right]$. Simar and Zelenyuk (2018) show that the conventional CLT is applied to τ_n. Specifically,

$$\frac{\sqrt{n}}{\sigma_{\tau}}\left(\tau_n - \tau\right) \xrightarrow{d} N(0, 1), \tag{35}$$

where

$$\sigma_{\tau}^{2} = \tau^{2}\left(\frac{\sigma_1^2}{\mu_1^2} + \frac{\sigma_2^2}{\mu_2^2} - 2\frac{\sigma_{12}}{\mu_1\mu_2}\right), \tag{36}$$

where $\sigma_1^2 = VAR\left[Z^{i\partial}\right]$, $\sigma_2^2 = VAR\left[Z^i\right]$, and $\sigma_{12} = COV\left[Z^{i\partial}, Z^i\right]$.

In practice, τ_n is not observable since the true efficiency scores are not observable, thus we need to estimate the population parameter and perform statistical inference about it based on the weighted mean of DEA/FDH efficiency scores, $\hat{\tau}_n$. $\hat{\tau}_n$ can also be rewritten as

$$\hat{\tau}_n = \frac{\hat{\mu}_{1,n}}{\hat{\mu}_{2,n}} = \frac{(1/n)\sum_{i=1}^{n} \hat{Z}^{i\partial}}{(1/n)\sum_{i=1}^{n} Z^i}, \tag{37}$$

where $\hat{Z}^{i\partial} = \widehat{OTE}\left(X^i, Y^i\right) pY^i$.

Simar and Zelenyuk (2018) extend the theory in Kneip et al. (2015) to develop the CLTs for the weighted mean. Specifically, under the regularity conditions specified in Kneip et al. (2015), Theorem 2 in Simar and Zelenyuk (2018) establishes that for $N + M \leq 5$ if DEA-CRS is used, for $N + M \leq 4$ if DEA-VRS is used, and for $N+M \leq 3$ if FDH is used, the CLTs for the weighted mean of DEA/FDH efficiency scores are given by

$$\frac{\sqrt{n}}{\hat{\sigma}_{\tau,n}} \left(\hat{\tau}_n - \hat{B}_{\hat{\tau}_n,\kappa} - \tau + R_{n,\kappa}\right) \xrightarrow{d} N(0, 1) \tag{38}$$

where $\hat{\sigma}_{\tau,n}$ is a consistent estimator of σ_τ given by

$$\hat{\sigma}_{\tau,n}^2 = \hat{\tau}_n^2 \left(\frac{\hat{\sigma}_{1,n}^2}{\hat{\mu}_{1,n}^2} + \frac{\hat{\sigma}_{2,n}^2}{\hat{\mu}_{2,n}^2} - 2\frac{\hat{\sigma}_{12,n}}{\hat{\mu}_{1,n}\hat{\mu}_{2,n}}\right), \tag{39}$$

with $\hat{\sigma}_{1,n}$, $\hat{\sigma}_{2,n}$, and $\hat{\sigma}_{12,n}$ are empirical versions of σ_1, σ_2, and σ_{12}, respectively. $\hat{B}_{\hat{\tau}_n,\kappa}$ is the generalized jackknife estimator of bias of $\hat{\tau}_n$, which can be obtained using a similar procedure as discussed for the case of the simple mean. Specifically,

$$\hat{B}_{\hat{\tau}_n,\kappa} = \frac{\hat{B}_{\hat{\mu}_{1,n},\kappa}}{\hat{\mu}_{2,n}} = \frac{L^{-1}\sum_{\ell=1}^{L} (2^\kappa - 1)^{-1}\left(\hat{\mu}_{1,n,\ell}^* - \hat{\mu}_{1,n}\right)}{\hat{\mu}_{2,n}}, \tag{40}$$

where $\hat{\mu}_{1,n,\ell}^*$ is analogous to $\bar{\theta}_{n,\ell}^*$ as defined in (25).

Alternatively, for $\kappa < 1/2$, the CLTs for the weighted mean of DEA/FDH efficiency scores are given by

$$\frac{\sqrt{n_\kappa}}{\hat{\sigma}_{\tau,n}} \left(\hat{\tau}_{n_\kappa} - \hat{B}_{\hat{\tau}_n,\kappa} - \tau + R_{n,\kappa}\right) \xrightarrow{d} N(0, 1), \tag{41}$$

where $\hat{\tau}_{n,\kappa}$ is a subsample version of $\hat{\tau}_n$, in the sense that the averages are taken over a random subsample $\mathcal{S}_{n_\kappa} \subset \mathcal{S}^n$ of size $n_\kappa = \lfloor n^{2\kappa} \rfloor$. Formally

$$\hat{\tau}_{n,\kappa} = \frac{n_{\kappa}^{-1} \sum_{\left\{i \middle| (X^i, Y^i) \in \mathcal{S}_{n_\kappa}\right\}} \hat{Z}^{i\partial}\left(\mathcal{S}^n\right)}{n_{\kappa}^{-1} \sum_{\left\{i \middle| (X^i, Y^i) \in \mathcal{S}_{n_\kappa}\right\}} Z^i}, \tag{42}$$

where $\hat{Z}^{i\partial}\left(\mathcal{S}^n\right) = \widehat{OTE}\left(X^i, Y^i \middle| \mathcal{S}^n\right) Z^i$.

Using the new CLTs, one can construct the $(1-\alpha)$ 100% confidence interval for the weighted mean in the population as

$$\left[\hat{\tau}_n - \hat{B}_{\hat{\tau}_n,\kappa} \pm z_{1-\alpha/2} \frac{\hat{\sigma}_{\tau,n}}{\sqrt{n}}\right], \tag{43}$$

when (38) can be applied, and

$$\left[\hat{\tau}_{n,\kappa} - \hat{B}_{\hat{\tau}_n,\kappa} \pm z_{1-\alpha/2} \frac{\hat{\sigma}_{\tau,n}}{\sqrt{n_\kappa}}\right], \tag{44}$$

when (41) can be applied.

As in the case of the simple mean, the bias of envelopment estimators is also a source of bias in estimating the variance of aggregate efficiency. To improve the accuracy of statistical inference in a finite sample, following Simar and Zelenyuk (2020), one can use the bias-corrected version of the estimator of variance

$$\tilde{\sigma}_{\tau,n}^2 = \hat{\tau}_n^2 \left(\frac{\tilde{\sigma}_{1,n}^2}{\hat{\mu}_{1,n}^2} + \frac{\hat{\sigma}_{2,n}^2}{\hat{\mu}_{2,n}^2} - 2\frac{\hat{\sigma}_{12,n}}{\hat{\mu}_{1,n}\hat{\mu}_{2,n}}\right), \tag{45}$$

where

$$\tilde{\sigma}_{1,n}^2 = \hat{\sigma}_{1,n}^2 + \hat{B}_{\hat{\mu}_{1,n},\kappa}^2. \tag{46}$$

The new CLTs also apply if $\tilde{\sigma}_{\tau,n}$ is used in place of $\hat{\sigma}_{\tau,n}$ in (38) and (41). Therefore, the $(1-\alpha)$ 100% confidence interval for the weighted mean in the population can also be obtained by

$$\left[\hat{\tau}_n - \hat{B}_{\hat{\tau}_n,\kappa} \pm z_{1-\alpha/2} \frac{\tilde{\sigma}_{\tau,n}}{\sqrt{n}}\right], \tag{47}$$

when (38) can be applied, and

$$\left[\hat{\tau}_{n,\kappa} - \hat{B}_{\hat{\tau}_n,\kappa} \pm z_{1-\alpha/2} \frac{\tilde{\sigma}_{\tau,n}}{\sqrt{n_\kappa}}\right], \tag{48}$$

when (41) can be applied.

As demonstrated by intensive Monte Carlo simulations in Simar and Zelenyuk (2020), for both the case of simple mean and weighted mean, the confidence intervals constructed using the bias-corrected version of the variance estimator have a better coverage and thus are more reliable.

4 Data and Variables

4.1 *Sample*

For this study we were able to obtain a fairly disaggregated dataset on public hospitals in Queensland that, among other things, consists of 153,472 data points for 704 different types of inpatient procedures across 109 hospitals in 2 years. To our understanding of the literature, it is a relatively rich dataset as researchers often have to use a much more aggregated data such as patient days (e.g., Sherman 1984; Bates et al. 2006; Hu et al. 2012) or the number of patients or admissions or discharges (e.g., Athanassopoulos et al. 1999; McCallion et al. 1999; Hao and Pegels 1994; Grosskopf et al. 2001; Zuckerman et al. 1994; Staat 2006; Tiemann and Schreyögg 2009; Besstremyannaya 2013). Indeed, despite many attempts we also were not able to obtain similar level of data for other States or Territories of Australia. So, while we cannot generalise our conclusions for other States, Territories or especially other countries, our results for Queensland might still be insightful, at least as a food for thought, for researchers across the globe.

Specifically, the sample for our analysis consists of 109 public acute hospitals in Queensland, Australia in the two financial years (FY) 2012/13 and 2013/2014.[19] The data is provided by the Queensland Department of Health, extracted from the two data collections which are the Financial and Residential Activity Collection (FRAC) and the Monthly Activity Collection (MAC).

4.2 *Variables*

The list of variables used in this study is shown in Table 2 (with their notations and descriptions). In the following subsections, we will discuss each variable in details.

4.2.1 Inpatient Output

The main variable of interest in this analysis is the inpatient output. In our data, inpatient output is recorded in terms of admitted patient episodes, which are categorised into 704 Diagnosis-Related Groups (DRGs).

[19]In Australia, a financial year starts on 1 July and ends on 30 June of the next calendar year.

Table 2 The description of variables

Variables	Description
Inputs	
FLABOUR	Labour input (PCA-based aggregation)
BEDS	Capital input (total beds)
DMSUP	Consumable input (drug and medical supply expenditure) (2012/13 constant price) ($1,000,000s)
Outputs	
OUT	Outpatient output (non-admitted occasions of service) (1000s)
WEPS1	Inpatient output (price-based aggregation using varying weights) (1000s)
WEPS2	Inpatient output (price-based aggregation using 2012/13 weights) (1000s)
WEPS3	Inpatient output (price-based aggregation using 2013/14 weights) (1000s)
FEPS	Inpatient output (PCA-based aggregation)
TEPS	Inpatient output (non-weighted aggregation) (1000s)

In principle, each DRG represents a separate type of output, yet they need to be aggregated to make the model feasible. Here, we consider two approaches to aggregate the inpatient output: (1) PCA-based approach and (2) Price-based approach. Moreover, for the price-based approach, we consider three different weight systems, which are: (1) varying DRG cost weights (different weights for different years), (2) constant cost weights using FY 2012/13 cost weights, and (3) constant cost weights using FY 2013/14 cost weights.[20] Besides these, we also compare the results from PCA-based and price-based approaches with those from a non-weighted aggregation approach, in which the inpatient output is measured by the raw count of the total number of episodes.

The relationship between the cost weights in FY 2012/13 and FY 2013/14 is shown in Fig. 1. We can see that there is a variation (but small in magnitude) in the cost weights across the years, and thus it is useful to assess the sensitivity of the results with respect to the different weight systems as was done in the simulation exercises in Zelenyuk (2020).

Figure 2 shows the total number of episodes of each DRG and the number of hospitals producing the DRG in 2013/14.[21] It can be seen from the figure that DRGs with high weights are produced in a small quantity and at few hospitals, whereas the DRGs that are produced in a massive quantity at many hospitals usually have low weights. The DRGs with high cost weights typically include highly specialised surgical procedures, such as Insertion of Ventricular Assist Device (A10Z), Liver

[20]It is worth mentioning here that under the activity-based funding system in Australia, each inpatient DRG has a single cost weight, which is updated annually based on historical costs incurred by hospitals in treating patients belonging to the DRG. An inpatient episode in each DRG is then funded at a flat rate based upon the DRG cost weight and the so-called national efficient price (NEP)–a fixed price paid per weighted unit.

[21]The patterns in the year 2012/13 are similar. To save space, we only present a graph for the year 2013/14.

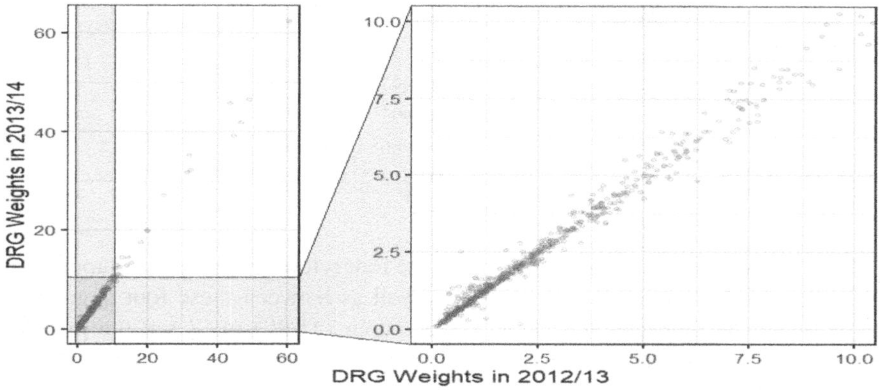

Fig. 1 Scatter plot of DRG weights in 2012/13 vs. 2013/14. The right panel is a zoomed part of the left panel

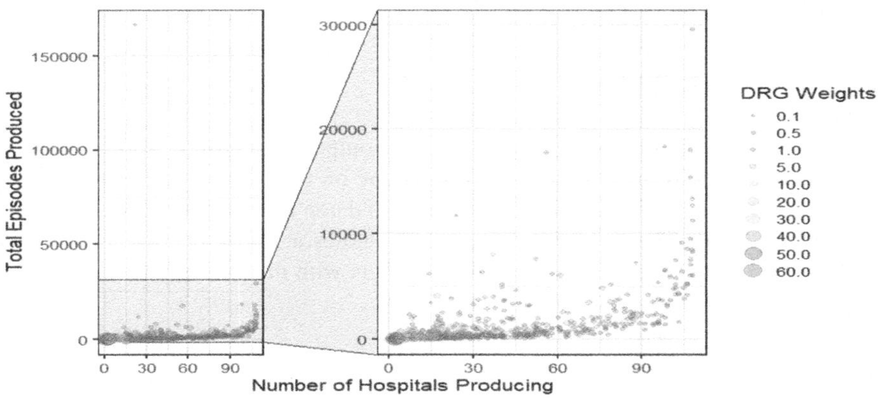

Fig. 2 Total number of episodes of each DRG and the number of hospitals producing the DRG in 2013/14. Each point represents a DRG. The colour and size of the points represent DRGs' weights. The right panel is a zoomed part of the left panel

Transplant (A01Z), and so on. Meanwhile, the ones with low cost weights are usually simple medical procedures, such as Electroconvulsive Therapy (U60Z), Haemodialysis (L61Z), and so on.

With regard to the PCA-based approach, the aggregate inpatient output (i.e., the first factorial variable) contains 65.39% information in all 704 DRGs as $\delta_1 = 0.6539$. Interestingly, in spite of the fact that the weights from the PCA-based approach are negatively correlated with the DRG cost weights,[22] there are

[22]E.g., pairwise Pearson linear correlation coefficients between PCA-based weights and DRG cost weights are around −0.33.

Table 3 Pairwise Pearson linear correlation coefficients among five aggregated inpatient outputs

	WEPS1	WEPS2	WEPS3	FEPS	TEPS
WEPS1	1				
WEPS2	0.9997	1			
WEPS3	0.9997	0.9995	1		
FEPS	0.9966	0.9958	0.9974	1	
TEPS	0.9813	0.9800	0.9828	0.9867	1

almost perfect positive correlations among the four relevant aggregated outputs (i.e., WEPS1, WEPS2, WEPS2 and FEPS), as well as between these four aggregated outputs and the non-weighted aggregated output (TEPS) (e.g., see the pairwise Pearson linear correlation coefficients in Table 3). The relationships among the aggregated outputs discussed here might make one conjecture that the nature of the dataset makes the results of analysis indifferent across all aggregation approaches regardless of whether they are weighted or non-weighted, but it is not necessarily the case and we will examine this conjecture in Sect. 5.

4.2.2 Outpatient Output

Another hospital output used in this study is outpatient output (OUT). Hospital outpatient output is measured by a raw count of the number of non-admitted occasions of service including both emergency and non-emergency services for non-admitted patients. The outpatient output also shows a strong positive correlation with each of the five aggregated inpatient outputs with the pairwise Pearson linear correlation coefficients being around 0.97–0.99.

4.2.3 Inputs

We use three inputs to model the production process of hospitals, which are: labour input (PCA-based aggregation of full time equivalent staff in six labour categories), capital input (proxied by the number of beds), and consumable input (proxied by drug, surgical and medical supply expenditure). On the input side, this is therefore a mixture of PCA-based approach and price-based approach.

The descriptive statistics of variables used in the study are shown in Table 4. As in any typical healthcare related data, we observe great variation and positive skewness for all variables in the working sample.

Table 4 The descriptive statistics of variables

Variables	Mean	Median	Q1	Q3	Std. Dev.	Min	Max
Inputs							
FLABOUR	1.01	0.12	0.07	0.42	2.43	0.02	15.56
BEDS	92.55	20.00	10.00	48.00	185.42	5.00	1024.00
DMSUP	10.12	0.42	0.15	2.37	27.45	0.03	167.35
Outputs							
OUT	88.60	19.88	6.31	52.23	166.02	1.31	954.90
WEPS1	9.93	0.87	0.31	4.05	23.05	0.03	132.61
WEPS2	9.94	0.83	0.29	3.93	23.18	0.03	132.61
WEPS3	9.90	0.88	0.31	4.05	22.87	0.03	129.36
FEPS	9.35	1.04	0.43	3.94	20.84	0.043	115.25
TEPS	9.07	1.02	0.38	4.58	18.37	0.05	95.58

Table 5 Model Specifications for the case of 3-inputs and 2-outputs models

	Inputs			Outputs					
Model	FLABOUR	BEDS	DMSUP	OUT	WEPS1	WEPS2	WEPS3	FEPS	TEPS
Model 1	✓	✓	✓	✓	✓				
Model 2	✓	✓	✓	✓		✓			
Model 3	✓	✓	✓	✓			✓		
Model 4	✓	✓	✓	✓				✓	
Model 5	✓	✓	✓	✓					✓

5 A Comparison of Estimated Efficiency Scores Among Different Aggregation Approaches

In the following analysis, we compare the DEA-CRS estimates of hospital efficiency across the five models in which three inputs and two outputs are utilised (see model specifications in Table 5).[23]

Table 6 shows the descriptive statistics on the central tendency and dispersion of the estimated efficiency scores from the five models. At first glance, we can see that the descriptive statistics for the estimated efficiency scores from Model 1, Model 2, and Model 3, where the inpatient outputs are aggregated using price-based approaches with different weight systems, are almost identical. To further investigate the relationship among the estimated efficiency scores of these three models, we look at their pairwise Spearman rank-order correlation coefficients. It is shown in Table 7 that the estimated efficiency scores are nearly perfectly

[23]As a sensitivity analysis, we also consider the case of 3-inputs and 1-output models, where we aggregate outpatient output and inpatient output using PCA-based aggregation, and the case of 1-input and 1-output models, where we further aggregate the three inputs into a single measure of hospital inputs using PCA-based aggregation. The results from the sensitivity analysis are qualitatively similar to the discussion in this section (see the results in the Appendix).

Table 6 Descriptive statistics of estimated efficiency scores using DEA-CRS estimator for the case of 3-inputs and 2-outputs models

Model	Mean	Median	Q1	Q3	Std. Dev.	Min	Max
Model 1	1.55	1.38	1.22	1.70	0.54	1.00	4.08
Model 2	1.54	1.37	1.21	1.70	0.53	1.00	4.05
Model 3	1.52	1.34	1.17	1.71	0.54	1.00	4.08
Model 4	1.61	1.43	1.23	1.80	0.58	1.00	4.68
Model 5	1.73	1.58	1.24	1.97	0.70	1.00	5.15

Table 7 Pairwise Spearman rank-order correlation coefficients of estimated efficiency scores using DEA-CRS estimator across models for the case of 3-inputs and 2-outputs models

	Model 1	Model 2	Model 3	Model 4	Model 5
Model 1	1.00				
Model 2	0.99	1.00			
Model 3	0.97	0.98	1.00		
Model 4	0.94	0.94	0.93	1.00	
Model 5	0.89	0.90	0.87	0.90	1.00

Table 8 95% confidence intervals for the simple mean of DEA-CRS efficiency scores for the case of 3-inputs and 2-outputs models

	Model 1	Model 2	Model 3	Model 4	Model 5
DEA estimate	1.55	1.54	1.52	1.61	1.73
Bias corrected	1.83	1.81	1.75	1.90	2.04
Std. error	0.06	0.06	0.06	0.07	0.08
Bias corrected std. error	0.07	0.07	0.07	0.08	0.09
LB of est. CI	1.69	1.68	1.62	1.73	1.84
UB of est. CI	1.94	1.92	1.87	2.00	2.15
LB of est. CI-improved	1.68	1.67	1.61	1.72	1.82
UB of est. CI-improved	1.95	1.94	1.88	2.02	2.17

LB Lower Bound, *UB* Upper Bound, *Est.* Estimated, *Std. Error* Standard Error, *CI* Confidence Interval

Computations are done in Matlab adopting the code from Simar and Zelenyuk (2020)

correlated with all the correlation coefficients being around 0.97–0.99, implying that the rankings of the individual hospitals based on the estimated efficiency scores from the three models are almost the same.

We then compare the 95% confidence intervals for the simple and weighted means of efficiency scores estimated from these three models. Table 8 reports the results for the simple mean and Table 9 presents the estimates for the weighted means.[24] We can see that for both the simple mean and the weighted mean, the

[24]It is worth mentioning here that to perform the analysis for the weighted mean, it requires information about the relative price between the outpatient output and the aggregate inpatient output. For the price-based aggregate inpatient output, we use the information about the relative

Table 9 95% confidence intervals for the weighted mean of DEA-CRS efficiency scores for the case of 3-inputs and 2-outputs models

	Model 1	Model 2	Model 3	Model 4	Model 5
DEA estimate	1.31	1.30	1.19	n/a	n/a
Bias corrected	1.58	1.55	1.28	n/a	n/a
Std. error	0.03	0.03	0.03	n/a	n/a
Bias corrected std. error	0.04	0.04	0.03	n/a	n/a
LB of est. CI	1.51	1.48	1.23	n/a	n/a
UB of est. CI	1.64	1.61	1.34	n/a	n/a
LB of est. CI-improved	1.49	1.46	1.23	n/a	n/a
UB of est. CI-improved	1.66	1.63	1.34	n/a	n/a

LB Lower Bound, *UB* Upper Bound, *Est.* Estimated, *Std. Error* Standard Error, *CI* Confidence Interval
Computations are done in Matlab adopting the code from Simar and Zelenyuk (2020)

biased corrected DEA estimates as well as the 95% confidence intervals are almost the same between Model 1 and Model 2. Compared to the first two models, the results from Model 3 are slightly different for the case of simple mean, and the difference is more pronounced when we consider the weighted mean. Specifically, the 95% confidence interval of the weighted mean from Model 3 is to the left and does not overlap with those from Model 1 and Model 2. However, the difference here might just be particular for the dataset used in this study. For this dataset, when using the fixed weights of FY2013/14, the quantity of the aggregate output of most (but not all) observations in FY2012/13 increases, but the estimated production frontier is moving inward due to a particular observation. This observation is on the estimated frontiers no matter whether WEPS1, WEPS2, or WEPS3 is used in the DEA models, and its value of WEPS3 is significantly smaller than WEPS1 and WEPS2. As a result, the estimated frontier of Model 3 around this observation is significantly lower than those of Model 1 and Model 2. Consequently, the estimated efficiency scores of other observations in this segment of the production space are smaller in Model 3 compared to Model 1 and Model 2. Moreover, the observation that we are referring to is relatively large (in terms of revenue), and thus so do its peers, and this is the reason why the difference is more pronounced when we compare the weighted mean. This is also a good real data example of the importance of accounting for the economic weight of each efficiency score in the averaging of the scores over observations of very different sizes.

Furthermore, we examine the whole distribution of estimated efficiency scores from the three models by comparing their estimated densities and performing the adapted Li test for the equality of the densities. From Fig. 3, we see that the estimated densities of estimated efficiency scores from Model 1 and Model 2 are

price in Nguyen and Zelenyuk (2021a). However, the price information is not available for PCA-based aggregate inpatient output as well as for the raw count of inpatient episodes, thus we focus only on the simple mean for Model 4 and Model 5.

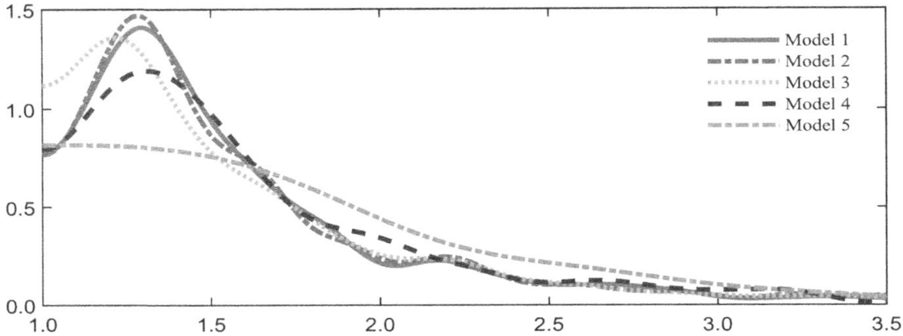

Fig. 3 Density estimates of the estimated efficiency scores using the DEA-CRS estimators for different models for the case of 3-inputs and 2-outputs models. Kernel-based with Silverman's (1986) reflection method: Gaussian kernel and bandwidth is selected by the method of Sheather and Jones (1991)

Table 10 The adapted Li test for equality of distributions of DEA-CRS efficiency scores by different models for the case of 3-inputs and 2-outputs models

	The Li test statistics	Bootstrap p-value	Decision (at 5% level of significance)
Model 1 vs. Model 2	−0.01	0.99	Do not reject H_0
Model 1 vs. Model 3	1.30	0.06	Do not reject H_0
Model 2 vs. Model 3	0.96	0.11	Do not reject H_0
Model 1 vs. Model 4	0.04	0.97	Do not reject H_0
Model 2 vs. Model 4	0.19	0.79	Do not reject H_0
Model 3 vs. Model 4	1.76	0.03	Reject H_0
Model 1 vs. Model 5	4.47	0.00	Reject H_0
Model 2 vs. Model 5	4.63	0.00	Reject H_0
Model 3 vs. Model 5	5.21	0.00	Reject H_0
Model 4 vs. Model 5	2.69	0.01	Reject H_0

Notes:

H_0 The densities of estimated efficiency scores of the two corresponding models are equal

Computations are done in Matlab adopting the code from Simar and Zelenyuk (2006), with 2000 bootstrap replications using Gaussian kernel, and Silverman (1986) robust rule of thumb bandwidth

almost overlapped. And, although the estimated densities from Model 3 is slightly to the left of those from Model 1 and Model 2, the adapted Li test shows that there is insufficient statistical evidence to conclude that the densities of the estimated efficiency scores from these three models are pairwise different (see Table 10). All in all, the analysis suggests that the efficiency scores estimated from the three models, in which inpatient outputs are aggregated using different weight systems, provide the same information about the overall efficiency level as well as the relative rankings of individual hospitals in our sample. This implies the robustness of results with respect to a slight change in relative prices.

Let us now look at the estimated efficiency scores from Model 4, where inpatient outputs are aggregated using the PCA-based approach and let us compare the estimates with those from Model 1, Model 2, and Model 3. We can see that although the simple mean of estimated efficiency scores from Model 4 are slightly higher (i.e., indicating overall less efficient) than those from the first three models, the 95% confidence intervals for the population mean constructed using the estimated efficiency scores from the four models almost overlap. Besides, the pairwise Spearman rank-order correlation coefficients between the estimates from Model 4 and each of those from the first three models are still very high (all around 0.93–0.94). Moreover, the estimated density of the estimated efficiency scores from Model 4 looks similar to those from the first three models, especially from Model 1 and Model 2. The adapted Li test supports the visualisation, showing no statistical pairwise difference between the densities of the estimated efficiency scores from Model 1, Model 2 and Model 4. The results suggest that the estimated efficiency scores from the DEA models are also robust with respect to the two different aggregation approaches.

To examine the conjecture discussed in Sect. 4 that due to the nature of the data, the results might be indifferent regardless of whether the weighted or non-weighted aggregation approaches are utilised, we compare the estimated efficiency scores from Model 5 with those from the first four models, focusing on their whole distribution. We can see from Fig. 3 that the estimated density of the efficiency scores obtained from Model 5 is very different from the estimated densities of those obtained from the first four models. Our visualisation is then confirmed by the adapted Li test, where the null hypotheses about the equality between densities of estimated efficiency scores from Model 5 and from each of the first four models are all rejected with very small p-values. The results highlight the importance of utilising the appropriate approaches for dimension reduction in the DEA context: estimated efficiency scores under PCA-based aggregation and price-based aggregation approaches are similar, but very different compared with those obtained under the naive non-weighted approach.

6 Concluding Remarks

Given the increasing attention to dimension reduction in the context of DEA with large dimensions for inputs and outputs, in this study we revisit the output aggregation in hospital efficiency analysis, where the challenge of big wide data for DEA has been present since the early 1980s. Although not cast explicitly, researchers in the field usually use price-based aggregation to aggregate a hospital's output to deal with the most challenging issue of DEA with big data—the 'curse of dimensionality'.

Using data on public hospitals in Queensland, Australia, we find that the choice of price systems (with small variation in prices) does not significantly affect the DEA estimates under the price-based aggregation approach. Moreover, the estimated efficiency scores from the DEA models are also robust with respect to the price-based aggregation and the PCA-based aggregation approaches. The robustness of the results suggests that the PCA-based aggregation can be viewed as a viable alternative for DEA practitioners who are unable to/or unwilling to use the price-based approach, e.g., due to unavailable or unreliable price information.

It is worth highlighting here that the results in this study are particular for the working sample, and should be interpreted with care. Even though, this paper can be viewed as a pilot study to remind DEA practitioners, especially those who work with hospital data, that price-based aggregation is just one approach (among others) to reduce the dimension of production space to deal with the 'curse of dimensionality'. Thus in applied works, several approaches should be applied to ensure the robustness of results.

Recently, the Least Absolute Shrinkage and Selection Operator (LASSO) has been applied into the DEA context as a promising technique for dimension reduction (see more discussions in Lee and Cai 2020; Chen et al. 2021, and references therein). As a result, a fruitful direction for future research is to empirically compare the DEA estimates based on the LASSO approach with those based on price-based and PCA-based aggregation approaches.

Acknowledgments We thank the Editors/referees for fruitful comments. We acknowledge the support from The University of Queensland and the financial support from the Australian Research Council (FT170100401). We also thank Zhichao Wang, Evelyn Smart, and Anne-Claire Bouton for their valuable feedback. We acknowledge and thank Queensland Health for providing part of the data that we used in this study. These individuals and organizations are not responsible for the views expressed in this paper.

Appendix

A *Results for the Case of 3-Inputs and 1-Output Models*

Here we aggregate outputs into a single measure of hospital outputs using the PCA-based approach and estimate 3-inputs and 1-output production models using DEA-CRS. Specifically, we aggregate the outpatient output with each of the four aggregated inpatient outputs (i.e., WEPS1, WEPS2, WEPS3, and FEPS) and denote the aggregated outputs as FOUT1, FOUT2, FOUT3, and FOUT4, respectively (see model specifications in Table 11 and the corresponding results in Fig. 4 and Tables 12, 13, 14, 15 and 16).

Table 11 Model specifications for the case of 3-inputs and 1-output models

	Inputs			Outputs				
Model	FLABOUR	BEDS	DMSUP	FOUT1	FOUT2	FOUT3	FOUT4	FOUT5
Model 6	✓	✓	✓	✓				
Model 7	✓	✓	✓		✓			
Model 8	✓	✓	✓			✓		
Model 9	✓	✓	✓				✓	
Model 10	✓	✓	✓					✓

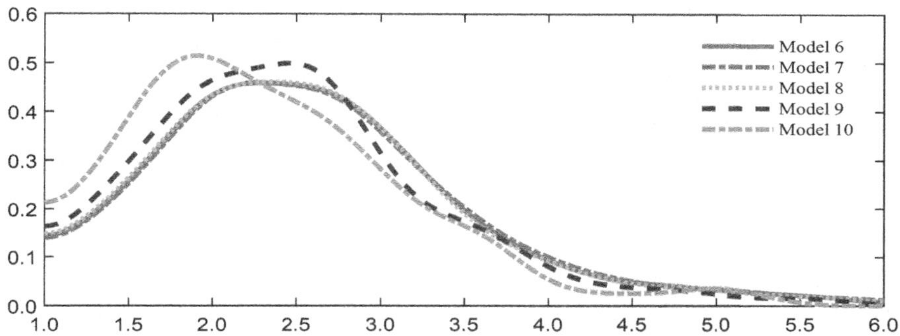

Fig. 4 Density estimates of the estimated efficiency scores using the DEA-CRS estimators for different models for the case of 3-inputs and 1-output models. Kernel-based with Silverman's (1986) reflection method: Gaussian kernel and bandwidth is selected by the method of Sheather and Jones (1991)

Table 12 Descriptive statistics of estimated efficiency scores using DEA-CRS estimator for the case of 3-inputs and 1-output models

Model	Mean	Median	Q1	Q3	Std. Dev	Min	Max
Model 6	2.61	2.49	1.98	3.07	1.01	1.00	8.43
Model 7	2.63	2.50	1.99	3.07	1.02	1.00	8.48
Model 8	2.60	2.49	1.95	3.07	1.00	1.00	8.41
Model 9	2.49	2.41	1.87	2.87	0.92	1.00	7.71
Model 10	2.36	2.22	1.74	2.81	0.93	1.00	8.10

B Results for the Case of 1-Input and 1-Output Models

Here we further aggregate three inputs into a single measure of input using the PCA-based approach, denoted as FINPUT, and estimate 1-input and 1-output production models using DEA-CRS (see model specifications in Table 17 and the corresponding results in Fig. 5 and Tables 18, 19, 20, 21 and 22).

Table 13 95% confidence intervals for the simple mean of DEA-CRS efficiency scores for the case of 3-inputs and 1-outputs model

	Model 6	Model 7	Model 8	Model 9	Model 10
DEA estimate	2.61	2.63	2.60	2.49	2.36
Bias corrected	3.43	3.46	3.41	3.26	3.00
Std. error	0.07	0.07	0.07	0.06	0.06
Bias corrected std. error	0.09	0.09	0.09	0.08	0.08
LB of est. CI	3.30	3.32	3.27	3.13	2.88
UB of est. CI	3.57	3.59	3.54	3.38	3.12
LB of est. CI-improved	3.26	3.28	3.24	3.10	2.85
UB of est. CI-improved	3.60	3.63	3.58	3.42	3.15

LB Lower Bound, *UB* Upper Bound, *Est.* Estimated, *Std. Error* Standard Error, *CI* Confidence Interval

Computations are done in Matlab adopting the code from Simar and Zelenyuk (2020)

Table 14 95% confidence intervals for the weighted mean of DEA-CRS efficiency scores for the case of 3-inputs and 1-output models

	Model 6	Model 7	Model 8	Model 9	Model 10
DEA estimate	2.66	2.66	2.65	2.65	2.54
Bias corrected	3.73	3.72	3.71	3.66	3.49
Std. error	0.05	0.05	0.05	0.06	0.07
Bias corrected std. error	0.09	0.09	0.09	0.09	0.10
LB of est. CI	3.63	3.62	3.61	3.55	3.34
UB of est. CI	3.83	3.82	3.81	3.77	3.63
LB of est. CI-improved	3.55	3.55	3.54	3.48	3.29
UB of est. CI-improved	3.90	3.90	3.88	3.83	3.68

LB Lower Bound, *UB* Upper Bound, *Est.* Estimated, *Std. Error* Standard Error, *CI* Confidence Interval

Computations are done in Matlab adopting the code from Simar and Zelenyuk (2020)

Table 15 Pairwise Spearman rank-order correlation coefficients of the estimated efficiency scores using the DEA-CRS estimator across models for the case of 3-inputs and 1-output models

	Model 6	Model 7	Model 8	Model 9	Model 10
Model 6	1				
Model 7	0.999	1			
Model 8	0.999	0.999	1		
Model 9	0.982	0.979	0.983	1	
Model 10	0.975	0.973	0.977	0.979	1

Table 16 The adapted Li test for the equality of distributions of DEA-CRS efficiency scores by different models for the case of 3-inputs and 1-output models

	The Li test statistics	Bootstrap p-value	Decision (at 5% level of significance)
Model 6 vs. Model 7	0.006	0.994	Do not reject H_0
Model 6 vs. Model 8	−0.001	0.998	Do not reject H_0
Model 7 vs. Model 8	0.022	0.974	Do not reject H_0
Model 6 vs. Model 9	0.417	0.573	Do not reject H_0
Model 7 vs. Model 9	0.404	0.595	Do not reject H_0
Model 8 vs. Model 9	0.378	0.620	Do not reject H_0
Model 6 vs. Model 10	2.131	0.016	Reject H_0
Model 7 vs. Model 10	2.221	0.013	Reject H_0
Model 8 vs. Model 10	2.012	0.018	Reject H_0
Model 9 vs. Model 10	1.229	0.071	Do not reject H_0

Notes:

H_0: The densities of estimated efficiency scores of the two corresponding models are equal

Computations are done in Matlab adopting the code from Simar and Zelenyuk (2006), with 2000 bootstrap replications using Gaussian kernel, and Silverman (1986) robust rule of thumb bandwidth

Table 17 Model Specifications for the case of 1-input and 1-output models

Model	Inputs	Outputs				
	FINPUT	FOUT1	FOUT2	FOUT3	FOUT4	FOUT5
Model 11	✓	✓				
Model 12	✓		✓			
Model 13	✓			✓		
Model 14	✓				✓	
Model 15	✓					✓

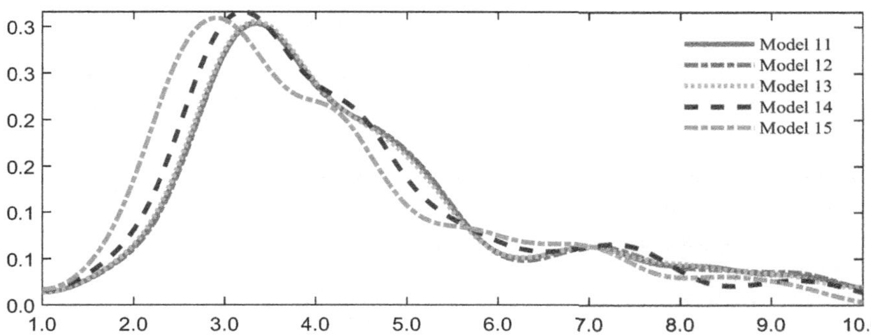

Fig. 5 Density estimates of the estimated efficiency scores using the DEA-CRS estimators for different models for the case of 1-input and 1-output models. Kernel-based with Silverman's (1986) reflection method: Gaussian kernel and bandwidth is selected by the method of Sheather and Jones (1991)

Table 18 Descriptive statistics of the estimated efficiency scores using the DEA-CRS estimator for the case of 1-input and 1-output models

Model	Mean	Median	Q1	Q3	Std. Dev	Min	Max
Model 11	4.82	4.09	3.21	5.46	2.37	1.00	14.23
Model 12	4.86	4.10	3.23	5.47	2.41	1.00	14.23
Model 13	4.78	4.05	3.19	5.42	2.33	1.00	13.75
Model 14	4.46	3.85	3.07	5.27	2.06	1.00	12.21
Model 15	4.28	3.69	2.79	5.14	2.11	1.00	12.76

Table 19 95% confidence intervals for the simple mean of DEA-CRS efficiency scores for the case of 1-input and 1-outputs model

	Model 11	Model 12	Model 13	Model 14	Model 15
DEA estimate	4.82	4.86	4.78	4.46	4.28
Bias corrected	5.61	5.64	5.55	5.15	4.94
Std. error	0.16	0.16	0.16	0.14	0.14
Bias corrected std. error	0.17	0.17	0.17	0.15	0.15
LB of est. CI	5.29	5.32	5.24	4.88	4.66
UB of est. CI	5.92	5.96	5.86	5.42	5.22
LB of est. CI-improved	5.28	5.31	5.22	4.86	4.65
UB of est. CI-improved	5.94	5.98	5.87	5.44	5.24

LB Lower Bound, *UB* Upper Bound, *Est.* Estimated, *Std. Error* Standard Error, *CI* Confidence Interval
Computations are done in Matlab adopting the code from Simar and Zelenyuk (2020)

Table 20 95% confidence intervals for the weighted mean of DEA-CRS efficiency scores for the case of 1-input and 1-output models

	Model 11	Model 12	Model 13	Model 14	Model 15
DEA estimate	4.18	4.18	4.17	4.15	3.99
Bias corrected	4.86	4.86	4.84	4.79	4.60
Std. Error	0.13	0.13	0.13	0.14	0.17
Bias corrected std. error	0.14	0.14	0.14	0.15	0.18
LB of est. CI	4.60	4.60	4.59	4.51	4.26
UB of est. CI	5.11	5.11	5.10	5.06	4.95
LB of est. CI-improved	4.59	4.59	4.57	4.50	4.25
UB of est. CI-improved	5.13	5.12	5.12	5.08	4.96

LB Lower Bound, *UB* Upper Bound, *Est.* Estimated, *Std. Error* Standard Error, *CI* Confidence Interval
Computations are done in Matlab adopting the code from Simar and Zelenyuk (2020)

Table 21 Pairwise Spearman rank-order correlation coefficients of the estimated efficiency scores using the DEA-CRS estimator across models for the case of 1-input and 1-output models

	Model 11	Model 12	Model 13	Model 14	Model 15
Model 11	1				
Model 12	0.999	1			
Model 13	0.999	0.999	1		
Model 14	0.988	0.987	0.990	1	
Model 15	0.981	0.980	0.982	0.985	1

Table 22 The adapted Li test for the equality of distributions of DEA-CRS efficiency scores by different models for the case of 1-input and 1-output models

	The Li test statistics	Bootstrap p-value	Decision (at 5% level of significance)
Model 11 vs. Model 12	−0.004	0.995	Do not reject H_0
Model 11 vs. Model 13	0.001	0.998	Do not reject H_0
Model 12 vs. Model 13	0.009	0.989	Do not reject H_0
Model 11 vs. Model 14	0.372	0.577	Do not reject H_0
Model 12 vs. Model 14	0.426	0.518	Do not reject H_0
Model 13 vs. Model 14	0.271	0.700	Do not reject H_0
Model 11 vs. Model 15	3.103	0.004	Reject H_0
Model 12 vs. Model 15	3.276	0.002	Reject H_0
Model 13 vs. Model 15	2.864	0.008	Reject H_0
Model 14 vs. Model 15	1.419	0.050	Reject H_0

Notes:

H_0 The densities of the estimated efficiency scores of the two corresponding models are equal

Computations are done in Matlab adopting the code from Simar and Zelenyuk (2006), with 2000 bootstrap replications using Gaussian kernel, and Silverman (1986) robust rule of thumb bandwidth

References

Adler, N., & Golany, B. (2001). Evaluation of deregulated airline networks using data envelopment analysis combined with principal component analysis with an application to Western Europe. *European Journal of Operational Research, 132*(2), 260–273.

Adler, N., & Golany, B. (2007). PCA-DEA: Reducing the curse of dimensionality. In J. Zhu & W. D. Cook (Eds.), *Modeling data irregularities and structural complexities in data envelopment analysis* (pp. 139–153). Springer US.

Aigner, D., Lovell, C. A. K., & Schmidt, P (1977). Formulation and estimation of stochastic frontier production function models. *Journal of Econometrics, 64*(6), 1263–1297.

Aragon, Y., Daouia, A., & Thomas-Agnan, C. (2005). Nonparametric frontier estimation: A conditional quantile-based approach. *Econometric Theory, 21*(2), 358–389.

Athanassopoulos, A., Gounaris, C., & Sissouras, A. (1999). A descriptive assessment of the production and cost efficiency of general hospitals in Greece. *Health Care Management Science, 2*(2), 97–106.

Bates, L. J., Mukherjee, K., & Santerre, R. E. (2006). Market structure and technical efficiency in the hospital services industry: A DEA approach. *Medical Care Research and Review, 63*(4), 499–524.

Besstremyannaya, G. (2013). The impact of Japanese hospital financing reform on hospital efficiency: A difference-in-difference approach. *The Japanese Economic Review, 64*(3), 337–362.

Cazals, C., Florens, J.-P., & Simar, L. (2002). Nonparametric frontier estimation: A robust approach. *Journal of Econometrics, 106*(1), 1–25.

Charles, V., Aparicio, J., & Zhu, J. (2019). The curse of dimensionality of decision-making units: A simple approach to increase the discriminatory power of data envelopment analysis. *European Journal of Operational Research, 279* (3), 929–940.

Charnes, A., Cooper, W. W., & Rhodes, E. (1978). Measuring the efficiency of decision making units. *European Journal of Operational Research, 2* (6), 429–444.

Chen, Y., Tsionas, M., & Zelenyuk, V. (2021). LASSO+DEA for small and big data. Omega, *91*, 102419.

Daouia, A., & Simar, L. (2007). Nonparametric efficiency analysis: A multivariate conditional quantile approach. *Journal of Econometrics, 140* (2), 375–400.

Daraio, C., & Simar, L. (2007). Economies of scale, scope and experience in the Italian motor-vehicle sector. In C. Daraio & L. Simar (Eds.), *Advanced robust and non-parametric methods in efficiency analysis: Methodology and applications* (pp. 135–165). Springer Science & Business Media.

Deprins, D., Simar, L., & Tulkens, H. (1984). Measuring labor efficiency in post offices. In M. G. Marchand, P. Pestieau, & H. Tulkens (Eds.), *The performance of public enterprises: Concepts and measurements* (pp. 243–267). Amsterdam, North-Holland.

Färe, R., & Grosskopf, S. (1985). A nonparametric cost approach to scale efficiency. *The Scandinavian Journal of Economics, 87*, 594–604.

Färe, R., Grosskopf, S., & Zelenyuk, V. (2004). Aggregation bias and its bounds in measuring technical efficiency. *Applied Economics Letters, 11*(10), 657–660.

Färe, R., & Primont, D. (1995). In R. Färe & D. Primont (Eds.), *Multi-output production and duality: Theory and applications*. New York: Kluwer Academic Publishers.

Färe, R., & Zelenyuk, V. (2003). On aggregate Farrell efficiencies. *European Journal of Operational Research, 146*(3), 615–620.

Fetter, R. B. (1991). Diagnosis related groups: Understanding hospital performance. *Interfaces, 21*(1), 6–26.

Grosskopf, S. (1996). Statistical inference and nonparametric efficiency: A selective survey. *Journal of Productivity Analysis, 7*(2–3), 161–176.

Grosskopf, S., Margaritis, D., & Valdmanis, V. (2001). Comparing teaching and non-teaching hospitals: A frontier approach (teaching vs. non-teaching hospitals). *Health Care Management Science, 4*(2), 83–90.

Hao, S. S., & Pegels, C. C. (1994). Evaluating relative efficiencies of Veterans Affairs medical centers using Data Envelopment, ratio, and multiple regression analysis. *Journal of Medical Systems, 18*(2), 55–67.

Härdle, W., & Simar, L. (2020). *Applied multivariate statistical analysis*. Springer.

Hollingsworth, B. (2003). Non-parametric and parametric applications measuring efficiency in health care. *Health Care Management Science, 6*(4), 203–218.

Hollingsworth, B. (2008). The measurement of efficiency and productivity of health care delivery. *Health Economics, 17* (10), 1107–1128.

Hu, H. H., Qi, Q., & Yang, C. H. (2012). Evaluation of China's regional hospital efficiency: DEA approach with undesirable output. *Journal of the Operational Research Society, 63*(6), 715–725.

Kneip, A., Park, B. U., & Simar, L. (1998). A note on the convergence of nonparametric DEA estimators for production efficiency scores. *Econometric Theory, 14* 783–793.

Kneip, A., Simar, L., & Wilson, P. W. (2008). Asymptotics and consistent bootstraps for DEA estimators in nonparametric frontier models. *Econometric Theory, 24*(6), 1663–1697.

Kneip, A., Simar, L., & Wilson, P. W. (2015). When bias kills the variance: Central limit theorems for DEA and FDH efficiency scores. *Econometric Theory, 31*(2), 394–422.

Kneip, A., Simar, L., & Wilson, P. W. (2016). Testing hypotheses in nonparametric models of production. *Journal of Business & Economic Statistics, 34*(3), 435–456.

Kohl, S., Schoenfelder, J., Fügener, A., & Brunner, J. O. (2019). The use of data envelopment analysis (DEA) in healthcare with a focus on hospitals. *Health Care Management Science, 22*(2), 245–286.

Lee, C.-Y., & Cai, J.-Y. (2020). LASSO variable selection in data envelopment analysis with small datasets. *Omega, 91*, 102019.

Li, Q. (1996). Nonparametric testing of closeness between two unknown distribution functions. *Econometric Reviews, 15*(3), 261–274.

McCallion, G., McKillop, D. G., Glass, J. C., & Kerr, C. (1999). Rationalizing Northern Ireland hospital services towards larger providers: Best-practice efficiency studies and current policy. *Public Money & Management, 19*(2), 27–32.

Meeusen, W., & van Den Broeck, J. (1977). Efficiency estimation from Cobb-Douglas production functions with composed error. *International Economic Review, 18*(2), 435–444.

Mouchart, M., & Simar, L. (2002). *Efficiency analysis of air controllers: First insights* (Consulting report #0202). Institut de Statistique, Université Catholique de Louvain. Belgium.

Nguyen, B. H., & Zelenyuk, V. (2021a). Aggregate efficiency of industry and its groups: The case of Queensland public hospitals. *Empirical Economics, 60*(6), 2795–2836.

Nguyen, B. H., & Zelenyuk, V. (2021b). Robust efficiency analysis of public hospitals in Queensland, Australia. In Daouia A., Ruiz-Gazen A. (Eds.), *Advances in contemporary statistics and econometrics* (pp. 221–242). New York: Springer.

O'Neill, L., Rauner, M., Heidenberger, K., & Kraus, M. (2008). A cross-national comparison and taxonomy of DEA-based hospital efficiency studies. *Socio-Economic Planning Sciences, 42*(3), 158–189.

Rosenblatt, M. (1956). Remarks on some nonparametric estimates of a density function. *The Annals of Mathematical Statistics, 27*(3), 832–837.

Sheather, S. J., & Jones, M. C. (1991). A reliable data-based bandwidth selection method for kernel density estimation. *Journal of the Royal Statistical Society. Series B (Methodological), 53*(3), 683–690.

Shephard, R. W. (1953). *Cost and production functions*. Princeton University Press.

Shephard, R. W. (1970). *Theory of cost and production functions*. Princeton University Press.

Sherman, H. D. (1984). Hospital efficiency measurement and evaluation: Empirical test of a new technique. *Medical Care, 22*(10), 922–938.

Sickles, R., & Zelenyuk, V. (2019). *Measurement of productivity and efficiency*. Cambridge University Press.

Silverman, B. W. (1986). *Density estimation for statistics and data analysis*. Chapman and Hall.

Simar, L., & Wilson, P. W. (1998). Sensitivity analysis of efficiency scores: How to boot-strap in Nonparametric Frontier Models. *Management Science, 44*(1), 49–61.

Simar, L., & Wilson, P. W. (2000). A general methodology for bootstrapping in non-parametric frontier models. *Journal of Applied Statistics, 27* (6), 779–802.

Simar, L., & Wilson, P. W. (2007). Estimation and inference in two-stage, semi-parametric models of production processes. *Journal of Econometrics, 136* (1), 31–64.

Simar, L., & Wilson, P. W. (2015). Statistical approaches for non-parametric frontier models: A guided tour. *International Statistical Review, 83*(1), 77–110.

Simar, L., & Wilson, P. W. (2020). Hypothesis testing in nonparametric models of production using multiple sample splits. *Journal of Productivity Analysis, 53*(3), 287–303.

Simar, L., & Zelenyuk, V. (2006). On testing equality of distributions of technical efficiency scores. *Econometric Reviews, 25*(4), 497–522.

Simar, L., & Zelenyuk, V. (2007). Statistical inference for aggregates of Farrell-type efficiencies. *Journal of Applied Econometrics, 22*(7), 1367–1394.

Simar, L., & Zelenyuk, V. (2011). Stochastic FDH/DEA estimators for frontier analysis. *Journal of Productivity Analysis, 36*(1), 1–20.

Simar, L., & Zelenyuk, V. (2018). Central limit theorems for aggregate efficiency. *Operations Research, 66*(1), 137–149.
Simar, L., & Zelenyuk, V. (2020). Improving finite sample approximation by central limit theorems for estimates from data envelopment analysis. *European Journal of Operational Research, 284*(3), 1002–1015.
Staat, M. (2006). Efficiency of hospitals in Germany: A DEA-bootstrap approach. *Applied Economics, 38*(19), 2255–2263.
Tiemann, O., & Schreyögg, J. (2009). Effects of ownership on hospital efficiency in Germany. *Academy of Management Proceedings, 2009*(1), 1–6.
Wilson, P. W. (2018). Dimension reduction in nonparametric models of production. *European Journal of Operational Research, 267*(1), 349–367.
Zelenyuk, V. (2020). Aggregation of inputs and outputs prior to data envelopment analysis under big data. *European Journal of Operational Research, 282*(1), 172–187.
Zhu, J. (1998). Data envelopment analysis vs. principal component analysis: An illustrative study of economic performance of Chinese cities. *European Journal of Operational Research, 111*(1), 50–61.
Zhu, J. (2020). DEA under big data: Data enabled analytics and network data envelopment analysis. *Annals of Operations Research*, in press
Zuckerman, S., Hadley J., & Lezzoni, L. (1994). Measuring hospital efficiency with frontier cost functions. *Journal of Health Economics, 13*(3), 255–280.

Parallel Processing and Large-Scale Datasets in Data Envelopment Analysis

Dariush Khezrimotlagh

1 Introduction

Charnes et al. (1978) developed data envelopment analysis (DEA). DEA is a boundary approach to estimate the production frontier and measure the performance of decision-making units (DMUs) with multiple inputs and multiple outputs. DEA determines the optimal inputs and outputs of DMUs to obtain efficiency. The survey of Emrouznejad and Yang (2018) of the last four decades shows that DEA has been used in many aspects, contexts, and activities. DEA is also a data enabled analytics tool (Zhu, 2020). From DEA data is enveloped, and one can (1) study the pattern of data, (2) find the best and worst practice DMUs, (3) suggest individual and common optimal weights corresponding to inputs and outputs, (4) suggest optimal inputs and outputs to achieve efficiency, (5) determine the reference sets for inefficient DMUs, and also (6) rank DMUs. For a background on DEA see Ray (2004), Cooper et al. (2007), Zhu (2014), Kao (2014), Toloo (2014), and Khezrimotlagh and Chen (2018).

DEA models are generally solved for one DMU at a time. When the number of DMUs (the number of inputs and outputs) increases, the elapsed time to evaluate all DMUs substantially increases. There are a few studies in the DEA literature to decrease the computational complexities of solving a DEA model. The main existing methods proposed by Barr and Durchholz (1997), Dulá (2011), Chen and Lai (2017), and Khezrimotlagh et al. (2019).

The elapsed time to evaluate DMUs depends on (at least) three factors, that is, cardinality, dimension, and density. Cardinality refers to the number of DMUs, where the dimension is the number of inputs and outputs. Density is also the percentage of efficient DMUs among all efficient and inefficient DMUs. As shown

D. Khezrimotlagh (✉)
Department of Mathematics and Computer Science, School of Science, Technology, and Engineering, Pennsylvania State University, Harrisburg, PA, USA
e-mail: dk@psu.edu

J. Zhu, V. Charles (eds.), *Data-Enabled Analytics*, International Series in Operations Research & Management Science 312, https://doi.org/10.1007/978-3-030-75162-3_6

in Khezrimotlagh and Zhu (2020), the production function, the distribution of inefficient DMUs, and the number of computers (or processors) in parallel can also affect the elapsed time to evaluate DMUs. The software used to run the models, different algorithms to solve a linear programming (LP), different versions of Windows, different preferences for Windows, and different way of programming an existing procedure may affect the elapsed time, too.

The strength of the existing methods to decrease the elapsed time of applying a DEA model has been studied according to cardinalities, dimensions, and densities of datasets. It is worth mentioning that, as a rule of thumb, when cardinality of a dataset is three times greater than the dimension of that dataset, it is less likely to observe a large number of efficient DMUs within the dataset (Banker et al. 1989). According to Toloo et al. (2015), almost all existing DEA applications have less than 7 dimensions. Thus, when cardinalities are more than 20 (and correspondingly more than 200, 2000, 20,000, and so on), we do not expect to have a large number of efficient DMUs within a dataset. For example, Khezrimotlagh et al. (2019) studied one of the largest available datasets in the DEA literature with 30,099 cardinalities and seven dimensions. The percentage of efficient DMUs in their dataset was only about 12.28%. Among these efficient DMUs, only 84 efficient DMUs are required to generate the corresponding PPS of DMUs (identified by Dulá's, 2011 method). This means that only 0.278% of DMUs were required to obtain their corresponding PPS. The larger the cardinalities, the smaller the densities and the larger the dimensions, the larger the densities. Khezrimotlagh and Zhu (2020) illustrate that when cardinalities are 50,000 (1,000,000) and dimensions are 10, using a uniform approach to generate a dataset, the total number of efficient DMUs within the sample is around 6% (1.1%). For larger cardinalities when dimensions are 10, it would be very rare to observe a dataset with higher than 10% densities. Nevertheless, in theory we may consider higher densities and dimensions. The performances of the existing methods are changed when densities, dimensions, and cardinalities are changed.

In this study, we illustrate the main existing methods to decrease the elapsed time of applying a DEA model to evaluate a large-scale dataset. We emphasize that when enough processors are used in parallel, the proposed method of Khezrimotlagh et al. (2019) is generally the fastest available technique to evaluate a large-scale dataset. However, when the number of processors is limited, the proposed methods by Dulá (2011) and Chen and Lai (2017) are more effective. When densities increase and parallel processing is available, the performance of Chen and Lai's (2017) method increases substantially; however, densities are known after evaluation. Therefore, it is not clear when we should use the method of Chen and Lai (2017). In this study, we first improve the method of Chen and Lai (2017). The enhanced method is normally two to three times faster than their original method. We then implement the enhanced method into the method of Khezrimotlagh et al. (2019) to further improve its speed when densities are extreme. We show that, overall, the existing methods are slower than the proposed combined method regardless of the cardinalities, dimensions, and densities. Several simulation experiments are explained, and the outcomes are depicted in several tables.

The rest of this paper is organized in five sections. Section 2 reviews the main existing methods. Section 3 presents how to select a method for a particular dataset. Section 4 illustrates necessary enhancements for the existing methods and proposes a new methodology by combination of the enhanced methods. Simulation experiments are also discussed in Sect. 4 and conclusions are given in Sect. 5.

2 The Existing Methodologies

In this section, we explain the existing methodologies that are the most effective techniques for evaluating a large dataset. Assume that there are n DMUs ($D_j, j = 1, 2, \ldots, n$) with m positive inputs (x_{ij}, $i = 1, 2, \ldots, m$) and s positive outputs (y_{rj}, $r = 1, 2, \ldots, s$).

Model (1) is the variable returns to scale (VRS) radial model (Banker et al. 1984). It is solved once for each DMU to measure its efficiency. It has $n + 1$ decision variables, $m + s + 1$ constraints, and n nonnegativity restrictions.

$$\max \varphi, \tag{1}$$

s.t.

$$\sum_{j=1}^{n} \lambda_j x_{ij} \leq x_{il}, \text{ for } i = 1, \ldots, m, \tag{1.1}$$

$$\sum_{j=1}^{n} \lambda_j y_{rj} \geq \varphi y_{rl}, \text{ for } r = 1, \ldots, s, \tag{1.2}$$

$$\sum_{j=1}^{n} \lambda_j = 1, \tag{1.3}$$

$$\lambda_j \geq 0, \text{ for } j = 1, \ldots, n. \tag{1.4}$$

The dual LP for model (1), as shown by model (2), includes $m + s + 1$ variables and $n + 1$ constraints, and $m + s$ nonnegativity restrictions.

$$\min \sum_{i=1}^{m} v_i x_{il} + w, \tag{2}$$

s.t.

$$\sum_{r=1}^{s} u_r y_{rl} = 1, \tag{2.1}$$

$$\sum_{i=1}^{m} v_i x_{ij} - \sum_{r=1}^{s} u_r y_{rj} + w \geq 0, \text{ for } j = 1, \ldots, n, \tag{2.2}$$

$$v_i \geq 0, \text{ for } i = 1, \ldots, m, \tag{2.3}$$

$$u_r \geq 0, \text{ for } r = 1, \ldots, s. \tag{2.4}$$

The PPS that generates by D_j 's is as follows:

$$P = \left\{ (\boldsymbol{x}, \boldsymbol{y}) \in \mathbb{R}_{++}^{m+s} : \sum_{j=1}^{n} \lambda_j x_j \leq \boldsymbol{x}, \sum_{j=1}^{n} \lambda_j y_j \geq \boldsymbol{y}, \sum_{j=1}^{n} \lambda_j = 1 \right\}. \tag{3}$$

Model (1) can also be written as model (4), when D_l ($l = 1, 2, \ldots, n$) is evaluated according to the generated PPS by $D'_{j'}$'s ($j' = 1, 2, \ldots, n'$). Model (4) can also be called as the super-efficiency DEA model.

$$\max \varphi', \tag{4}$$

s.t.

$$\sum_{j'=1}^{n} \lambda\prime_j x\prime_{ij\prime} \leq x_{il}, \text{ for } i = 1, \ldots, m, \tag{4.1}$$

$$\sum_{j'=1}^{n} \lambda\prime_j y\prime_{rj\prime} \geq \varphi' y_{rl}, \text{ for } r = 1, \ldots, s, \tag{4.2}$$

$$\sum_{j'=1}^{n} \lambda\prime_{j\prime} = 1, \tag{4.3}$$

$$\lambda\prime_{j\prime} \geq 0, \text{ for } j' = 1, \ldots, n'. \tag{4.4}$$

The performance of models (1) and (2) might be different on different datasets. For some datasets, model (2) might perform faster, however overall, model (1) performs faster. As density increases model (1) also performs faster than model (2).

2.1 The Standard Approach

The standard approach evaluates each DMU independently by one of models (1) and (2), where the entire dataset is considered to generate the corresponding PPS. Using parallel processing, each DMU can be evaluated by one of the processors. When 16 processors are available, this means that 16 DMUs are evaluated simultaneously by one of models (1) and (2). For example, one may expect that the elapsed time to evaluate 16,000 DMUs using 16 processors is approximately equal to the elapsed time to evaluate 1000 DMUs using one processor. This advantage of the standard approach to allow parallel processing should not be neglected when the standard approach is compared with other approaches. Parallel processing is an approach that splits a large and complex task into smaller and simpler subtasks in which they can simultaneously and independently be run on multiple computers and CPUs, and thus decrease the elapsed time to complete the original task. For example, Zhu et al. (2018) uses 10 computers in parallel to apply the standard approach. When there is only one processor, the standard approach is one of the weakest existing approaches. As the number of processors increases, the standard approach performs faster. Using a computer with CPU i7 with 8 processors in parallel, the standard approach usually performs very well for small cardinalities. Considering the same resources, the larger the cardinality the weaker the standard approach. The standard approach, in general, is significantly weaker than the main existing methods proposed by Barr and Durchholz (1997), Dulá (2011), Chen and Lai (2017), and Khezrimotlagh et al. (2019).

2.2 The Hierarchical Decomposition (HD) Procedure

Barr and Durchholz (1997) propose a hierarchical decomposition (HD) procedure to decrease the elapsed time of applying a DEA model to a large-scale dataset. They design a parallel processing approach to firstly find the efficient DMUs, and secondly use the efficient DMUs to evaluate inefficient DMUs. The steps of implementing HD are illustrated as follows.

The n observed DMUs are considered as the main block. The first iteration of the procedure is started by dividing the main block to several subblocks of size p. As suggested by Khezrimotlagh et al. (2019), the size of the subblocks can be measured by Eq. (5).

$$p = \min\left(\text{int}\left(\sqrt{(m+s)\,n}\right), \text{int}\,(n/2)\right) + 1. \tag{5}$$

The symbol 'int' shows the integer part of a real number. Assume that $n = pQ + R$, where Q is the number of subblocks of size p. When $R > 0$, there are $Q + 1$ subblocks. We assume $R = 0$ in our discussion to ease illustrations.

Using parallel processing, DMUs in each subblock are simultaneously evaluated; for example, by 8, 12, 16, 32, or more processors (computers) in parallel. Instead of solving an LP with size $n + 1$, Q LPs with size $p + 1$ are evaluated. This technique not only decreases the elapsed time to evaluate DMUs by decreasing the size of LPs, but it also decreases the workload for each processor and thus decreases the elapsed time.

Assume that $\beta = 0.9$ and $\gamma = 1.5$, as suggested by Barr and Durchholz (1997) and Dulá (2011). The parameter β is for terminating the procedure and the parameter γ is for increasing the size of the subblocks in each iteration. Different selections for these two parameters may result different elapsed times.

When the first iteration is finished, the subblocks' best-practice DMUs are aggregated into one single block (inefficient DMUs are excluded). Assume the size of this single block is n_1. If $n_1/n \geq \beta = 0.9$, the DMUs in the single block of size n_1 are evaluated by one of models (1) and (2) to find the efficient DMUs and complete HD's first phase. In phase 2, the excluded inefficient DMUs are evaluated using model (4) to complete the entire task.

When n_1/n is less than $\beta = 0.9$, the second iteration is started by dividing the found single block of size n_1 (from the first iteration) into subblocks of size $\gamma \times p = 1.5 \times p$. The same approach is applied to find the best-practice DMUs for these subblocks and to aggregate them into one single block of size n_2. The third iteration is started if n_2/n_1 is less than $\beta = 0.9$, and so on. The entire procedure is terminated after a few iterations.

Using enough computers (processors) in parallel, when the density is not very large, HD is one of the fastest available approaches to evaluate DMUs. As density and dimension increase where the used recourses in parallel are the same, HD becomes weaker than the other main models proposed by Dulá (2011), Chen and Lai (2017), and Khezrimotlagh et al. (2019).

2.3 *Build Hull (BH)*

Dulá (2011) proposes an approach, called build hull (BH), to find the minimum number of efficient DMUs to generate the corresponding production possibility set (PPS) of DMUs. From his approach, the necessary efficient DMUs are found one at a time and the corresponding PPS is gradually generated. If only one single processor exists to evaluate DMUs, BH could be the fastest available technique, as it requires the least number of calculations in comparison with the existing techniques. However, when the number of processors increases, BH becomes weaker. Parallel processing is not applicable for BH as its subtasks are dependent on each other. Khezrimotlagh and Zhu (2020) show that BH is slower than their method, and when an i9 CPU is used (in comparison with an i5 or i7 CPU) BH can even be slower than DH. Their findings emphasize the importance of the parallel processing that should not be neglected when two methods are compared. Khezrimotlagh and Zhu (2020) show that the strength of their procedure in comparison with BH is the use of parallel processing, as the elapsed time for their method decreases when the number of processors increases.

To apply BH, Ali's (1993) theorem is used to construct an initial set of size 1. The theorem says "a DMU with the largest (smallest) value for the first output (input) lies on the efficient frontier". Dulá (2011) proposes the following additive model (where slacks and surpluses are equal with negative signs) to evaluate DMUs one at a time.

$$\min \alpha, \tag{6}$$

s.t.

$$\sum\nolimits_{j=1}^{n} \lambda_j x_{ij} - \alpha \le x_{il}, \text{ for i} = 1, \ldots, m, \tag{6.1}$$

$$\sum\nolimits_{j=1}^{n} \lambda_j y_{rj} + \alpha \ge y_{rl}, \text{ for } r = 1, \ldots, s, \tag{6.2}$$

$$\sum\nolimits_{j=1}^{n} \lambda_j = 1, \tag{6.3}$$

$$\lambda_j \ge 0, \text{ for } j = 1, \ldots, n', \tag{6.4}$$

$$\alpha \ge 0. \tag{6.5}$$

In the first iteration, there is only one DMU to generate the PPS (that is a rectangle). A DMU is then selected among the remaining DMUs and model (6) is applied to evaluate that DMU. The primal and dual optimal solutions of model (6) are calculated by software (using initial and optimal tableaus of the simplex method). If the primal optimal objective, α^*, is equal to 0, this means that the evaluated DMU belongs to the generated PPS and the DMU is flagged as an inefficient DMU. A flagged DMU may lie on the efficient frontier, but it does not necessary include such a flagged DMU to generate the PPS. When $\alpha^* > 0$, the evaluated DMU is an exterior DMU to the PPS. In this case, the dual optimal solution is used to calculate the following expression, for any other DMUs.

$$\sum\nolimits_{r=1}^{s} u_r^* y_{rj} - \sum\nolimits_{i=1}^{m} v_i^* x_{ij} + w^*. \tag{7}$$

A DMU with the maximum positive value of the expression (7) is then added to the initial set to start the second iteration. Dulá (2011) shows that when there is a unique DMU with the maximum value of expression (7), that DMU is an efficient DMU. The similar technique to use the dual optimal solutions for examining exterior DMUs of a PPS is also used in Chen and Cho (2009).

The second iteration uses the found two DMUs from the first iteration to generate the corresponding PPS. Similar to the first iteration, a DMU is selected from the remaining DMUs and model (6) is solved to find the primal and dual optimal solutions. A similar procedure is continued for the next iterations until there are no remaining DMUs. The number of iterations is not greater than the number of DMUs. After the final iteration, the first phase of BH is completed, and in the second phase, the efficient DMUs are used to evaluate all flagged DMUs.

Since iteration $k + 1$ cannot be started until the iteration k is completed ($k \geq 1$), parallel processing is not applicable. As cardinality increases, BH performs weaker than the existing methodologies where an adequate number of processors are used in parallel. When there is only one processor to run the model, BH is one of the fastest available techniques to evaluate a large-scale dataset.

2.4 Finding Reference Sets (FRS)

Chen and Lai (2017) propose an approach, called finding reference sets (FRS), that is useful to handle a large-scale dataset when the resources are limited. From their approach, one can evaluate a big dataset using the Microsoft Excel solver with a limitation of 300 variables. Unlike the previous methods, FRS includes one stage to evaluate all DMUs and finds the efficient DMUs after the entire evaluation. Their method includes selecting an initial subsample of a size greater than the number of inputs and outputs, applying the standard DEA model, checking the optimality condition (proposed by Chen and Cho, 2009), possible resampling, and repeating the process to find the appropriate reference sets for evaluating a DMU. FRS becomes stronger as densities and dimensions increase. Nevertheless, before evaluation, the percentage of efficient DMUs within a dataset is unknown. In this study, we propose a new methodology to remove this issue.

In order to illustrate FRS, assume that a DMU is selected. To start FRS, an initial subsample with a size greater than $m + s$ should be selected. A larger size for the initial subsample may increase the elapsed time to evaluate the DMU. The initial subsample includes the DMU itself and the remaining DMUs can be chosen randomly or using the following equation, as suggested by Chen and Lai (2017) for the output orientated approach. DMUs with higher values of $d(j, l)$ are added to the initial subsample. In Eq. (8), the symbol $\langle .,. \rangle$ is the dot product of two vectors and the symbol $\|.\|$ is the 2-norm of a vector.

$$d(j, l) = \arccos\left(\frac{\langle(\mathbf{0}, \mathbf{y}_l), (\mathbf{x}_j - \mathbf{x}_l, \mathbf{y}_j - \mathbf{y}_l)\rangle}{\|\mathbf{0}, \mathbf{y}_l\| \, \|(\mathbf{x}_j - \mathbf{x}_l, \mathbf{y}_j - \mathbf{y}_l)\|}\right). \tag{8}$$

Assume that an initial subsample including $m + s + 1$ DMUs is selected. The corresponding model (1) or (2) for this set is formulated and the DMU is evaluated. According to the dual or primal optimal solution, expression (7) is calculated for any other DMUs (Chen and Cho 2009). DMUs are then sorted in ascending order according to their corresponding values for expression (7). As suggested by Chen and Lai (2017, pp. 152), the DMU in the first position of the initial subsample is dropped, and the DMU in the last position of the remaining DMUs is added to the initial subsample. The next iteration can now be started by the same approach. The resampling will stop until expression (7) is negative for all remaining DMUs. In this case, the DMU is evaluated. The similar process is continued to evaluate all DMUs.

2.5 The Enhanced Hierarchical Decomposition (EHD) Procedure

Khezrimotlagh et al. (2019) improved the HD procedure, called enhanced HD (EHD), and proposed the fastest available technique, in general, to find efficient DMUs as well as inefficiency scores of all DMUs. They only used one computer with several processors to run parallel processing and comparisons; nevertheless, the impact of their approach is noticeable when more than one computer is used in parallel. The larger the number of processors (computers) in parallel, the stronger EHD to decrease the elapsed time of solving a DEA model. When densities are extreme, EHD becomes weaker. For example, when densities are 100%, HD and EHD become weaker than the standard approach because the first iteration in their first phase is an additional step to the standard approach. Nevertheless, before evaluation, the percentage of efficient DMUs within a dataset is unknown. It is impossible to declare the density of a dataset which is not yet evaluated. Density of a dataset is known after evaluation. Khezrimotlagh and Zhu (2020), however, suggest a technique in the first iteration of their method to quickly estimate the percentage of efficient DMUs within a dataset. In this case, after the first iteration in the first phase of EHD, one can decide which technique is more effective to evaluate DMUs and decrease the elapsed time to find efficient DMUs. In this study, we use their technique and propose a new technique to increase the performance of EHD when densities are extreme. The steps to implement EHD are illustrated as follows.

Khezrimotlagh et al. (2019) suggest selecting one particular single block to evaluate any other blocks in the first phase of HD. To find the particular block, they suggest an algorithm that chooses a subsample of DMUs which are potential for being efficient DMUs. Assume that the subsample size (the size of blocks) is calculated by Eq. (5).

To find the subsample, they first use the Ali's (1993) theorems to select $m + s$ DMUs (or more); such that, DMU_l is located on the VRS frontier if one of the following equations holds: $x_{il} = \min \{x_{ij} | j = 1, \ldots, n\}$, for $i = 1, \ldots, m$, and $y_{rl} = \max \{y_{rj} | j = 1, \ldots, n\}$, for $r = 1, \ldots, s$. After that, the percentiles of each input and each output are found. And for each unselected DMU_j a pre-score is assigned from the following algorithm.

$$
\begin{Bmatrix}
u = 0, \\
\quad \begin{Bmatrix}\text{For each } i \text{ and } k_i, \\ \quad \text{If } x_{ij} \le k_i - \text{th percentile of } \boldsymbol{x}^i \text{ then} \\ \quad\quad u = u + 1 \end{Bmatrix} \\
\quad \begin{Bmatrix}\text{For each } r \text{ and } k'_r, \\ \quad \text{If } y_{rj} \ge k\prime_r - \text{th percentile of } \boldsymbol{y}^r \text{ the} \\ \quad\quad u = u + 1 \end{Bmatrix} \\
\text{pre\_score}\, j = u
\end{Bmatrix}
\tag{9}
$$

DMUs are then sorted in descending order corresponding to the assigned pre-scores. The higher the pre-scores, the more likely to have an efficient DMU. The remaining DMUs (to construct a subsample of size p) are selected from those DMUs with higher pre-scores.

The rest of the procedure can be programmed similar to that of HD. The difference between HD and EHD is that blocks in the first iteration of HD are evaluated according to the generated PPS of the selected subsample; that is, model (4) or its dual is used to evaluate DMUs in each block.

Khezrimotlagh et al. (2019) and Khezrimotlagh and Zhu (2020) showed that when the density is very small, it is more likely to complete the entire evaluation in the above first iteration.

If there are some exterior DMUs corresponding to the PPS of the selected subsample, the next iteration is started by aggregating the exterior DMUs and the subsample's DMUs to construct the second subsample. Here, users can estimate the density of the dataset and make decisions to use an appropriate procedure. The efficient DMUs in the second subsample are found and the rest of the DMUs are evaluated according to the generated PPS of efficient DMUs.

3 Selecting an Appropriate Method

None of the existing methods can be considered as a complete package to evaluate DMUs with different cardinalities, dimensions, and densities. When resources are limited (for example, when only one processor is used), BH is the choice. If the standard version of the Microsoft Excel Solver is used, FRS is the choice. As number of processors in parallel increases, EDH is the choice. When densities and dimensions are rare and extreme, and resources are limited, enhanced FRS is the choice, and so on.

The meaning of extreme is changed as the number of processors in parallel is changed. When there are 4 processors in parallel, 4000 cardinalities might be considered as an extreme cardinality. As number of processors in parallel increases to 12 or 16 processors, 4000 cardinalities can be considered as a small cardinality. For small cardinalities, the standard approach can be more powerful than the existing methods. For example, using 16 processors in parallel, the standard approach can be the fastest available technique to evaluate a dataset including 1000 DMUs with 20 dimensions and 100% densities.

Using the standard version of the Microsoft Excel Solver with the limitation of 200 variables, 2% (10%) densities among 10,000 (2000) DMUs can be considered an extreme density. In contrast, when we use 16 processors in parallel using MATLAB software, 2% (10%) densities among 10,000 (2000) DMUs are a small density. We also note that 10% densities do not necessarily mean that 10% of DMUs are required to generate their corresponding PPS. BH finds the minimum number of efficient DMUs to generate the PPS of DMUs, followed by EDH, whereas the standard approach and HD uses all DMUs to generate the PPS. For instance, in the example of Khezrimotlagh et al. (2019), the number of efficient DMUs among 30,099 DMUs is 2545, where BH (EHD) only used 84 (457) of these efficient DMUs to generate the PPS. In this case, HD used all 2545 DMUs to generate their corresponding PPS.

As a result, it is not clear "what is an extreme density, dimension, or cardinality?" without measuring the potential of the available resources. Even if, for a particular number of processors in parallel, we find a boundary for extreme densities or extreme dimensions, the meaning of extreme will be changed as the number of resources in parallel is changed. Therefore, users should be aware of their potential recourses before choosing a method to apply a DEA model.

4 Proposed Methods

4.1 Introducing the Enhanced FRS (EFRS)

In this section, we first improve the method of Chen and Lai (2017) to become two to three times faster in general. It is worth mentioning that, from the model, a reference set to evaluate a DMU may be dropped from the initial subsample in one iteration and added again to the initial subsample in another iteration. This results in unnecessary calculations and increases the elapsed time to evaluate the DMU. It is also possible that the number of iterations approaches to infinity. For example, when the size of the initial subsample is not appropriately selected, there is a chance to observe a large number of iterations to find the necessary reference sets for evaluating a DMU. To enhance the elapsed time of applying FRS, we suggest using Eq. (5) for the size of the initial subsample. We also suggest a desired number of iterations to evaluate DMUs; for example, $m + s + 1$ iterations at the most. The higher the upper bound for possible interactions, the larger the elapsed time in general. In the final iteration, we then aggregate all DMUs with positive values for expression (7) and evaluate the DMU. We ignore dropping DMUs from the

Table 1 Comparing FRSs and EFRSs

Cardinality	Dimension	Density	FRS1	FRS2	EFRS1	EFRS2	EFRS
3000	4	10.40%	94.38	66.91	83.07	62.09	43.55
		60.17%	99.65	71.99	90.14	66.37	48.83
	20	47.40%	145.55	132.45	145.19	145.93	79.89
		73.07%	178.91	152.7	162.18	150.46	136.93
10,000	4	10.12%	554.81	389.93	454.69	342.46	156.39
		60.00%	590.46	462.65	527.43	383.98	427.32
	20	34.18%	1034.38	974.08	1037.3	963.27	666.86
		68.35%	1670.17	1413.22	1483.44	1230.69	1239.55

initial subsample and only add potential candidates to the initial subsample in each iteration. We call this approach, the enhanced FRS (EFRS).

Ali's (1993) theorems can be added to the initial subsample to apply FRS. This might decrease the elapsed time for some DMUs and at the same time increase the elapsed time for some other DMUs. Since FRS aims to find the corresponding reference sets for evaluating a DMU, the smaller initial subsample (including all reference sets of the evaluated DMU) results in decreasing the elapsed time to evaluate that DMU. Thus, we only select the initial subsample using Eq. (8). For an input-oriented approach, we also use the similar equation discussed in Chen and Lai (2017). For other returns to scale technologies, we consider both equations for input-orientation and output-orientation.

We now consider the following scenarios.

- FRS1: FRS with an initial subsample of size $m + s + 1$ and an upper bound $2 \times (m + s + 1)$ for iterations.
- FRS2: FRS with an initial subsample of size $m + s + 1$ and an upper bound $m + s + 1$ for iterations.
- EFRS1: FRS with an initial subsample of size $m + s + 1$ and an upper bound $2 \times (m + s + 1)$ for iterations without dropping a DMU in each iteration.
- EFRS2: FRS with an initial subsample of size $m + s + 1$ and an upper bound $m + s + 1$ for iterations without dropping a DMU in each iteration.
- EFRS: FRS with an initial subsample of size p (using Eq. 5) and an upper bound $m + s + 1$ for iterations without dropping a DMU in each iteration.

We use a computer with Windows 10 and CPU i7 with 8 identical processors in parallel to examine the strength of these changes. The same conditions are considered for all different scenarios. As can be seen in Table 1, EFRS performs faster than the other scenarios. The pros of the suggested enhancements can be seen in the table. As the upper bound for the number of iterations increases, FRS becomes weaker, unless densities are extremely large (e.g., above 80%) and in this case, EFRS1 performs faster. In overall, FRS becomes faster when we do not drop a DMU in each iteration. For larger densities, the strengths of the scenarios are similar; however, EFRS is generally faster than other scenarios.

Table 2 Comparing EHD and EFRS

Cardinality	Dimension	Density	EHD		EFRS
			Phase 1	Phase 2	
2000	2	0.30%	2.87	2.87	4.13
	4	1.65%	2.29	3.39	3.37
	6	7.90%	2.23	3.33	4.49
	8	16.20%	2.23	3.29	4.95
	10	24.40%	2.45	3.53	5.17
64,000	2	0.03%	45.47	79.63	1346.72
	4	0.15%	42.55	77.51	1600.42
	6	0.82%	44.64	87.01	1869.65
	8	2.31%	52.70	141.17	2126.87
	10	5.08%	106.05	382.85	2477.96
128,000	2	0.01%	84.01	84.01	4207.86
	4	0.11%	81.71	151.36	6967.86
	6	0.55%	86.57	181.41	7932.09
	8	1.48%	105.63	343.85	9668.67
	10	3.75%	268.58	1394.32	11,373.46

4.2 *Comparing EHD and EFRS*

In the previous section, we enhanced the FRS method. We now compare EFRS with the method of Khezrimotlagh et al. (2019), EHD, where datasets are generated by a uniform distribution without controlling densities. We use a computer with Windows 10 and CPU i9 with 16 identical processors in parallel and the same conditions to compare the methods. The following table illustrates the elapsed times of these methods where cardinalities are 2000, 64,000, and 128,000, and dimensions are 2, 4, 6, 8, and 10. As cardinality increases, the density decreases, and when dimension increases, the density increases.

When density is 3.75%, cardinality 128,000, and dimension is 10, EHD is eight times faster than EFRS. In this case, EHD finds the set of efficient DMUs in 5 minutes, where it takes 3 hours and 15 minutes for EFRS to find the set of efficient DMUs. EHD finds the efficiency of the entire dataset in 23 minutes only.

When densities are 0.01%, 0.11%, and 0.55%, cardinality 128,000, and dimension is 2, 4, and 6, EHD is 50, 46, and 43 times faster than EFRS, respectively (Table 2).

As discussed, densities are measured after evaluations. However, in theory we can control densities. When densities are increased, the strength of EHD is decreased. In contrast, EFRS becomes stronger for larger densities. Thus, since densities are unknown before evaluation, we may choose one method that is not actually appropriate to evaluate that dataset. In the next subsection, we combine EFRS and EHD to introduce a new method with the properties of both EHD and EFRS.

4.3 Implementing EFRS into EHD

Suppose that a large-scale dataset is given with cardinality n (e.g., above 10,000 DMUs). Assume that the dimension is $m + s$ and $t > 1$ is the number of processors in parallel. We note that densities are unknown before evaluation of DMUs and we are not able to assume preferences for densities without applying an appropriate method. Here, we propose the following method to deal with unknown densities situations. In order to evaluate the given dataset with cardinality n,

(i) Calculate the subsample's size, p, using Eq. (5).
(ii) Find the first subsample of size, p, using Ali's (1993) theorems and the algorithm 9.
(iii) Divide DMUs into $Q + 1$ blocks at the most, where $n = pQ + R$.
(iv) Using model (4), evaluate the DMUs in each block according to the PPS of the first subsample.
(v) Find the exterior DMUs to the PPS of the first subsample.
(vi) If there are no exterior DMUs, the entire dataset is evaluated, and the task is completed.
(vii) If there are some exterior DMUs, aggregate all the exterior DMUs and the efficient DMUs of the first subsample into one single subsample, called second subsample.
(viii) Apply EFRS on the second subsample to find the efficient DMUs to generate the PPS of all DMUs. In this step, all inefficient DMUs within the second subsample are evaluated, and density is known.
(ix) Apply the enhanced FRS on the remaining DMUs where the initial subsample to evaluate each DMU is chosen from the found efficient DMUs in the previous step.

We note that, in general, the size of the second subsample is very small. For example, using a uniform approach to generate datasets, the average size for the second subsample when cardinality is 4000 and dimension is 6 is about 161.23 with the standard deviation 30.75. When cardinality is 32,000 and dimension is 6, the average size for the second subsample is 452.37 with the standard deviation 54.77. Similarly, when cardinality is 128,000 and dimension is 10, the average size for the second subsample is 6351.30 with the standard deviation 364.51. Nevertheless, the second subsamples' sizes increase as densities increase. When densities are extreme, using EFRS to find efficient DMUs within the second subsample decreases the elapsed time substantially.

Table 3 illustrates the performance of the enhanced methods when densities are controlled. A computer with Windows 10, CPU i9, and 16 processors in parallel is used. For extreme dimensions and densities, the strength of EHD substantially decreases, whereas EFRS and EHD-EFRS become more powerful. When cardinalities are 10,000 and 30,000, dimension is 6, and density is about 5%, EHD performs faster than EFRS, but slower than EHD-EFRS. When density increases to 55%, EHD becomes very slower than EHD-EFRS. In this case, EHD-

Table 3 Comparing the enhanced methods

Cardinality	Dimension	Density	EFRS	EHD		EHD-EFRS	
				Phase 1	Phase 2	Phase 1	Phase 2
3000	6	5%	10.85	3.78	5.03	5.19	6.55
		55%	17.56	9.14	10.43	17.43	19.64
	18	40%	20.67	11.7	13	18.63	20.51
		60%	32.64	16.76	17.63	30.52	31.87
10,000	6	5%	31.155	13.225	15.15	11.915	13.98
		55%	102.61	155.98	179.95	92.12	96.97
	18	30%	193.37	274.17	287.54	128.65	137.79
		60%	280.34	614.27	634.81	260.15	269.45
30,000	6	5%	447.77	29.26	54.63	25.78	48.54
		55%	1259.70	3923.69	4707.76	904.18	987.82
	18	20%	1437.36	9910.39	10,116.57	695.28	758.69
		60%	2713.10	37,193.15	38,074.31	2493.39	2539.85

EFRS is four times faster than EHD to find the efficient DMUs. When cardinality is 30,000, dimension is 18, and density is about 60%, both EHD-EFRS and EFRS become about 15 times faster than EHD. When densities are extremely large (e.g., above 70%), the differences between EFRS and EHD-EFRS become negligible. When densities are small, the differences between EHD and EHD-EFRS are also negligible. As a result, the combination of EHD and EFRS is suggested to evaluate a large-scale dataset. This outcome suggests a single methodology for applying DEA models regardless of densities, cardinalities, and dimensions of datasets.

5 Conclusion

In this study, we show that the strengths of the existing methods are affected when cardinality, dimension, and density are changed. We emphasize that density is unknown before evaluation, but raise the concern that, if density is large, different methods should be used. We conclude that the method of Khezrimotlagh et al. (2019), EHD, becomes weak for extreme densities. For this reason, the method of Chen and Lai (2017), FRS, is improved, called EFRS, and then it is implemented into the method of Khezrimotlagh, called EHD-EFRS. Using simulation experiments, it is determined that EHD-EFRS removes the concern of different possible densities of datasets effecting a reliable final outcome. EHD-EFRS includes both properties of EHD and EFRS, and regardless the densities of datasets, it generally performs faster than all existing methods.

References

Ali, A. I. (1993). Streamlined computation for data envelopment analysis. *European Journal of Operational Research, 64*(1), 61–67.
Banker, R. D., Charnes, A., & Cooper, W. W. (1984). Some models for estimating technical and scale inefficiencies in data envelopment analysis. *Management Science, 30*(9), 1078–1092.
Banker, R. D., Charnes, A., Cooper, W. W., Swarts, J., & Thomas, D. (1989). An introduction to data envelopment analysis with some of its models and their uses. *Research in Governmental and Nonprofit Accounting, 5*(1), 125–163.
Barr, R. S., & Durchholz, M. L. (1997). Parallel and hierarchical decomposition approaches for solving large-scale data envelopment analysis models. *Annals of Operations Research, 73*, 339–372.
Charnes, A., Cooper, W. W., & Rhodes, E. (1978). Measuring the inefficiency of decision-making units. *European Journal of Operational Research, 2*(6), 429–444.
Chen, W. C., & Cho, W. J. (2009). A procedure for large-scale DEA computations. *Computers & Operations Research, 36*(6), 1813–1824.
Chen, W. C., & Lai, S. Y. (2017). Determining radial efficiency with a large data set by solving small-size linear programs. *Annals of Operations Research, 250*(1), 147–166.
Cooper, W. W., Seiford, L. M., & Tone, K. (2007). *Data envelopment analysis: a comprehensive text with models, applications, references and DEA-solver software (Vol. 2)*. New York: Springer.
Dulá, J. H. (2011). A method for data envelopment analysis. *INFORMS Journal on Computing, 23*(2), 284–296.
Emrouznejad, A., & Yang, G. L. (2018). A survey and analysis of the first 40 years of scholarly literature in DEA: 1978–2016. *Socio-Economic Planning Sciences, 61*, 4–8.
Kao, C. (2014). *Network data envelopment analysis* (Vol. 10, pp. 26–33). Springer.
Khezrimotlagh, D., & Chen, Y. (2018). *Decision making and performance evaluation using data envelopment analysis*. Springer.
Khezrimotlagh, D., & Zhu, J. (2020). A faster method to Deal with big data in data envelopment analysis. In V. Charles, J. Aparicio, & J. Zhu (Eds.), *Data science and productivity analytics*. Springer. ISBN:978-3-030-43383-3.
Khezrimotlagh, D., Zhu, J., Cook, W., & Toloo, M. (2019). Data envelopment analysis and big data. *European Journal of Operational Research, 274*(3), 1047–1054.
Ray, S. C. (2004). *Data envelopment analysis: Theory and techniques for economics and operations research*. Cambridge university Press.
Toloo, M. (2014). *Data envelopment analysis with selected models and applications* (Vol. 30). VSB-TU Ostrava.
Toloo, M., Barat, M., & Masoumzadeh, A. (2015). Selective measures in data envelopment analysis. *Annals of Operations Research, 226*(1), 623–642.
Zhu, J. (2014). *Quantitative models for performance evaluation and benchmarking: Data envelopment analysis with spreadsheets* (Vol. 213). Springer.
Zhu, J. (2020). DEA under big data: Data enabled analytics and network data envelopment analysis. *Annals of Operations Research*, (In Press).
Zhu, Q., Wu, J., & Song, M. (2018). Efficiency evaluation based on data envelopment analysis in the big data context. *Computers & Operations Research, 98*, 291–300.

Network DEA and Big Data with an Application to the Coronavirus Pandemic

Hirofumi Fukuyama and William L. Weber

1 Introduction

In 2001, Doug Laney (2001) proposed three characteristics defining big data: volume, velocity, and variety. Volume refers to the increasingly large amounts of data that businesses and governments acquire due to innovations in data collection such as Universal Product Codes (UPC), electronic data interchange (EDI), point of sale (POS) tracking, smart devices (phones, watches and wearable health trackers), videos and social media. Moreover, data are becoming available more frequently (Velocity) increasing the ability of decision-makers to respond to changing conditions in a more timely manner. As firms merge and new partnerships are formed, data management confronts problems with the Variety of data available, since different organizations code and process data differently. More recently, veracity, variability, and value have been added to the "Vs" defining big data. Veracity refers to the ability of data users to verify the accuracy of data and trust the sources generating the data. Variability refers to the fact that the data may be collected/observed at different times and frequencies. This variability in data collection means that missing data requires some variables be interpolated across time. Finally, the ability of data analytics to decipher and decode the data into useful information for decision makers is necessary for Big Data to add value to a firm's profits or other stakeholders' objectives.

Econometric tools for use with big data have been developed and Varian (2014) provides a good overview of research that has investigated data manipulation,

H. Fukuyama (✉)
Fukuoka University, Fukuoka, Japan
e-mail: fukuyama@fukuoka.u.ac.jp

W. L. Weber
Southeast Missouri State University, Cape Girardeau, MO, USA
e-mail: wlweber@semo.edu

J. Zhu, V. Charles (eds.), *Data-Enabled Analytics*, International Series in Operations Research & Management Science 312, https://doi.org/10.1007/978-3-030-75162-3_7

variable selection, estimation techniques, and prediction. Less research has been devoted to big data in DEA (Data Envelopment Analysis) and this paper reviews some of that research and overviews how NDEA (Network DEA) might be used as a tool for evaluating big data. In addition, we provide an empirical example of how NDEA can be used with big data by examining the coronavirus pandemic that began in early 2020. Our NDEA model estimates US state performance in minimizing deaths from Covid. The model incorporates undesirable spillover effects of Covid, i.e. contagion, across time and between states. In addition, we incorporate the number of tests for Covid as an input that can be used to avert deaths from Covid while maintaining some level of economic activity. Our dynamic method investigates whether intertemporal reallocations of Covid tests could have helped reduce Covid deaths.

2 Recent Work on DEA and Big Data

The DEA (Data Envelopment Analysis) approach to data analysis assumes that producers use inputs, $x \in R_+^N$, to produce outputs $y \in R_+^M$. Data on inputs and outputs are observed for $k = 1, \ldots, K$ DMUs (decision-making units) in periods $t = 1, \ldots, T$.

The technology coefficient matrix (Koopmans, 1951) is given by

$$\mathbf{T}^t = \begin{pmatrix} -x_{11}^t & \cdots & -x_{1K}^t \\ \vdots & \ddots & \vdots \\ -x_{N1}^t & \cdots & -x_{NK}^t \\ y_{11}^t & \cdots & y_{1K}^t \\ \vdots & \ddots & \vdots \\ y_{M1}^t & \cdots & y_{MK}^t \end{pmatrix}, \qquad t = 1, \ldots, T, \tag{1}$$

with $N + M$ rows and K columns where the kth column of $\mathbf{T}^t$ represents the inputs used and outputs produced by DMU k in period t and is given by the vectors $-\mathbf{x}_k^t$ and $\mathbf{y}_k^t$.

With respect to Big Data in the DEA context; "volume" is represented by the number of columns (DMUs) in the technology matrix, $\mathbf{T}^t$, that is, the number of DMUs used to construct the technology; "velocity", refers to the idea that data are collected and available in an increasing number of time periods, so one observes an increasing number of technology matrices in the sequence $t = 1, \ldots, T, \ldots$; "variety" is seen in the number of inputs used and outputs produced, with greater variety increasing the number of rows in the technology matrix. This change in the number of rows was discussed by Coale (1951) in his comments to a paper by Simon (1951) on technological change, where the values of the cells in the coefficient matrix change. Furthermore, Bellman (1957, 1961) introduced the term "curse of dimensionality" and in the context of DEA, this curse means that adding inputs/outputs to $\mathbf{T}^t$ can reduce the discriminatory power of DEA to measure the performance of DMUs.

Post-multiplying the technology matrices by a $K \times 1$ vector of intensity variables, $\boldsymbol{\lambda}^t = (\lambda_1{}^t, \ldots, \lambda_K{}^t)'$, yields the transformed technology matrix

$$\mathbf{T}^t\boldsymbol{\lambda}^t = \begin{pmatrix} -\sum_{k=1}^{K} x_{1k}^t \lambda_k^t \\ \vdots \\ -\sum_{k=1}^{K} x_{Nk}^t \lambda_k^t \\ \sum_{k=1}^{K} y_{1k}^t \lambda_k^t \\ \vdots \\ \sum_{k=1}^{K} y_{Mk}^t \lambda_k^t \end{pmatrix}, \quad t = 1, \ldots, T. \tag{2}$$

Given $\mathbf{T}^t\boldsymbol{\lambda}^t$, the constant returns to scale DEA technology set for period t is

$$\begin{aligned} &\left\{ \begin{pmatrix} \mathbf{x}^t \\ \mathbf{y}^t \end{pmatrix} \middle| \; T^t\boldsymbol{\lambda}^t \geq \begin{pmatrix} -\mathbf{x} \\ \mathbf{y} \end{pmatrix}, \; \boldsymbol{\lambda}^t \geq \mathbf{0} \right\} \\ &= \left\{ \begin{pmatrix} \mathbf{x}^t \\ \mathbf{y}^t \end{pmatrix} \middle| \; \sum_{k=1}^{K} \mathbf{x}_k^t \lambda^t{}_k \leq \mathbf{x}^t, \; \sum_{k=1}^{K} \mathbf{y}_k^t \lambda_k^t \geq \mathbf{y}^t, \; \boldsymbol{\lambda}^t \geq \mathbf{0} \right\}, \quad t = 1, \ldots, T. \end{aligned} \tag{3}$$

Constraining the intensity variables to be non-negative allows the technology to exhibit constant returns to scale. Non-increasing returns to scale is imposed by adding the intensity variable constraint $\sum_{k=1}^{K} \lambda_k^t \leq 1$ to (3). Variable returns to scale is modeled by adding the constraint $\sum_{k=1}^{K} \lambda_k^t = 1$ to (3). Variable returns to scale allows the technology to exhibit increasing, constant, or decreasing returns to scale.

Input or output efficiency measures can be defined on the technology. The variable returns to scale output technical efficiency index of DMU "o" is estimated using DEA as

$$\begin{aligned} \widehat{OTE}^t\left(\mathbf{x}_o^t, \mathbf{y}_o^t\right) &= \max_{\theta,\lambda} \quad \theta \quad \text{subject to} \\ &\sum_{k=1}^{K} \lambda_k^t x_{kn} \leq x_{on}^t, \quad n = 1, \ldots, N, \\ &\sum_{k=1}^{K} \lambda_k^t y_{km} \geq \theta y_{om}^t, \quad m = 1, \ldots, M, \\ &\sum_{k=1}^{K} \lambda_k^t = 1, \\ &\lambda_k^t \geq 0, \quad k = 1, \ldots, K, \quad t = 1, \ldots, T. \end{aligned} \tag{4}$$

The index gives the maximum proportion expansion in outputs given inputs. Efficient DMUs have $\widehat{OTE}^t\left(\mathbf{x}_o^t, \mathbf{y}_o^t\right) = 1$ and inefficient producers have $\widehat{OTE}^t\left(\mathbf{x}_o^t, \mathbf{y}_o^t\right) > 1$ with higher values indicating greater inefficiency.

Input technical efficiency is estimated as

$$
\begin{aligned}
\widehat{ITE}\left(\mathbf{y}_o^t, \mathbf{x}_o^t\right) = \min_{\phi,\lambda} \quad & \phi \text{ subject to} \\
& \sum_{k=1}^{K} \lambda_k^t x_{kn} \leq \phi x_{on}^t, \quad n = 1, \ldots, N, \\
& \sum_{k=1}^{K} \lambda_k^t y_{km} \geq y_{om}^t, \quad m = 1, \ldots, M, \\
& \sum_{k=1}^{K} \lambda_k^t = 1, \\
& \lambda_k^t \geq 0, \quad k = 1, \ldots, K, \quad t = 1, \ldots, T.
\end{aligned} \tag{5}
$$

The input technical efficiency index gives the proportional contraction in inputs, given outputs. Efficient DMUs have $\widehat{ITE}\left(\mathbf{y}_o^t, \mathbf{x}_o^t\right) = 1$. Inefficient DMUs have $\widehat{ITE}\left(\mathbf{y}_o^t, \mathbf{x}_o^t\right) < 1$ with smaller values indicating lower efficiency.

Let Ψ^t represent the true technology such that

$$
\Psi^t = \{(\mathbf{x}, \mathbf{y}) : \quad \mathbf{x} \text{ can produce } \mathbf{y} \text{ in } t\}, \quad t = 1, \ldots, T. \tag{6}
$$

The true, but unobserved measures of output and input technical efficiencies for DMU "o" are

$$
\widehat{OTE}^t\left(\mathbf{x}_o^t, \mathbf{y}_o^t\right) = \max\left\{\theta : \quad \left(\mathbf{x}_o^t, \theta \mathbf{y}_o^t\right) \in \Psi^t\right\}. \tag{7}
$$

and

$$
\widehat{ITE}^t\left(\mathbf{y}_o^t, \mathbf{x}_o^t\right) = \min\left\{\phi : \quad \left(\phi \mathbf{x}_o^t, \mathbf{y}_o^t\right) \in \Psi^t\right\}, \quad t = 1, \ldots, T. \tag{8}
$$

Zelenyuk (2019) showed that the differences between the DEA estimates of output and input efficiencies and the true measure of output and input efficiency are given by $\widehat{OTE}\left(\mathbf{x}_o^t, \mathbf{y}_o^t\right) - \widehat{OTE}\left(\mathbf{x}_o^t, \mathbf{y}_o^t\right) = O_p\left(K^{-r}\right)$ and $\widehat{ITE}\left(\mathbf{y}_o^t, \mathbf{x}_o^t\right) - \widehat{ITE}\left(\mathbf{y}_o^t, \mathbf{x}_o^t\right) = O_p\left(K^{-r}\right)$, where $O_p(K^{-r})$ is the order of magnitude and $r = 2/(N + M)$ for the constant returns to scale technology and $r = 2/(N + M + 1)$ for the variable returns to scale technology. Thus, one challenge for using DEA to analyze Big Data is that as the variety of data increases, the number of rows in the technology matrix $\mathbf{T}^t$ increases, resulting in a slower convergence of the DEA estimator of efficiency to the true measure of efficiency.

In a Monte Carlo simulation, Zelenyuk (2019) investigated various aggregation methods for reducing the number of rows in the technology matrix to speed the convergence of the DEA estimator to the true estimator. These methods included aggregating inputs to total costs and aggregating outputs to total revenues and then using costs (revenues) in the DEA measure of input (output) technical efficiency. This aggregation reduces the number of rows in the technology matrix from $N + M$ to $M + 1$ when aggregating inputs to costs to estimate *ITE*. Similarly, aggregating outputs to revenues to estimate *OTE* reduces the number of rows from $N + M$ to $N + 1$. The simulation found a high positive and significant correlation between the true measures of efficiency and the DEA estimates of efficiency using cost/revenue aggregates of inputs/outputs.

To aggregate outputs to revenues, let $\mathbf{p}' = (p_1, \ldots, p_M) \in R_+^M$ be a price vector which is common to all DMUs. Then, revenues are $r_k = \mathbf{p}'\mathbf{y}_k$, $k = 1, \ldots, K$. Given revenues, an alternative DEA revenue-based output technical efficiency measure for DMU "o" is

$$
\begin{aligned}
&\widetilde{OTE}(\mathbf{x}, r) = \max_{\lambda} \quad \frac{1}{r_o^t} \sum_{k=1}^{K} \lambda_k^t r_k^t \\
&\sum_{k=1}^{K} \lambda_k^t x_{kn}^t \leq x_{on}^t, \quad n = 1, \ldots, N, \\
&\sum_{k=1}^{K} \lambda_k^t = 1, \\
&\lambda_k^t \geq 0, \quad k = 1, \ldots, K.
\end{aligned}
\tag{9}
$$

Comparing efficiency as measured by (4) with that measured by (9) we have $\widetilde{OTE}(\mathbf{x}, r) \geq OTE(\mathbf{x}, \mathbf{y})$. Thus, under VRS, the aggregation of outputs to revenues using a common price vector allows the number of constraints to be reduced from $N + M + 1$ to $N + 1$ and results in a greater potential expansion of a DMU's revenues relative to the revenue frontier than the potential expansion of outputs to the output frontier.

A similar procedure can be used to aggregate inputs to costs to construct a cost-based measure of input technical efficiency. Let costs be represented by $c_k = \mathbf{w}'\mathbf{x}_k^t$ where $\mathbf{w}' = (w_1, \ldots, w_N) \in R_+^N$ is the input price vector common to all DMUs. Given costs, an alternative DEA input technical efficiency measure is

$$
\begin{aligned}
&\widetilde{ITE}(\mathbf{y}, c) = \min_{\lambda} \quad \frac{1}{c_o^t} \sum_{k=1}^{K} \lambda_k^t c_k^t \\
&\sum_{k=1}^{K} \lambda_k^t y_{km}^t \geq y_{om}^t, \quad m = 1, \ldots, M, \\
&\sum_{k=1}^{K} \lambda_k^t = 1, \\
&\lambda_k^t \geq 0, \quad k = 1, \ldots, K.
\end{aligned}
\tag{10}
$$

Comparing input efficiency measured by (5) with that measured by (10) we have $\widetilde{ITE}(\mathbf{y}, c) \leq ITE(\mathbf{y}, \mathbf{x})$. Thus, aggregating inputs to costs reduces the number of constraints from $N + M + 1$ to $M + 1$ and allows a larger potential contraction in actual costs relative to minimum (frontier) costs than the potential contraction of actual inputs relative to minimum inputs needed to produce a given output.

Regarding the sample size needed for a given number of inputs and outputs various rules of thumb suggest that $K > 3(N + M)$ or $K \geq \max(N \times M,\ 3(N + M))$ (Cooper et al., 2007). When the number of DMUs is too small to effectively avoid the curse of dimensionality, Charles et al. (2019) suggest simple indexes to aggregate outputs/inputs or subsets of outputs/inputs. For instance, the Benefit-of-the-Doubt (BOD) method developed by Cherchye et al. (2007) is one way of constructing a DEA composite indicator of outputs without reference to inputs. This index is formulated as

$$\begin{aligned} \tilde{CI}_o &= \min_{\mu_k} \sum_{k=1}^{K} \mu_k \cdot 1 \\ s.t. \quad & \sum_{k=1}^{K} \mu_k y_{mk} \geq y_{mo} \qquad m = 1, \ldots, M \\ & \mu_k \geq 0, \qquad k = 1, \ldots, K. \end{aligned} \tag{11}$$

A value of $\tilde{CI}_o = 1$ indicates that DMU "o" produces on the BOD frontier. To improve the discriminatory power of the BOD method, the Super-Efficiency index of Andersen and Petersen (1993) takes the form

$$\begin{aligned} CI_o &= \min_{\mu_k} \sum_{k=1}^{K} \mu_k \cdot 1 \\ s.t. \quad & \sum_{k \neq o}^{K} \mu_k y_{mk} \geq y_{ko} \qquad m = 1, \ldots, M \\ & \mu_k \geq 0 \qquad k = 1, \ldots, k. \end{aligned} \tag{12}$$

In (12), a DMU can have $CI_o > 1$ because its outputs are not included on the left-hand side of the $m = 1, \ldots, M$ constraints. Charles et al. (2019) suggest that aggregate output indexes, such as $\tilde{CI}_o$ or CI_o, could be used to measure input technical efficiency as in (5). Similarly, aggregate input indexes could be constructed using the BOD or Super-Efficiency methods with those indexes then used to measure output technical efficiency as in (4).

Zhu (2020) found 38 studies that combined Big Data with DEA and suggests that network DEA can be used to deal with the "value" characteristic of big data. To illustrate, Zhu examined transportation and logistics' systems as networks. Kaffash et al. (2021) review big data algorithms that have been used in ITS (intelligent transportation systems) to predict traffic flows and times, traffic signs and traffic hazards. They suggest that although DEA has not been widely used in ITS, various DEA models might be capable of analyzing Big Data in ITS.

Various algorithms have been proposed for reducing the computational time necessary to find the DMUs that define the technological frontier (Zhu et al., 2018; Khezrimotlagh et al., 2018). These algorithms basically take subsamples of DMUs from all available DMUs and find the best-practice DMUs in the subsample.

Here, we note that these best-practice DMUs might not be efficient for the entire sample. Then, using the best-practice sample DMUs, all other DMUs are evaluated relative to the best-practice subsample technology. The DMUs that are outside the technology are then combined with the best-practice subsample DMUs and the best-practice technology for the entire sample of DMUs is then determined. These types of algorithms reduce the computational time for DEA considerably for big data sets with respect to "volume", i.e., the number of DMUs.

3 Network DEA

Black-box DEA models have been widely employed to measure DMU performance. In these types of models, inputs enter the black box and outputs exit with no information provided on the inner workings of the box. Network models are an attempt to "open" the black box by accounting for intermediate products that are outputs/inputs within the black box. Opening the black box provides greater information to managers and policy analysts about the production process and can potentially improve decision-making. For instance, network models typically include intermediate products that are outputs of one stage of production and are inputs to another stage of production. These intermediate products can flow between stages and across time, and can include undesirable by-products such as polluting by-products of electric utilities, non-performing loans by banks, or infections that occur due to surgical procedures in hospitals. Network DEA (NDEA) models specify the production technology to account for linkages between stages/divisions of a company, across time, locations, or feedback effects between different producers (Färe & Grosskopf, 1996a, b, 2000). Applications of NDEA have been offered for banks (Akther et al. 2013; Fukuyama & Weber, 2015), primary and secondary schools (Grosskopf et al., 2015), manufacturing firms generating pollutants (Bostian et al., 2018), knowledge production in the form of patents, Ph.D. students and publications by universities (Fukuyama et al., 2016), and knowledge production in the form of scientific publications by countries (Bostian et al., 2020a, b). Cantor and Poh (2020) derived the primal and dual LP problems for a general network system allowing overall efficiency to be decomposed into divisional/stage efficiencies with an application to electric utilities. Fukuyama et al. (2020) analyzed the production of desirable final outputs with three separate technologies accounting for intermediate products and undesirable outputs. In one technology, energy and other inputs are used to produce an intermediate product which is used in a second technology to produce the final desirable output. The third technology accounted for the costly disposability of undesirable outputs that are generated from fossil fuel combustion. In this model, the costly disposability assumption requires production to take place within the intersection of the technologies so that reductions in undesirable outputs (carbon emissions) requires reductions in energy use. In turn, less energy reduces the intermediate products, which finally reduces desirable outputs.

Time substitution models are another form of network DEA. In these models, intertemporal resource reallocation can occur so as to take advantage of technological change or scale economies/diseconomies. Time substitution optimizes production to maximize desirable outputs and minimize undesirable outputs from given inputs, or, to minimize input usage given target outputs. Färe et al. (2012) and Bostian et al. (2018) provide time substitution models when pollution is a joint byproduct of desirable output production. Dynamic models account for contemporaneous production decisions to affect future production possibilities as when investment decisions taken in one period affect the capital stock in subsequent periods (Färe et al., 2018; Bostian et al. 2020a, b) or when the timing of loans impacts subsequent levels of nonperforming loans (Fukuyama & Weber, 2017a, b).

4 A Network Model of Covid

4.1 The NDEA Model and Performance Measures

The coronavirus pandemic of 2020 continues to impose large costs in terms of deaths and lost economic activity. In the US, the pandemic has cycled through the states with Washington and Massachusetts the first states to experience pandemic deaths followed by deaths in all 50 states and the District of Columbia (DC). As the pandemic has proceeded, states varied their responses from strict lockdowns to business as usual. To examine the performance of states in dealing with the pandemic, we develop and estimate a dynamic network model. Data from *The Covid Tracking Project* are updated daily and these data have various Big Data characteristics. For instance, the *volume* and *velocity* of data in *The Covid Tracking Project* include the 50 states plus DC for 300+ days and counting. The data include a *variety* of 39 variables including the number of deaths, hospitalizations, people in intensive care units (ICUs) of hospitals, the number of tests given to detect coronavirus, whether the tests were positive or negative, and the number of recoveries. In addition, the *veracity* of the data collected is monitored with each state's daily data receiving a quality grade (A+, A, B, C, or D). Furthermore, the data has the *variability* characteristic of Big Data as we combine daily Covid data with quarterly real state GDP data that are interpolated to daily data. Our goal in this section is to construct a dynamic network model of the pandemic that allows for time substitution of Covid tests to optimize production, i.e., minimize Covid deaths and maximize real state GDP. Network Data Envelopment Analysis (NDEA) is used to construct the pandemic technology and to add *value* by identifying states and time periods with high and low performance.

Comparing indicators such as the number of deaths or recoveries from coronavirus across states suffers from omitted variable bias as those indicators do not account for things like population, loss of economic activity, and the potential for viral infections/deaths in one state to spillover and impose costs on other states.

States like New York or California, with large populations and large travel/tourism industries have already suffered greater numbers of infections and deaths than smaller states with fewer out-of-state visitors like North Dakota. However, as mentioned earlier, controlling for too many variables in an NDEA model leads to the curse of dimensionality.

To avoid the curse of dimensionality we specify a relatively parsimonious NDEA model. At the beginning of the pandemic ($t = 0$) each state and DC has a population, $x_{1k}^0, \quad k = 1, \ldots, K$. In periods $t = 1, \ldots, T$, states begin testing for the virus with some individuals testing positive and other individuals testing negative. The total number of tests is denoted x_{2k}^t. In our dynamic model, tests (x_{2k}^t) will be allowed to be reallocated across time. We use total tests as the relevant input impacting deaths, rather than only positive tests. Given that false positive and false negatives both occur during testing, the total number of tests is likely to be a better measure of resource usage. Some individuals who become infected with coronavirus die, b_k^t, while others experience varying degrees of recovery. We use the lagged number of own deaths, b^{t-1}, as a link between production in two periods. Lagged own deaths are likely to be positively correlated with infections and so increases in deaths (infections) in one period can affect production possibilities in a subsequent period.

In addition to a state's own deaths, deaths (infections) in other states can spillover and affect production possibilities. We model this potential spillover as follows. First, we calculated the great circle distance from the population center of each state to the population center of every other state.[1] Let the distance between states j and k be represented by $Dist_{k,j} = Dist_{j,k}$. Second for each state we calculated the number of deaths per unit of distance summed over all states and let that sum be the spillover, $B_k^{t-1} = \sum_{j \neq k}^{K} \frac{b_j^{t-1}}{Dist_{j,k}}$. Again, we assume that the external spillover, B_k^{t-1}, affects production possibilities in the next period. The index, B_k^{t-1}, means that a state like New York will receive a greater spillover from close populous states like New Jersey, Connecticut, and Pennsylvania, than it will receive from far, less populous states like Wyoming. Using NDEA, Fukuyama et al. (2016) and Bostian et al. (2020a, b) incorporated spillover indexes to model knowledge production where knowledge is a public good which, once produced, becomes available to all other producers in a subsequent period. Given initial population, x_{1k}^0, the subsequent populations are $x_{1k}^s = x_{1k}^0 - \sum_{t=1}^{s} b_k^{t-1}, \quad t = 1, \ldots, s, \ldots, T$. We set $B_k^0 = 0$ in $t = 0$. Finally, each state generates economic activity (real state GDP) represented by y_k^t. This economic activity is positively related to population and the number of tests.

The static technology in each period is given as

[1] We use the Haversine formula to calculate the great circle distance between two states j and k as $Dist_{jk} = a \times \cos^{-1}\{\cos\delta_j \cos \delta_k \cos (\lambda_j - \lambda_k) + \sin \delta_j \sin \delta_k\}$ where $a=$ radius of the earth in miles (2853.222 miles), δ_j and δ_k are the latitudes of j and k in radians and λ_j and λ_k are the longitudes of j and k in radians.

$$T^t = \left\{ \left(x^t, b^{t-1}, B^{t-1}, y^t, b^t \right) : \left(x^t, b^{t-1}, B^{t-1} \right) \text{ can produce } \left(y^t, b^t \right) \right\} \quad t=1, \ldots, T, \tag{13}$$

where $x^t = \left(x_1^t, x_2^t \right)$ and $B^{t-1} = \sum_{j \neq k}^{K} \frac{b_j^{t-1}}{Dist_{k,j}}$.

The properties of the technology include the following:

$$\begin{aligned}
&i. \quad \left(x^t, b^{t-1}, B^{t-1}, y^t, b^t \right) \in T^t \text{ and } x^{t'} \geq x^t \text{ then } \left(x^{t'}, b^{t-1}, B^{t-1}, y^t, b^t \right) \in T^t,\\
&ii. \quad \left(x^t, b^{t-1}, B^{t-1}, y^t, b^t \right) \in T^t \text{ and } y^{t'} \leq y^t \text{ then } \left(x^t, b^{t-1}, B^{t-1}, y^{t'}, b^t \right) \in T^t,\\
&iii. \quad \left(x^t, b^{t-1}, B^{t-1}, y^t, b^t \right) \in T^t \text{ and } b^{t-1'} \geq b^{t-1} \text{ then } \left(x^t, b^{t-1'}, B^{t-1}, y^t, b^t \right) \in T^t,\\
&iv. \quad \left(x^t, b^{t-1}, B^{t-1}, y^t, b^t \right) \in T^t \text{ and } B^{t-1'} \geq B^{t-1} \text{ then } \left(x^t, b^{t-1}, B^{t-1'}, y^t, b^t \right) \in T^t,\\
&v. \quad \left(x^t, b^{t-1}, B^{t-1}, y^t, b^t \right) \in T^t \text{ then for } 0 \leq \theta^t \leq 1, \left(x^t, b^{t-1}, B^{t-1}, \theta^t y^t, \theta^t b^t \right) \in T^t.
\end{aligned} \tag{14}$$

Properties i. through iv. are monotonicity properties of the technology that impose strong disposability for inputs (x^t), desirable outputs (y^t), own previous deaths (b^{t-1}), and spillover deaths from other states (B^{t-1}). Property (v) imposes joint weak disposability between the desirable output (y^t) and the undesirable output (b^t). Joint weak disposability of outputs means that reductions in the undesirable output along the frontier of the technology must be accompanied by proportional reductions in the desirable output. In essence, given (x^t, b^{t-1}, B^{t-1}), reducing the number of deaths from Covid imposes a cost in terms of foregone economic activity.[2]

We specify a dynamic technology that allows for reallocation of tests for the virus across periods. Paul Romer (https://paulromer.net/) was an advocate (March 23, 24, 25, 2020) of early and numerous tests for coronavirus followed by quarantine of people testing positive. His simulation showed that, relative to random isolation of people, if people who test positive, including false positives, isolate, the proportion of the isolated population who are actually infected is higher, and fewer people need to isolate.

Our dynamic network model uses time substitution methods to investigate the optimal timing of virus tests. Time substitution holds the number of tests constant across all periods, but allows when the tests are given to be chosen optimally to reduce Covid deaths. Let $\tilde{X}_{2k} = \sum_{t=1}^{T} x_{2k}^t$ represent the number of the total number of tests given across all periods. The dynamic technology takes the form

[2]One could also impose joint weak input disposability between inputs (x^t) and the two spillovers, b^{t-1} and B^{t-1} (see Fukuyama & Weber, 2017a, b). Joint weak input disposability would allow proportional expansions in inputs and spillovers.

$$DT = \left\{ \left(x_1^0, \tilde{X}_2, b^{t-1}, B^{t-1}, y^t, b^t\right) : \quad \left(x_1^0 - \sum_{t=1}^{t} b^{t-1}, x_2^t, b^{t-1}, B^{t-1}\right) \text{ can produce } \left(y^t, b^t\right) \text{ and } \sum_{t=1}^{T} x_2^t \leq \tilde{X}_2, \quad t = 1, \ldots, T \right\}. \tag{15}$$

Figure 1 depicts the network technology where $x_c^t = (x_{1c}^t, x_{2c}^t)$ and $x_d^t = (x_{1d}^t, x_{2d}^t)$ and are updated as $x_c^{t+1} = \left(x_{1c}^t - b_c^t, x_{2c}^{t+1}\right)$ and $x_d^{t+1} = \left(x_{1d}^t - b_d^t, x_{2d}^{t+1}\right)$. Fig. 1 shows desirable final outputs (y) exiting the technology in each period, but undesirable outputs from one period spillover and affect the technology in the subsequent period in two ways: first, there is the own effect as undesirable outputs b^t produced by one DMU affects the subsequent period technology for that same DMU, and second, undesirable outputs produced by one DMU, say DMU "*c*" spillover and affect the technology for other DMUs, i.e., in Fig. 1, DMU "*d*".

We use DEA (Data Envelopment Analysis) to represent the dynamic network technology. As its name implies, DEA envelops observed inputs and outputs to form the best-practice technology. This technology consists of piecewise linear constraints such that a DMU can produce no more output using no less input than a linear combination of all DMUs' inputs and outputs. The static DEA technology associated with (13) is given as

$$\begin{aligned} T_{DEA,VRS}^t = \Big\{ & \left(x^t, b^{t-1}, B^{t-1}, y^t, b^t\right) : \\ & x_1^t \geq \sum_{k=1}^{K} \lambda_k^t x_{1k}^t, \quad x_2^t \geq \sum_{k=1}^{K} \lambda_k^t x_{2k}^t, \quad b^{t-1} \geq \sum_{k=1}^{K} \lambda_k^t b_k^{t-1}, \quad B^{t-1} \geq \sum_{k=1}^{K} \lambda_k^t B_k^{t-1}, \\ & y^t \leq \sum_{k=1}^{K} \theta_k^t \lambda_k^t y_k^t, \quad b^t = \sum_{k=1}^{K} \theta_k^t \lambda_k^t b_k^t, \quad \lambda_k \geq 0, \quad k = 1, \ldots, K, \\ & 0 \leq \theta_k^t \leq 1, \quad k = 1, \ldots, K, \quad \sum_{k=1}^{K} \lambda_k^t = 1 \Big\}. \end{aligned} \tag{16}$$

The intensity variables, λ_k^t, and the output scaling factors, θ_k^t, are used to envelop the data. The output scaling factors, θ_k^t, are associated with joint weak disposability between (y^t, b^t) (Kuosmanen, 2005; Färe & Grosskopf, 2009). Under joint weak disposability of outputs, the outputs y and b can be reduced proportionally, but the b output is not freely disposable. That is, there is an opportunity cost of reducing b along the frontier of the technology; less of the y output must also be produced.

The Kuosmanen (2005) transformation can be used to get a reduced form for the product of the intensity variables and the scaling factors for outputs, $\left(\theta_k^t \lambda_k^t\right)$. We follow Fukuyama and Weber (2017a, b) who estimated a dynamic network model with y and b outputs using the Kuosmanen (2005) transformation to make (16) into a system of linear equations. Let $\gamma_k^t = \theta_k^t \lambda_k^t$ and let $u_k^t = \left(1 - \theta_k^t\right) \lambda_k^t$. Then $\lambda_k^t = \left(u_k^t + \gamma_k^t\right)$. Upon substitution, the DEA technology is

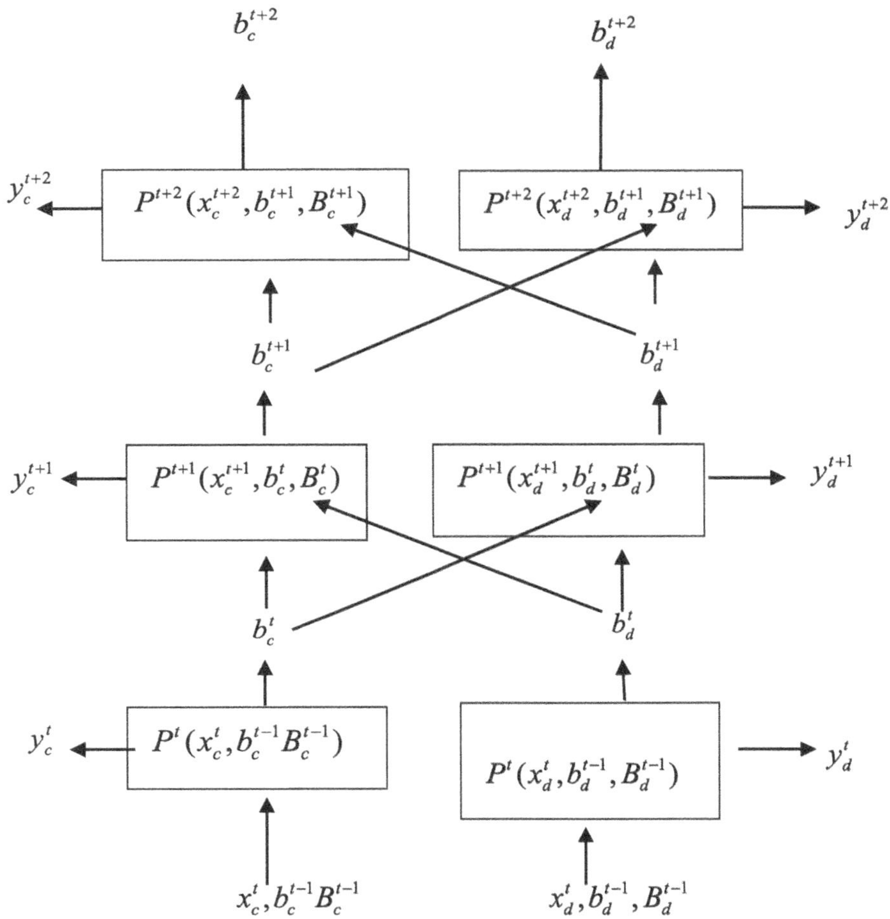

Fig. 1 A dynamic network model

$$
\begin{aligned}
T^t_{DEA,VRS} = \Big\{ & \left(x^t, b^{t-1}, B^{t-1}, y^t, b^t\right) : \\
& x_1^t \geq \sum_{k=1}^{K} \left(u_k^t + \gamma_k^t\right) x_{1k}^t, \quad x_2^t \geq \sum_{k=1}^{K} \left(u_k^t + \gamma_k^t\right) x_{2k}^t, \\
& b^{t-1} \geq \sum_{k=1}^{K} \left(u_k^t + \gamma_k^t\right) b_k^{t-1}, \quad B^{t-1} \geq \sum_{k=1}^{K} \left(u_k^t + \gamma_k^t\right) B_k^{t-1}, \\
& y^t \leq \sum_{k=1}^{K} \gamma_k^t y_k^t, \quad b^t = \sum_{k=1}^{K} \gamma_k^t b_k^t, \\
& u_k^t \geq 0, \quad \gamma_k^t \geq 0, \quad k = 1, \ldots, K, \quad \sum_{k=1}^{K} \left(u_k^t + \gamma_k^t\right) = 1 \Big\}.
\end{aligned}
\tag{17}
$$

The directional output distance function seeks the maximum expansion in desirable outputs and contraction in undesirable outputs for the directional vector $g = (g_y, g_b)$. For the static DEA technology (16) this distance function is estimated as

$$\begin{aligned}\vec{D}_o^t\left(x, b^{t-1}, B^{t-1}, b^t, y^t; g\right) = \max_{\beta,\gamma,u}\Big\{\beta^t : \quad & b_o^t - \beta^t g_b = \sum_{k=1}^K \gamma_k^t b_k^t, \quad y_o^t + \beta^t g_y \le \sum_{k=1}^K \gamma_k^t y_k^t, \\ & x_{1o}^t \ge \sum_{k=1}^K \left(u_k^t + \gamma_k^t\right) x_{1k}^t, \quad x_{2o}^t \ge \sum_{k=1}^K \left(u_k^t + \gamma_k^t\right) x_{2k}^t, \\ b_o^{t-1} \ge \sum_{k=1}^K \left(u_k^t + \gamma_k^t\right) b_k^{t-1}, \quad & B_o^{t-1} \ge \sum_{k=1}^K \left(u_k^t + \gamma_k^t\right) B_k^{t-1}, \\ u_k^t \ge 0, \quad \gamma_k^t \ge 0, \quad k = 1, \ldots, K, \quad & \sum_{k=1}^K \left(u_k^t + \gamma_k^t\right) = 1\Big\}, \quad t = 1, \ldots, T.\end{aligned} \tag{18}$$

In each period, the static model chooses the intensity variables, $\lambda_k^t = u_k^t + \gamma_k^t$, and β^t to simultaneously minimize b^t and maximize y_k^t, given the directional vector $g = (g_b, g_y)$. Problem (18) is solved $K \times T$ times. When $\vec{D}_o^t(\cdot) = 0$ the DMU produces on the frontier of the static technology. Values of $\vec{D}_o^t(\cdot) > 0$ indicate inefficiency, with larger values indicating greater inefficiency.

For the dynamic technology we impose joint weak disposability between (y^t, b^t) as in the static model, but we also allow for an intertemporal reallocation of tests $\left(x_2^t\right)$. Using the Kuosmanen transformation the dynamic DEA technology is

$$\begin{aligned} DT = \Big\{ & \left(x_1^0, \tilde{X}_2, b^{t-1}, B^{t-1}, y^t, b^t; g\right) : \\ & x_1^t \ge \sum_{k=1}^K \left(u_k^t + \gamma_k^t\right) x_{1k}^t, \quad x_{1k}^t = x_{1k}^0 - \sum_{t=1}^t b_k^{t-1}, \quad t = 1, \ldots, T, \\ & x_2^t \ge \sum_{k=1}^K \left(u_k^t + \gamma_k^t\right) x_{2k}^t, \quad \sum_{t=1}^T x_2^t \le \tilde{X}_2, \quad t = 1, \ldots, T, \\ & b^{t-1} \ge \sum_{k=1}^K \left(u_k^t + \gamma_k^t\right) b_k^{t-1}, \quad B^{t-1} \ge \sum_{k=1}^K \left(u_k^t + \gamma_k^t\right) B_k^{t-1}, \quad t = 1, \ldots, T \\ & y^t \le \sum_{k=1}^K \gamma_k^t y_k^t, \quad b^t = \sum_{k=1}^K \gamma_k^t b_k^{t-1}, \quad t = 1, \ldots, T, \\ & u_k^t \ge 0, \quad \gamma_k^t \ge 0, \quad k = 1, \ldots, K, \quad \sum_{k=1}^K \left(u_k^t + \gamma_k^t\right) = 1, \quad t = 1, \ldots, T\Big\}.\end{aligned} \tag{19}$$

We estimate a directional output distance function to measure the distance from observed outputs, $\left(y_o^t, b_o^t\right)$, to the frontier of the dynamic technology. This distance function projects outputs onto the dynamic frontier for the directional vector $g = (g_b, g_y)$.

The optimization problem for the dynamic model is

$$\overrightarrow{DD} = \max_{\delta^t, \gamma^t, u^t, x_2^t} \left\{ \begin{array}{l} \sum_{t=1}^{T} \delta^t : b_o^t - \delta^t g_b = \sum_{k=1}^{K} \gamma_k^t b_k^t, \quad y_o^t + \delta^t g_y \leq \sum_{k=1}^{K} \gamma_k^t y_k^t, \quad t = 1, \ldots, T, \\ \quad x_{1o}^t \geq \sum_{k=1}^{K} \left(u_k^t + \gamma_k^t\right) x_{1k}^t, \quad x_{1o}^t = x_{1o}^0 - \sum_{t=1}^{t} \left(b_o^{t-1} - \delta^{t-1} g_b\right), \quad t = 1, \ldots, T, \\ \quad x_2^t \geq \sum_{k=1}^{K} \left(u_k^t + \gamma_k^t\right) x_{2k}^t, \quad \sum_{t=1}^{T} x_2^t \leq \tilde{X}_{2o}, \quad t = 1, \ldots, T, \\ \quad b_o^{t-1} \geq \sum_{k=1}^{K} \left(u_k^t + \gamma_k^t\right) b_k^{t-1}, \quad B_o^{t-1} \geq \sum_{k=1}^{K} \left(u_k^t + \gamma_k^t\right) B_k^{t-1}, \quad t = 1, \ldots, T, \\ \quad u_k^t \geq 0, \quad \gamma_k^t \geq 0, \quad k = 1, \ldots, K, \quad \sum_{k=1}^{K} \left(u_k^t + \gamma_k^t\right) = 1, \quad t = 1, \ldots, T. \end{array} \right\} \tag{20}$$

The dynamic optimization problem gives the maximum contraction in b and expansion in y summed over all periods allowing an optimal intertemporal reallocation of x_2. This problem is solved K times. Since the status quo, $x_{2o}^t, \quad t = 1, \ldots, T$ is a feasible, but not necessarily optimal solution in (20), we have $\overrightarrow{DD} = \sum_{t=1}^{T} \delta^{t*} \geq \sum_{t=1}^{T} \beta^{*t}$ where $\beta^{*t}, \quad t = 1, \ldots, T$ are the optimal expansion factors for the series of static problems (18), and $\delta^{*t}, \quad t = 1, \ldots, T$, are the optimal expansion factors in (20). Comparing the actual number and timing of tests given, x_{2o}^t, with the optimal number and timing of tests, x_2^{*t}, will provide evidence in favor or against Romer's hypothesis. We note that in (20), the population variable, x_{1o}^t, in each period t = 2, ...,T, depends on the initial population, deaths, and deaths averted in all past periods due to an optimal reallocation in the number of test given, i.e., $x_{1o}^t = x_{1o}^0 - \sum_{t=1}^{t} \left(b_o^{t-1} - \beta^{t-1} g_b\right)$. Since averting deaths in previous periods expands the potential population in subsequent periods, the monotonicity property in (14) iii. means that the technology has the potential to expand in future periods.

4.2 Data and NDEA Performance Estimates

The data on deaths from Covid are from *The Covid Tracking Project* and are updated daily and there are day of week effects regarding when deaths occur or are reported. Weekdays typically have more reported deaths than weekends and so we take a seven day moving average of deaths as the undesirable output, (b^t). We do the same for positive and negative tests. Figures 2 and 3 graph the 7 day moving averages of deaths, total tests (positive plus negative tests) for the 50 states and DC. The data are current to January 13, 2021.

The individual states vary in the magnitude and the timing of their peaks. Figure 4 graphs the 7 day moving average for the five largest states. Peak deaths occur first in New York during March 2020, then Texas and Florida in late July and early August 2020, Pennsylvania in December 2020, and finally California in January 2021.

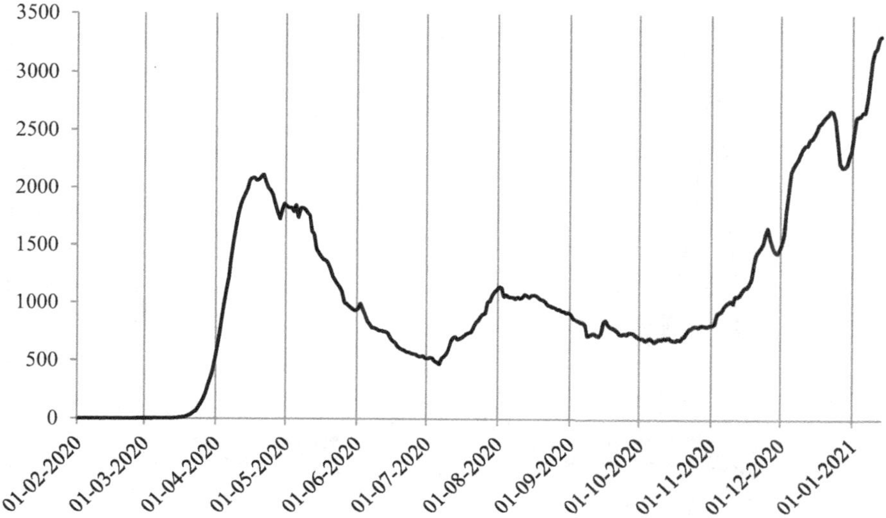

Fig. 2 7 day moving average deaths (b) in US to 1-13-2021

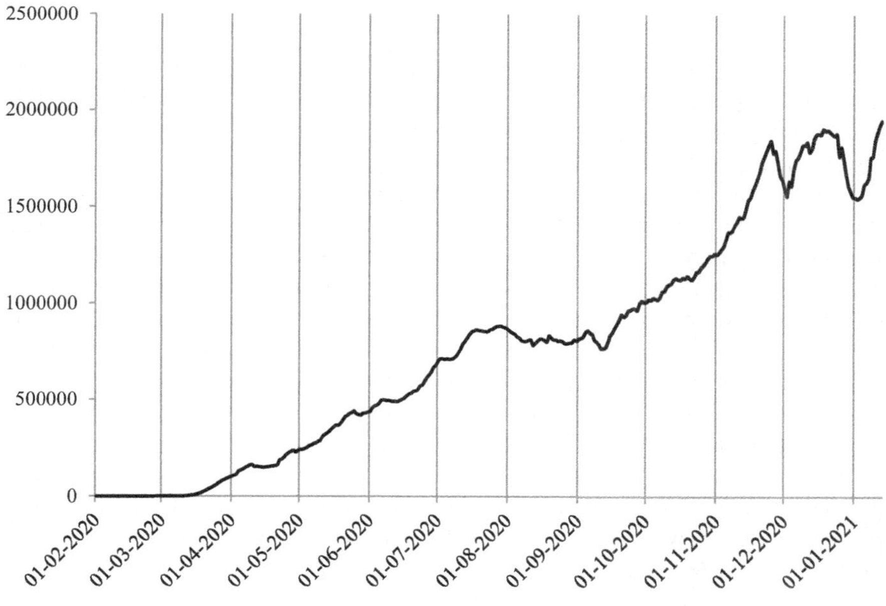

Fig. 3 7 day moving average tests (x_2) for Covid, US to 1-13-2021

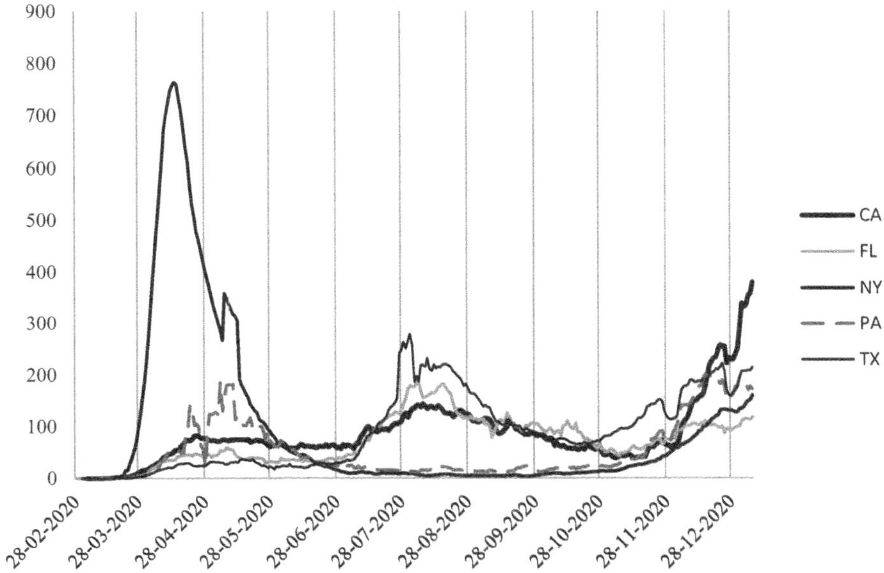

Fig. 4 7 day moving average deaths in 5 largest states

Figure 5 shows Covid deaths in medium sized states where Connecticut and Louisiana experienced peak deaths in April 2020, but then rebounded slightly in December and January 2021. The other three states, Oregon, Kentucky, and Oklahoma experienced peak deaths in November/December.

Data on state populations come from the Census Bureau for 2019. Data on each state's contribution to real GDP come from the Bureau of Economic Analysis. Real state GDP is reported quarterly but enters our model on a daily basis. Therefore, we calculated the quarter to quarter growth rate in each state's real GDP and converted that growth rate to a daily growth rate. Then, we used the compounding formula $y^{t+1} = y^t \exp(r)$ where r equals the daily growth rate to calculate state real GDP on each day in the quarter. State real GDP is reported up to the third quarter of 2020 with the fourth quarter data scheduled to be released March 26, 2021. For the fourth quarter of 2020 and up to January 7, 2021 we assumed that each state's GDP equals that state's GDP at the end of the third quarter. The state centers of population in 2010 were derived from the Census Bureau and the distance between population centers was used to calculate the spillover variable. As mentioned earlier, the quality of data varies by state. Twelve states[3] reported negative values or zero values for the seven day moving average of deaths and were dropped from the

[3]States with 0 or negative values for 7 day moving average deaths were AK, HI, ID, ME, MT, ND, NH, SD, VT, WV, WY, and WA. States with negative values for seven day moving average tests were AL, KY, NC, and SC.

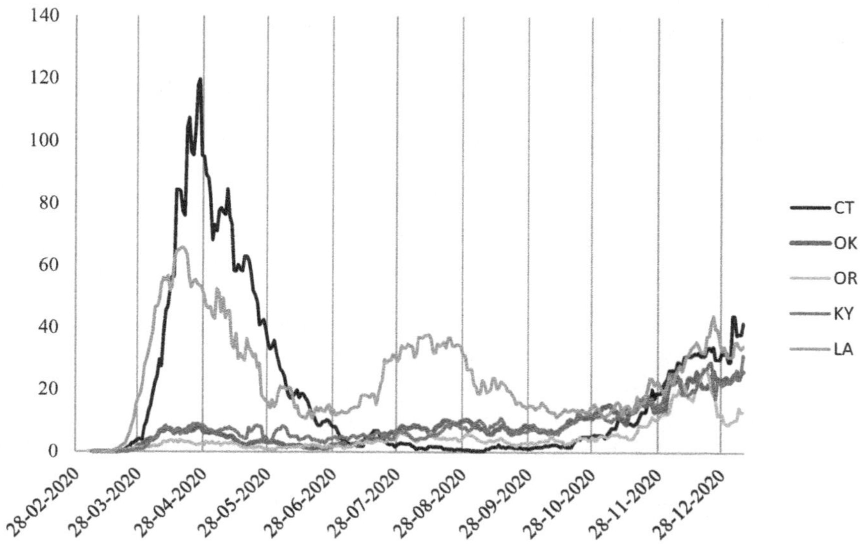

Fig. 5 7 day moving average deaths in medium sized states to 1-13-2021

Table 1 Descriptive Statistics, K = 35 DMUs and T = 272 days. (4-11-2020 to 1-7-2021)

	mean	Std.dev.	Min.	Max.
Population= x_1^0	8118923	8192633	712816	39368078
Deaths= b	31	28	3	103
Real GDP= y (millions)	397221	532355	26079	2785188
Tests= x_2	23626	25792	3359	124516
Spillover= B	1.78	1.47	0.26	12.84

technology. An additional four states that reported a negative seven day moving average of tests were dropped. This left 34 states and the District of Columbia. The model is estimated for the 272 days beginning April 11, 2020 and ending January 7, 2021. We chose April 11 as the starting date because by then all states in our sample had experienced some Covid deaths and had a positive 7-day moving average for deaths.

Table 1 reports descriptive statistics of the variables used in the model. The 34 states and DC have a larger average population than the average of all 50 states. On average, the 34 states and DC experienced about 31 deaths per day, gave 23,600 tests per day and experienced a spillover from other states of about 1.78 deaths per mile of distance. Average annual real GDP was approximately $400 billion.

To estimate the static and dynamic models we choose directional vectors equal to the daily mean values of real state GDP and the seven-day average of Covid deaths: $g_y = \overline{y}$ and $g_b = \overline{b}$ as reported in Table 1. Given these directional vectors, the number of deaths that could have been averted on a given day if the state operated on the frontier of the technology will equal $\beta^t \overline{b}$ using the static model (18) estimates and $\delta^t \overline{b}$ using the dynamic model (20) estimates. In addition to these averted deaths, moving to the best practice static frontier could increase real GDP by $\beta^t \overline{y}$ (static) or by $\delta^t \overline{y}$ by moving to the best-practice dynamic frontier. Since $\sum_{t=1}^{T} \delta^t \geq \sum_{t=1}^{T} \beta^t$, we can derive an indicator of the potential deaths averted via an optimal intertemporal reallocation of tests: $Gains = \left(\sum_{t=1}^{T} \delta^t - \sum_{t=1}^{T} \beta^t \right) \overline{b}$, given the directional scaling vector $g_b = \overline{b}$.

Figure 6 graphs average daily static inefficiency found as β^t from the solution to (18) and average daily dynamic inefficiency found as δ^t from the solution to (20). It is possible for $\beta^t \geq \delta^t$ on any given day. This might occur if one or more states are closer to the dynamic frontier than the static frontier because they are reducing their use of tests on that day and saving them for use at a different time, when more y and less b could be produced. However, since the static DEA technology is contained within the dynamic DEA technology it must be the case that $\sum_{t=1}^{T} \delta^t \geq \sum_{t=1}^{T} \beta^t$. For the 272 days, the averages, by state, were $\sum_{t=1}^{T} \delta^t = 144.1$ and $\sum_{t=1}^{T} \beta^t = 134.1$. Given these estimates and $g_b = \overline{b} = 31$, the average deaths averted are $(144.1 - 134.1) \times 31 = 310$ per state, or by 10,850 in all 35 states. Dynamic inefficiency reaches two early peaks on April 25 and May 16 and then declines to its lowest point on July 7. After that, dynamic inefficiency increases to August 2, then moves sideways to October 16 and then upward to its peak on December 9. From December 9 to December 30 dynamic inefficiency declines but increases thereafter. Static inefficiencies follow a similar pattern.

Four states—CA, DE, NJ, and OK – plus DC are on the dynamic frontier every day. Other states are on the frontier on some, but not all days. In fact, 25% of the 9520 state/day observations were on the dynamic frontier one day or another. The same four states and DC are on the static frontier every day, while 7 states are never on the frontier. In total, 40.9% (1849/9520) of the 9520 state/day observations reside on the static frontier at least one day. On 16 out of the 28 days from July 3 to July 28 static inefficiency was greater than dynamic inefficiency because the solution to the dynamic model indicates that the actual number of tests given would have been reallocated to other days when more deaths could have been averted. A similar phenomenon—static inefficiency greater than dynamic inefficiency—occurred on 5 of 6 days from September 20 to September 26 and on 8 of 13 days from November 10 to November 22. Again, tests would have been reallocated to other days to avert more deaths.

By January 7, 2021, the 35 states in our sample had given an average of 5.15 million tests for Covid from April 11, 2020 to January 7, 2021. Figure 7 graphs the cumulative proportion of actual tests and the cumulative proportion of optimal tests given during that period. Cumulative optimal tests are less than cumulative

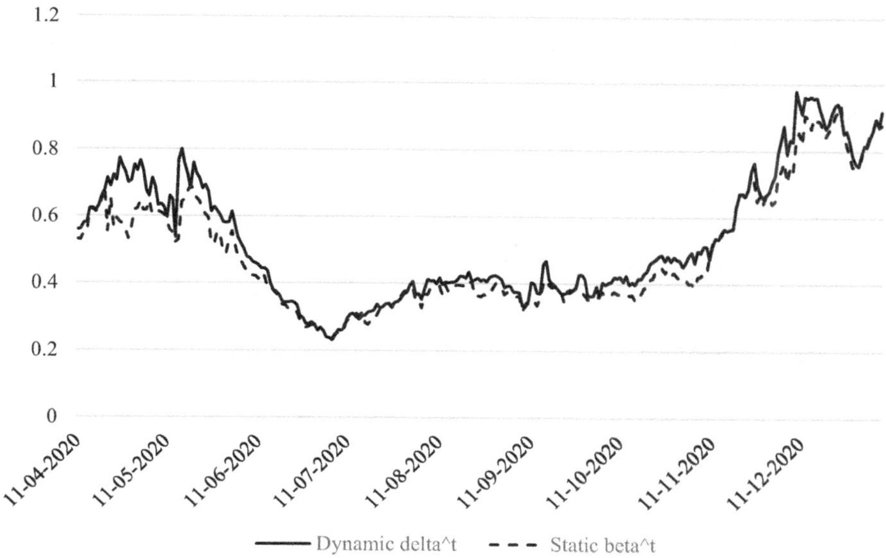

Fig. 6 Average daily mean inefficiency (β^t = static, δ^t =dynamic)

actual tests until November 24, 2020. On that day, the cumulative optimal proportion of tests exceeds the cumulative number of actual tests and that inequality remains until January 7, 2011 when the two cumulative proportions equal 1. However, there were 5 out of 6 days from April 23 to April 28 when average optimal tests were greater than average actual tests. This cluster occurs during the early two peaks in average dynamic inefficiency that occurred April 25 and May 16 and lends support to Romer's hypothesis.

The average daily gains (deaths averted) from reallocating tests across the 272 days are shown in Fig. 8. As shown earlier, these gains (deaths averted) are calculated as $Gains = \left(\delta^t - \beta^t\right)\overline{b}$.

These estimates offer some, but not overwhelming support of Romer's hypothesis that tests should be given early. The largest number of deaths averted from an optimal reallocation of tests occurred mid April to mid May, October to November 10, and from December 3 to December 21.

We urge caution in interpreting these results and offer several caveats. First, with six inputs/outputs and only 35 DMUs, the results of Zelenyuk (2019) suggest the potential for large deviations between the true technology and the DEA technology. One possible remedy would be to adjust the intensity variables so that a constant technology across all periods is assumed. This would reduce the number of intensity variables $\left(\lambda_k^t\right)$ from $K \times T = 35 \times 272$ to $K = 35$. Of course, this reduction would rule out any innovations that occurred during the pandemic. Second, our spillover variables, b^{t-1} and B^{t-1}, were ad hoc and alternative spillover variables might yield different results. Other variables that track spillovers between states might

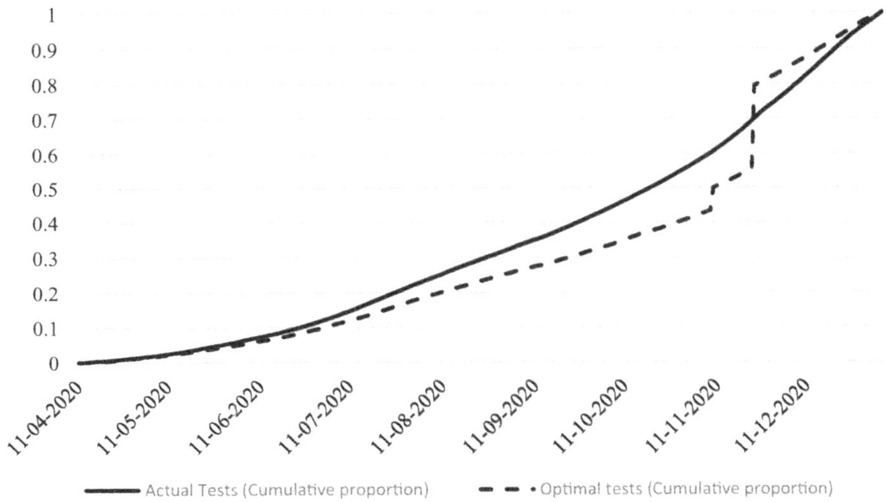

Fig. 7 Cumulative proportions of actual tests and optimal tests

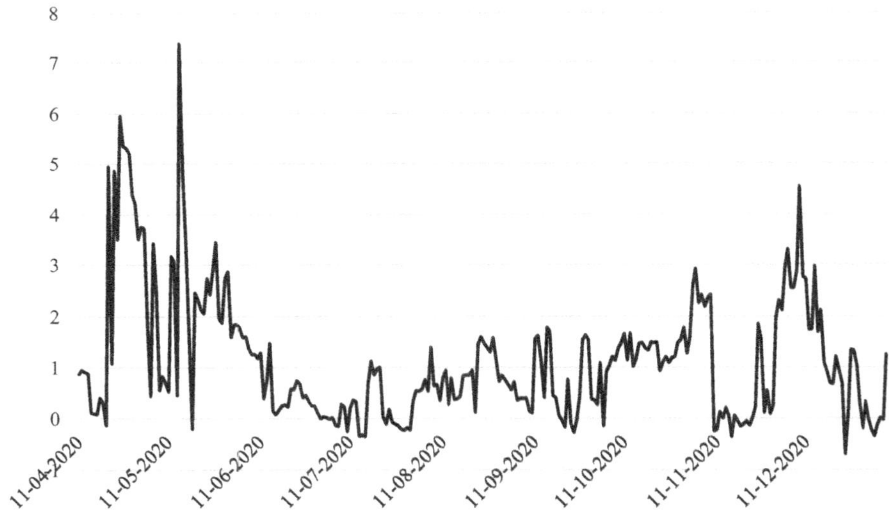

Fig. 8 Average daily deaths averted due to optimal intertemporal reallocation of tests

also include the size of the travel and tourism sector from each state, the number of workers who reside in one state but commute to work in another state, and the number of out-of-state students in higher educational institutions. To avoid the curse of dimensionality, the many potential spillover variables might be aggregated using the super-efficiency benefit of the doubt method shown in (12). Third, the data on real GDP were available on a quarterly basis and we converted those numbers to daily estimates assuming a constant rate of growth or decline from one quarter to another. It is likely that real GDP did not grow smoothly through each quarter. In addition, the fourth quarter state GDP figures were unavailable so we use the end of the third quarter GDP figures for the estimates in the fourth quarter 2020 and the first week of January 2021. Finally, we assumed strong disposability of the inputs and spillover variables in our NDEA model. This assumption might be further relaxed by adapting a joint weak disposability assumption between inputs and the two undesirable spillover variables. We leave these caveats for future research.

5 Summary

In this chapter we studied the characteristics of Big Data in the context of network DEA models. The characteristics of Big Data include volume, velocity, variety, veracity, value, and variability of the data, all of which must be accounted for in network DEA applications. In addition to studying how NDEA models can accommodate big data, we estimated a dynamic network model of the coronavirus pandemic. This model assumed that states seek to simultaneously maximize real GDP and minimize deaths from Covid given inputs. Comparing the dynamic performance of states with their static performance allowed us to estimate the number of Covid deaths that could have been averted due to an optimal intertemporal reallocation of tests for Covid. If the 35 states in our sample had all adopted the best-practice performance of states on the static DEA frontier approximately 145,000 deaths could have been averted, instead of the more than 280,000 deaths that actually occurred. In addition, an optimal reallocation of tests for Covid could have further reduced deaths by 10,850.

Acknowledgments We thank Shawna Grosskopf and Rolf Färe for helpful comments on an earlier draft of this paper.

References

Akther, S., Fukuyama, H., & Weber, W. L. (2013). Estimating two-stage network slacks-based inefficiency: An application to Bangladesh banking. *Omega: The International Journal of Management Science, 41*, 88–96.

Andersen, P., & Petersen, N. C. (1993). A procedure for ranking efficient units in data envelopment analysis. *Management Science, 39*, 1261–1264.

Bellman, R. E. (1957). *Dynamic programming*. Princeton University Press.

Bellman, R. E. (1961). *Adaptive control processes*. Princeton University Press.

Bostian, M. B., Färe, R., Grosskopf, S., Lundgren, T., & Weber, W. L. (2018). Time substitution for environmental performance: The case of Swedish manufacturing. *Empirical Economics, 54*(1), 129–152.

Bostian, M. B., Daraio, C., Färe, R., Grosskopf, S., Izzo, M. G., Leuzzi, L., Ruocco, G., & Weber, W. L. (2020a). Reconstructing nonparametric productivity networks. *Entropy, 22*(12), 1401. https://doi.org/10.3390/e22121401

Bostian, M., Daraio, C., Grosskopf, S., Ruocco, G., & Weber, W. L. (2020b). Sources and uses of knowledge in a dynamic network technology. *International Transactions in Operational Research, 27*, 1821–1844.

Bureau of Economic Analysis. GDP by State. https://www.bea.gov/data/gdp/gdp-state

Cantor, V. J. M., & Poh, K. L. (2020). Efficiency measurement for general network systems: A slacks-based measure model. *Journal of Productivity Analysis, 54*, 43–57.

Charles, V., Aparicio, J., & Zhu, J. (2019). The curse of dimensionality of decision-making units: A simple approach to increase the discriminatory power of data envelopment analysis. *European Journal of Operational Research, 279*, 929–940.

Cherchye, L., Moesen, W., Rogge, N., & van Puyenbroeck, T. (2007). An introduction to "benefit of the doubt" composite indicators. *Social Indicators Research, 82*, 111–145.

Coale, A. (1951). Comments on Simon. In T. C. Koopmans (Ed.), *Activity analysis of production and allocation: Proceedings of a conference*. John Wiley and Sons, Inc/Chapman and Hall, Limited.

Cooper, W. W., Seiford, L. M., & Zhu, J. (2007). *Data envelopment analysis: A comprehensive text with models, applications, references and DEA-solver software* (Second ed.). Springer.

Färe, R., & Grosskopf, S. (1996a). *Intertemporal production Frontiers: With dynamic DEA*. Kluwer Academic Publishers.

Färe, R., & Grosskopf, S. (1996b). Productivity and intermediate products: A frontier approach. *Economics Letters, 50*, 65–70.

Färe, R., & Grosskopf, S. (2000). Network DEA. *Socio-Economic Planning Sciences, 34*(1), 35–49.

Färe, R., & Grosskopf, S. (2009). A comment on weak disposability in nonparametric production analysis. *American Journal of Agricultural Economics, 91*, 535–538.

Färe, R., Grosskopf, S., Margaritis, D., & Weber, W. L. (2012). Technological change and timing reductions in greenhouse gas emissions. *Journal of Productivity Analysis, 37*, 205–216.

Färe, R., Grosskopf, S., Margaritis, D., & Weber, W. L. (2018). Dynamic efficiency and productivity: Modeling and an illustration. In C. A. K. Lovell, E. Grifell-Tatje, & R. Sickles (Eds.), *Oxford handbook of productivity* (pp. 183–209). Oxford University Press.

Fukuyama, H., & Weber, W. L. (2015). Measuring Japanese Bank performance: A dynamic network DEA approach. *Journal of Productivity Analysis, 44*(3), 249–264.

Fukuyama, H., & Weber, W. L. (2017a). Measuring bank performance with a dynamic network Luenberger indicator. *Annals of Operations Research, 250*(1), 85–104.

Fukuyama, H., & Weber, W. L. (2017b). Japanese bank productivity, 2007-2012: A dynamic network approach. *Pacific Economic Review, 22*(4), 649–676.

Fukuyama, H., Weber, W. L., & Xia, Y. (2016). Time substitution and network effects with an application to Nanobiotechnology policy for US universities. *Omega: The International Journal of Management Science., 60*, 34–44.

Fukuyama, H., Hashimoto, A., & Weber, W. L. (2020). Environmental efficiency, energy efficiency and aggregate well-being of Japanese prefectures. *Journal of Cleaner Production, 271*. https://doi.org/10.1016/j.jclepro.2020.122810

Grosskopf, S., Hayes, K., Taylor, L. L., & Weber, W. (2015). Centralized or decentralized control of school resources? A network model. *Journal of Productivity Analysis, 43*, 139–150.

Kaffash, S., Nguyen, A. T., & Zhu, J. (2021). Big data algorithms and applications in intelligent transportation system: A review and bibliometric analysis. *International Journal of Production Economics, 231*. https://doi.org/10.1016/j.ijpe/2020.107868

Khezrimotlagh, D., Zhu, J., Cook, W. D., & Toloo, M. (2018). Data envelopment analysis and big data. *European Journal of Operational Research, 274*. https://doi.org/10.1016/j.ejor.2018.10.044

Koopmans, T. C. (1951). Chapter 3: Analysis of production as an efficient combination of activities. In T. C. Koopmans (Ed.), *Activity analysis of production and allocation: Proceedings of a conference*. John Wiley and Sons, Inc/Chapman and Hall, Limited.

Kuosmanen, T. (2005). Weak disposability in nonparametric production with undesirable outputs. *American Journal of Agricultural Economics, 87*, 1077–1082.

Laney, D. (2001). *3D data management: Controlling data volume, velocity, and variety*. META Group.

Romer, P. https://paulromer.net/ Simulating Covid-19: Part 1. March 23, 2020.

Romer, P. https://paulromer.net/ Simulating Covid-19: Part 2. March 24, 2020.

Romer, P. https://paulromer.net/ Even a bad Covid test can help guide the decision to isolate. Simulating Covid: Part 3. March 25, 2020.

Simon, H. A. (1951). Chapter 15. Effects of technological change in a linear model. In T. C. Koopmans (Ed.), *Activity analysis of production and allocation: Proceedings of a conference*. John Wiley and Sons, Inc/Chapman and Hall, Limited.

The Covid Tracking Project at *The Atlantic*. https://covidtracking.com/data

United States Census Bureau. Annual Estimates of the Resident Population for the United States, Regions, States, and Puerto Rico: April 1, 2010 to July 1, 2019 (NST-EST2019-01) https://www.census.gov/data/tables/time-series/demo/popest/2010s-state-total.html

United States Census Bureau. Centers of Population. https://www2.census.gov/geo/docs/reference/cenpop2010/CenPop2010_Mean_ST.txt

Varian, H. R. (2014). Big data: New tricks for econometrics. *Journal of Economic Perspectives, 28*(2), 3–28.

Zelenyuk, V. (2019). *Data envelopment analysis and business analytics: The big data challenges and some solutions*. Centre for Efficiency and Productivity Analysis Working Paper Series No. WP07/2019.

Zhu, J. (2020). DEA under big data: Data enabled analytics and network data envelopment analysis. *Annals of Operational Research*https://doi.10.1007/s10479-020-03668-8

Zhu, Q., Wu, J., & Song, M. (2018). Efficiency evaluation based on data envelopment analysis in the big data context. *Computers and Operations Research, 98*, 291–300. https://doi.org/10.1016/j.cor.2017.06.017

Hierarchical Data Envelopment Analysis for Classification of High-Dimensional Data

Ming-Miin Yu, Kok Fong See, and Bo Hsiao

1 Introduction

Big data refers to large volumes of structured, semistructured, and unstructured data that cannot be processed and stored using traditional data systems within a specified period. The prevalence of big data has intensified the need for efficient tools and platforms for data mining, machine learning, predictive modeling, and other advanced analytics applications (Chen et al., 2014). Although the general attributes of big data are often characterized by the 3 Vs, namely, volume, velocity, and variety, several researchers, including Demchenko et al. (2013) and Loshin (2013), have considered two additional dimensions, value and veracity, for a more comprehensive and broader definition. Data science is the field of cleaning, preparing, and analyzing data regardless of whether it is structured or unstructured. Big data analytics involves a set of advanced technologies designed to efficiently operate and maintain data that are high in volume, variety and velocity. An algorithmic or mechanical process is used in data analytics to derive insights. The ability to utilize multicriteria decision-making (MCDM) for performance evaluation is one of the most critical needs in every society. The emergence and prevalence of data science in MCDM research can enable and expedite research outputs. Providing useful information for

M.-M. Yu (✉)
Department of Transportation Science, National Taiwan Ocean University, Keelung, Taiwan

K. F. See
Economics Programme, School of Distance Education, Universiti Sains Malaysia, Pulau Pinang, Malaysia

Manning School of Business, University of Massachusetts, Lowell, MA, USA

B. Hsiao
Department of Information Management, Chang Jung Christian University, Tainan, Taiwan

J. Zhu, V. Charles (eds.), *Data-Enabled Analytics*, International Series in Operations Research & Management Science 312, https://doi.org/10.1007/978-3-030-75162-3_8

decision-making by integrating big data, data science and data analytics is essential for policy makers.

The analytical hierarchy process (AHP), presented by Saaty (1980), is a typical MCDM method that provides a priori information about attribute significance; both offer readily justifiable, discerning and deliberate approaches to the MCDM strategy. Although the AHP and conjoint analysis are capable of finding the optimal solution for high-dimensional problems with simple and manual calculations, both methods determine the relative importance of attributes by aggregating the views of multiple decision-makers. For example, the AHP is based on pairwise comparison; this characteristic obliges decision-makers to consider their previous judgments and judge in a way that reflects these judgments. Overall, decision-makers cannot be completely subjective. Human emotions are obscure, and decision-makers may not be able to value them with the precise numeric quantities needed to make explicit judgments. The weights obtained from the data envelopment analysis (DEA) model have been adopted in the literature to represent the relative importance of attributes, wherein the more favorable attributes usually have larger associated weights. The weight, to some extent, represents the marginal contribution of the respective attribute (Kao & Hung, 2008).

Charnes et al. (1978) introduced the DEA model (known as the CCR DEA model), derived from productive efficiency, and it is used to measure the efficiency of decision-making units (DMUs) using multiple inputs and outputs, a concept first discussed by Farrell in 1957. One of the major issues in the DEA method is determining whether ratio measures are admissible as input and output data. Ratio measures are frequently utilized as context-related variables. For example, the mortality rate is commonly used to represent the quality of a country's health care system. Emrouznejad and Amin (2009) showed some of the difficulties of using standard DEA models given input and/or output ratios. Following this, Krivonozhko et al. (2015) compared ratio analysis with the DEA model and showed that using ratio analysis implies that one multidimensional space is projected onto other subspaces many times. Olesen et al. (2015) provided a positive answer to the long-standing debate of whether such data could be used in DEA.

Performance in the presence of multiple different criteria can be evaluated using DEA, but the results heavily depend on the selection, values, and weights assigned to the criteria (Popović et al., 2020). Hierarchical data envelopment analysis (H-DEA) is explicitly related with the value dimension of big data. It is considered to be "data enabled analytics" for solving complex problems with a hierarchical structure of criteria (Zhu, 2020). A hierarchical structure is used to configure the overall efficiency evolution problem, in which weighting schemes continually generate a large quantity of information, such as the marginal contribution of corresponding dimensions, categories, and indicators, for every DMU. This information impacts the different indicators used to represent the achievement of the goal; because it is nearly impossible to provide the necessary and appropriate information to policy makers with such complex and voluminous data, these data must be classified appropriately for convenience and ease of access and presented to the user. Classification is essential to the study of performance evaluation: H-DEA is used to classify high-dimensional data. Our goals in this chapter are twofold: first, to reassess the performance of 110 countries in terms of global food security standards using the H-DEA model; second, to determine how the 110 countries differ concerning the efficiency contributions of the 3 dimensions (i.e., affordability, availability, quality and safe) and to identify areas of improvement for each country.

The current chapter starts by adopting the approach of Cook and Kress (1990), which studies DEA models for aggregating preference rankings. We then present in Section "Four-level H-DEA" the works of Chen et al. (2019), who apply an H-DEA approach to general hierarchical structures. In Section "Cluster Approach", we present a cluster approach for classifying the available hierarchical structure weighting schemes and show that groups of DMUs with similar characteristics can use the weighting and component score schemes of the hierarchical structure and cluster analysis. The H-DEA results for the 2014 Global Food Security Index (GFSI) across 110 countries as presented in Chen et al. (2019) are reproduced in Section "The Results from the H-DEA Model". Chen et al. (2019) use two standard classifications, income level and geographical region, to analyze country group performance in food security. To avoid repeating the researcher's judgment, we present country group analysis using the cluster approach in Section "Country Grouping Using Multi-level K-Means Cluster Analysis". The conclusion and remarks are included in Section "Conclusions".

2 Methodology

2.1 Single-Level DEA

Initially, the DEA model for measuring the efficiency level of a DMU, denoted by DMUo, as proposed by Charnes et al. (1978) is expressed as follows:

$$\text{Max} \qquad h_o = \frac{\sum_{r=1}^{s} u_r y_{ro}}{\sum_{i=1}^{m} v_i x_{io}} \tag{1}$$

Subject to

$$\frac{\sum_{r=1}^{s} u_r y_{rj}}{\sum_{i=1}^{m} v_i x_{ij}} \leq 1.j = 1, \ldots, n$$

$$u_r \geq 0, \quad r = 1, \ldots, s$$

$$v_i \geq 0, \quad i = 1, \ldots., m$$

where j is the DMU index, $j = 1, \ldots, n$, r is the output index, $r = 1, \ldots, s$, i is the input index, $i = 1, \ldots, m$, y_{rj} is the value of the r-th output for the j-th DMU, x_{ij} is the value of the i-th input for the j-th DMU, u_r is the weight given to the r-th output, v_i is the weight given to the i-th input, and h_o is the relative efficiency of DMUo, the DMU under evaluation.

Suppose that the DMUs have the unit input and the index outputs y_{rj}, and the CCR DEA model is expressed as follows:

$$\text{Max} \qquad h_o = \frac{\sum_{r=1}^{s} u_r y_{ro}}{\sum_{i=1}^{m} v_i} \tag{2}$$

Subject to

$$\frac{\sum_{r=1}^{s} u_r y_{rj}}{\sum_{i=1}^{m} v_i} \leq 1. j = 1, \ldots, n$$

$$u_r \geq 0, \quad r = 1, \ldots, s$$

$$v_i \geq 0, \quad i = 1, \ldots, m$$

Taking the new weights, one obtains

$$\text{Max} \qquad h_o = \sum_{r=1}^{s} \mu_r y_{ro} \tag{3}$$

Subject to

$$\sum_{r=1}^{s} \mu_r y_{rj} \leq 1. \quad j = 1, \ldots, n$$

$$u_r \geq 0, \quad r = 1, \ldots, s$$

This is exactly the same as the model that was first discussed by Cook and Kress (1990) and then by Fernandez-Castro and Smith (1994), shown as model (4).

$$\text{Max} \qquad h_o = \sum_{i=1}^{m} \sum_{r=1}^{s} w_{ir} e_{ir}^{o} \tag{4}$$

Subject to

$$\sum_{i=1}^{m}\sum_{r=1}^{s} w_{ir}e_{ir}^{j} \leq 1.j = 1,\ldots,n$$

where $e_{ir}^{j} = \frac{y_r^j}{x_i^j}$, x_i^j $(i = 1,\ldots.,m)$ are the inputs, and y_r^j $(r = 1,\ldots,s)$ are the outputs of DMU j $(j = 1,\ldots,n)$.

2.2 *Four-Level H-DEA*

The GFSI was established by the Economist Intelligence Unit (EIU), and a total of 110 countries are included in the 2014 GFSI exercise. Determining weights in order to aggregate several attributes from the indicators is one of the foremost problems in constructing a composite index. The H-DEA framework can be classified into four levels. The framework goal and dimensions are the first and second levels, respectively. Each dimension has several categories, which are the third level, and each category has several detailed indicators, which comprise the fourth level. Because each dimension has different numbers of criteria and each criterion has a different number of detailed indicators, determining the importance of the weight for each detailed indicator under these dimensions and categories is likely to be extremely complex. To simplify the list, the relationship from the second level to the fourth level is assumed to have a strict hierarchical structure.

The overall performance of DMUo is computed using the following steps. First, a detailed indicator score is combined with the total weighted score for each category. Second, the category score is combined with the total weighted score for each dimension and is expressed as follows:

$$y_{o'} = \max\sum_{i=1}^{I} u_i\left(\sum_{j_i=1}^{J_i} u_{j_i}\sum_{k_{ji}=1}^{K_{ji}} u_{k_{ji}} y_{k_{ji}o'}\right) = \sum_{i=1}^{I}\sum_{j_i=1}^{J_i}\sum_{k_{ji}=1}^{K_{ji}} u_i u_{j_i} u_{k_{ji}} y_{k_{ji}o'} \quad (5)$$

where $y_{o'}$ indicates the total performance score for DMUo';

$y_{k_{ji}o'}$ denotes the score for detailed indicator k_{ji} of category j of dimension i of DMUo';

$u_{k_{ji}}$ denotes the weight of detailed indicator k_{ji} of category j under dimension i;

u_{j_i} denotes the weight of category j_i of dimension i;

u_i denotes the weight of dimension i;

O denotes the number of DMUs ($o = 1,\ldots,O.$);

I denotes the number of dimensions ($i = 1,\ldots,I.$);

j_i denotes the number of categories ($j_i = 1, \cdots, J_i.\ i = 1,\ldots,I.$);

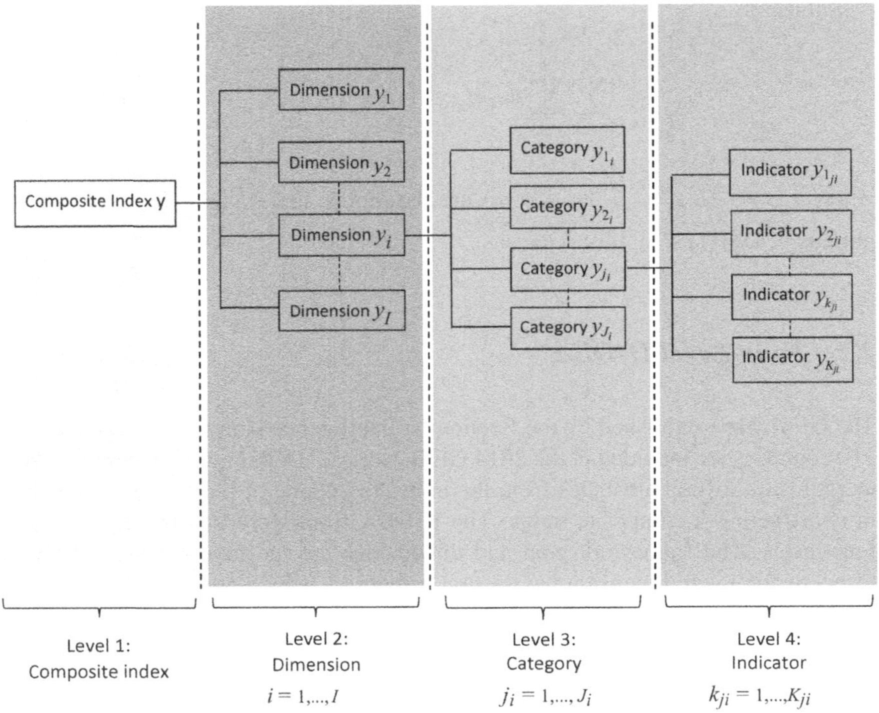

Fig. 1 The four-level H-DEA structure

k_{ji} denotes the number of detailed indicators ($k_{ji} = 1, \cdots, K_{ji}, j_i = 1, \cdots, J_i$. $i = 1, \cdots, I$).

Figure 1 provides a graphical representation of the hierarchical structure with subindicators. To obtain the composite index, the weights are required to aggregate all the indicators and subindicators at each level. Then, the final weighted score for the overall composite index for each DMU $o' = 1,...,O$ can be formulated as follows:

$$y_{o'} = \max \sum_{i=1}^{I} \sum_{j_i=1}^{J_i} \sum_{k_{ji}=1}^{K_{ji}} u_i u_{j_i} u_{k_{ji}} y_{k_{ji}o'} \tag{6}$$

DMUo' is assigned the highest possible efficiency score ($0 \leq y_{o'} \leq 1$) that constraints allow from the available data by choosing the optimal weight for each detailed indicator, category and dimension. If DMUo receives the maximal value $y_{o'} = 1$, then it is efficient, but if $y_{o'} < 1$, it is inefficient. The goal is to obtain a value for the index between zero and one, and each indicator, category and dimension also requires an index value between zero and one. The problem is solved subject to the following constraints for each DMUo:

Category level:

$$y_{j_i o} = \sum_{k_{ji}=1}^{K_{ji}} u_{k_{ji}} y_{k_{ji} o} \leq 1, j_i = 1, \cdots, J_i, i = 1, \cdots, I, o = 1, \cdots, O \tag{6.1}$$

Dimension level:

$$y_{io} = \sum_{j_i=1}^{J_i} u_{j_i} y_{j_i o} \leq 1, i = 1, \cdots, I, o = 1, \cdots, O \tag{6.2}$$

Goal level:

$$y_o = \sum_{i=1}^{I} u_i y_{io} \leq 1, \quad o = 1, \cdots, O. \tag{6.3}$$

The weights associated with the relative importance of each dimension in level 2, each category in level 3 and each indicator in level 4 must sum to one at each level:

$$\sum_{k_{ji}=1}^{K_{ji}} u_{k_{ji}} = 1, j_i = 1, \cdots, J_i, i = 1, \cdots, I, \tag{6.4}$$

$$\sum_{j_i}^{J_i} u_{j_i} = 1, \quad i = 1, \cdots, I. \tag{6.5}$$

$$\sum_{i=1}^{I} u_i = 1. \tag{6.6}$$

$$u_i, u_{j_i}, u_{k_{ji}} \geq \varepsilon, i = 1, \cdots, I, j_i = 1, \cdots, J_i, k_{ji} = 1, \cdots, K_{ji}.$$

where u_i, u_{j_i} and $u_{k_{ji}}$ are the weights of the indexes associated with dimension i and its category j_i and indicator k_{ji}, respectively, which are determined by each DMU such that the highest score will be produced for that DMU. Constraints (6.1)–(6.3) require that the aggregation values should not exceed unity at the three different levels.

This problem is clearly nonlinear in the sense that u_i, u_{j_i} and $u_{k_{ji}}$ are decision variables. If we implement a change in the variables, $u_{j_i} u_{k_{ji}} = \hat{u}_{k_{ji}}$, $u_i u_{j_i} u_{k_{ji}} = u_i \hat{u}_{k_{ji}} = \hat{\hat{u}}_{k_{ji}}$, model (6) can be converted to a linear programming model by integrating constraint (6.1) into (6.2) and substituting the nonlinear terms of $u_{j_i}u_{k_{ji}}$ with $\hat{u}_{k_{ji}}$ ($i = 1, \ldots, I$, $j_i = 1, \ldots, J_i$), thus obtaining (6.7):

$$y_{io} = \sum_{j_i=1}^{J_i} u_{j_i} \sum_{k_{ji}=1}^{K_{ji}} u_{k_{ji}} y_{k_{ji}o} = \sum_{j_i=1}^{J_i} \sum_{k_{ji}=1}^{K_{ji}} u_{j_i} u_{k_{ji}} y_{k_{ji}o} = \sum_{j_i=1}^{J_i} \sum_{k_{ji}=1}^{K_{ji}} \hat{u}_{k_{ji}} y_{k_{ji}o} \leq 1,$$
$$i = 1, \cdots, I, o = 1, \cdots, O \tag{6.7}$$

By further integrating constraint (6.7) into (6.3) and making a change in the variables $u_i \hat{u}_{k_{ji}} = \hat{\hat{u}}_{k_{ji}}$, (6.3) reduces to (6.8).

$$\begin{aligned} y_o &= \sum_{i=1}^{I} u_i y_{io} = \sum_{i=1}^{I} u_i \sum_{j_i=1}^{J_i} \sum_{k_{ji}=1}^{K_{ji}} \hat{u}_{k_{ji}} y_{k_{ji}o} \\ &= \sum_{i=1}^{I} \sum_{j_i=1}^{J_i} \sum_{k_{ji}=1}^{K_{ji}} \hat{\hat{u}}_{k_{ji}} y_{k_{ji}o} \leq 1, \quad o = 1, \cdots, O. \end{aligned} \tag{6.8}$$

To link (6.4), (6.5) and (6.6), we further multiply u_{j_i} by both sides of constraint (6.4) and u_i by both sides of constraint (6.5) to obtain the following relationship:

$$u_{j_i} \sum_{k_{ji}=1}^{K_{ji}} u_{k_{ji}} = u_{j_i}, j_i = 1, \cdots, J_i, i = 1, \cdots, I, \tag{6.4-1}$$

$$\sum_{k_{ji}=1}^{K_{ji}} \hat{u}_{k_{ji}} = u_{j_i}, j_i = 1, \cdots, J_i, i = 1, \cdots, I, \tag{6.4-2}$$

$$u_i \sum_{j_i}^{J_i} u_{j_i} = u_i, \quad i = 1, \cdots, I. \tag{6.5.1}$$

$$u_i \sum_{j_i}^{J_i} \sum_{k_{ji}=1}^{K_{ji}} \hat{u}_{k_{ji}} = u_i, \quad i = 1, \cdots, I. \tag{6.5-2}$$

$$\sum_{j_i}^{J_i} \sum_{k_{ji}=1}^{K_{ji}} \hat{\hat{u}}_{k_{ji}} = u_i, \quad i = 1, \cdots, I. \tag{6.5-3}$$

Model (6) can be re-expressed as follows:

$$y_{o'} = \max \sum_{i=1}^{I} \sum_{j_i=1}^{J_i} \sum_{k_{ji}=1}^{K_{ji}} \hat{\hat{u}}_{k_{ji}} y_{k_{ji}o'} \tag{7}$$

Subject to
Category level:

$$y_{j_i o} = \sum_{k_{ji}=1}^{K_{ji}} u_{k_{ji}} y_{k_{ji}o} \le 1,\ j_i = 1, \cdots, J_i, i = 1, \cdots, I, o = 1, \cdots, O \tag{7.1}$$

Dimension level:

$$y_{io} = \sum_{j_i=1}^{J_i} \sum_{k_{ji}=1}^{K_{ji}} \hat{u}_{k_{ji}} y_{k_{ji}o} \le 1, i = 1, \cdots, I, o = 1, \cdots, O \tag{7.2}$$

Goal level:

$$y_o = \sum_{i=1}^{I} \sum_{j_i=1}^{J_i} \sum_{k_{ji}=1}^{K_{ji}} \hat{\hat{u}}_{k_{ji}} y_{k_{ji}o} \le 1, \quad o = 1, \cdots, O. \tag{7.3}$$

Weight restrictions:

$$\sum_{k_{ji}=1}^{K_{ji}} \hat{u}_{k_{ji}} = u_{j_i}, j_i = 1, \cdots, J_i, i = 1, \cdots, I, \tag{7.4}$$

$$\sum_{j_i}^{J_i} \sum_{k_{ji}=1}^{K_{ji}} \hat{\hat{u}}_{k_{ji}} = u_i, \quad i = 1, \cdots, I. \tag{7.5}$$

$$\sum_{i=1}^{I} u_i = 1. \tag{7.6}$$

At optimality, for the composited indicator $y_{o'}$ $\left(o' = 1, \cdots, O\right)$, dimension i scores $y_{io'}$ $\left(i = 1, \cdots, I, o' = 1, \cdots, O\right)$ and category j scores $y_{j_i o}$ $(j_i = 1, \cdots, J_i, i = 1, \cdots, I, o' = 1, \cdots, O)$ are calculated as follows using the optimal $u_i^*, \hat{\hat{u}}_{k_{ji}}^*, \hat{u}_{k_{ji}}^*, u_{ji}^*, u_{k_{ji}}^*, i = 1, \ldots, I, j_i = 1, \ldots, J_i, k_{ji} = 1, \ldots, K_{ji}$

$$y_{j_io} = \sum_{k_{ji}=1}^{K_{ji}} u_{k_{ji}}^* y_{k_{ji}o},\ j_i = 1, \cdots, J_i, i = 1, \cdots, I, o = 1, \cdots, O \tag{8}$$

$$y_{io} = \sum_{j_i=1}^{J_i} \sum_{k_{ji}=1}^{K_{ji}} u_{j_i}^* u_{k_{ji}}^* y_{k_{ji}o},\ i = 1, \cdots, I, o = 1, \cdots, O \tag{9}$$

$$y_o = \sum_{i=1}^{I} \sum_{j_i=1}^{J_i} \sum_{k_{ji}=1}^{K_{ji}} u_i^* u_{j_i}^* u_{k_{ji}}^* y_{k_{ji}o}, \quad o = 1, \cdots, O. \tag{10}$$

Model (6) allows each DMUo to select the optimal weights that simultaneously maximize its composite indicator and prevent the composite indicators of all computed DMUs from exceeding a value of one.

2.3 Cluster Approach

Data clustering is used to identify the natural groupings of set objects in various disciplines that involve the analysis of multivariate data. This method of analysis uses one of the most popular and easiest partitional algorithms: the K-Means algorithm. This algorithm was discovered by several researchers, namely, Steinhaus (1956), Lloyd (1982), Ball and Hall (1965) and MacQueen (1967), and serves various scientific fields. The K-Means algorithm has grown to become one of the most widely used algorithms for clustering because it is simple, efficient, and easy to implement, achieving empirical success. Mathematically, the clusters are computed using the following expression:

Let $\mathbf{x}_o = \{x_i, i = 1, \cdots, n\}$, $o = 1, \cdots, O$ be a set of O observations, where each observation is an n-dimensional real vector that will be clustered into $K(K < O)$ sets of clusters $c_k = \{\mathbf{x}_o | o \in cluster\ k\}$, $k = 1, \cdots, K)$, $C = \{c_k, k = 1, \cdots, K\}$. The expression shows that K-Means is an algorithm for finding the partition that minimizes the squared error between the empirical cluster means and the cluster points. This is further expressed as follows:

Let μ_k be the mean of the observations in c_k. Each observation is assigned to the cluster with the nearest distance that has the least sum of squared Euclidean distance. This approach tries to keep the intra-cluster observations as similar as possible while also making the clusters as different (far) as possible. The sum of the squared Euclidean distance between μ_k and all the observations belonging to cluster c_k will be defined as Eq. (11):

$$J(c_k) = \sum_{\mathbf{x}_o \in c_k} \left\| \mathbf{x}_o - \mu_k \right\|^2 \tag{11}$$

It can be observed from the expression that the goal of the K-Means algorithm is to minimize the total sum of squared Euclidean distance over all clusters of K. This is shown in Eq. (12).

$$J(C) = \sum_{k=1}^{K} \sum_{\mathbf{x}_o \in c_k} \left\| \mathbf{x}_o - \mu_k \right\|^2 \tag{12}$$

The minimization of the problem is computationally difficult and is usually termed an NP-hard problem, even for $k = 2$. Hence, the K-Means algorithm, being a greedy algorithm, has a local minimum convergence. However, research indicates that there is a substantial probability that the K-Means algorithm can also converge to a global optimum, especially when there is a separation of the clusters.

The K-Means algorithm usually begins with an original partition that has K clusters, and the patterns are assigned to clusters to provide the greatest reduction in the sum of the squared distance between the data points and all centroids. The squared error usually decreases when there is an increase in the number of clusters K, where the number of clusters K (with $J(C) = 0$ when $K = O$) is minimized for K pre-defined distinct non-overlapping clusters.

The major steps in the K-Means algorithm are as follows:

Step 1: Choose an original partition containing K clusters. Repeat steps 2 and 3 until the membership in the clusters is stable, i.e., the assignment of data points to clusters is no longer changing.

Step 2: Create a new partition by allocating or assigning every pattern to the cluster with the closest mean. Assign each data point to the closest cluster (centroid).

Step 3: Calculate the newly created cluster centers.

Solving the K-Means algorithm requires user-specified parameters. The parameters include the number of clusters K, the initialization of clusters, and the metric distance. Given these parameters, the most critical step is choosing K, and the K-Means algorithm helps by running different independent K values so that the generated partition will be meaningful to the selected domain expert. However, different initializations could result in different final clusters because the K-Means only converges to local minima. However, the local minima can be overcome by running the K-Means algorithm for a given K that has various multiple partitions and involves selecting a smaller sum of the squared distance between data points and all centroids.

The K-Means algorithm has been expanded in various ways. These extensions include additional heuristics that involve clusters that are minimal in size and clusters that are merged and split. In the K-Means method, the Euclidean metric is most often used when calculating the distance between points and cluster centers. As such, the K-Means method is useful for finding clusters in data that are spherical and ball-shaped. When computing the K-Means using the Mahalanobis distance metric, it can help in the detection of a hyperellipsoidal shape. However, carrying out this computation can be costly. Furthermore, the K-Means variant can be used to calculate vector quantization using the Itakura-Saito distance metric, in which the K-Means distance is denoted by the L1-norm (Kashima et al., 2003). The following subsections provide a brief explanation of how the cluster approach can be extended to the multi-level K-Means algorithm and applied to grouping analysis after obtaining the results of the H-DEA model.

2.3.1 Multi-level Clustering Measurement

To provide a better understanding of the clustering approach, the GFSI results from Chen et al. (2019) are used as an example in the following discussion. Under the scoring clustering approach, we apply the K-Means algorithm to group 110 countries by global food security ranking. There are four-level H-DEA scores, with 1 variable at the goal level, I variables at the dimension level, J variables $\left(\sum_{i=1}^{I} J_i = J\right)$ at the category level, and K variables $\left(\sum_{i=}^{I}\sum_{j_i=1}^{J_i} K_{j_i} = K\right)$ at the indicator level. For each o-th DMU ($o = 1, \ldots, O$), we defined $\mathbf{y}_o^1 = \{y_o\}$ at the goal level, $\mathbf{y}_o^2 = \{y_{io}, i = 1, \cdots I\}$ at the dimension level, $\mathbf{y}_o^3 = \{y_{j_i o}, i = 1, \cdots I, j_i = 1, \cdots, J_i\}$ at the category level, and $\mathbf{y}_o^4 = \{y_{k_{ji} o}, i = 1, \cdots I, j_i = 1, \cdots, J_i, k_{ji} = 1, \cdots, K_{ji}\}$ at the indicator level. Without loss of generality, different numbers of clusters are assigned for different levels. We defined, $c_p^1\,(p = 1, \cdots, P)$, $c_{pq}^2\,(p = 1, \cdots, P, q = 1, \cdots, Q)$, $c_{pqr}^3\,(p = 1, \cdots, P, q = 1, \cdots, Q, r = 1, \cdots, R)$, and $c_{pqrs}^4\,(p=1, \cdots, P, q = 1, \cdots, Q, r = 1, \cdots, R, s = 1, \cdots, S)$ to represent the cluster sets for levels 1, 2, 3 and 4, respectively, which satisfy the following requirements, as shown in Eqs. (13,14,15,16).

$$c_p^1 = \left\{\mathbf{y}_o^1 : \left\|\mathbf{y}_o^1 - \boldsymbol{\mu}_p^1\right\|^2 \le \left\|\mathbf{y}^1 - \boldsymbol{\mu}_h^1\right\|^2; \forall h, 1 \le h \le P\right\} \tag{13}$$

where each $\mathbf{y}_o^1 = \{y_o\}$ is assigned to exactly one cluster $c_p^1\,(p = 1, \cdots, P)$, and $\boldsymbol{\mu}_\mathbf{p}^\mathbf{1}$ is the mean (centroid) of observations in cluster c_p^1 at the goal level.

For each cluster set $c_p^1\,(p = 1, \cdots, P)$, we further partition the observations belonging to cluster c_p^1 into Q sets, $c_{pq}^2\,(q = 1, \cdots, Q)$. c_{pq}^2 is expressed as follows:

$$c_{pq}^2 = \left\{\mathbf{y}_o^2 : \left|\mathbf{y}_o^2 - \boldsymbol{\mu}_{pq}^2\right\|^2 \le \left\|\mathbf{y}_o^2 - \boldsymbol{\mu}_{ph}^2\right\|^2; \forall h, 1 \le h \le Q\right\} \tag{14}$$

where $\mathbf{y}_o^2 = \{y_{io}, i = 1, \cdots I\}$, and $\boldsymbol{\mu}_{pq}^\mathbf{2}$ is the mean (centroid) of observations in cluster c_{pq}^2 at the dimension level.

For each cluster set c_{pq}^2 $(p = 1, \cdots, P, q = 1, \cdots, Q)$, we further partition the observations belonging to cluster c_{pq}^2 into R sets, c_{pqr}^3 $(r = 1, \cdots, R)$. c_{pqr}^3 is expressed as follows:

$$c_{pqr}^3 = \left\{ \mathbf{y}_o^3 : \left\| \mathbf{y}_o^3 - \boldsymbol{\mu}_{pqr}^3 \right\|^2 \leq \left\| \mathbf{y}_o^3 - \boldsymbol{\mu}_{pqh}^3 \right\|^2 ; \forall h, 1 \leq h \leq R \right\} \tag{15}$$

where $\mathbf{y}_o^3 = \left\{ y_{j_i o}, i = 1, \cdots I, j_i = 1, \cdots, J_i \right\}$, and $\boldsymbol{\mu}_{pqr}^{\mathbf{3}}$ is the mean (centroid) of observations in cluster c_{pqr}^3 at the category level.

For each cluster set c_{pqr}^3 $(p = 1, \cdots, P, q = 1, \cdots, Q, r = 1, \cdots, R)$, we further partition the observations belonging to cluster c_{pqr}^3 into S sets, c_{pqrs}^4 $(s = 1, \cdots, S)$, where c_{pqrs}^4 is expressed as follows:

$$c_{pqrs}^4 = \left\{ \mathbf{y}_o^4 : \left\| \mathbf{y}_o^4 - \boldsymbol{\mu}_{pqrs}^4 \right\|^2 \leq \left\| \mathbf{y}_o^4 - \boldsymbol{\mu}_{pqhs}^4 \right\|^2 ; \forall h, 1 \leq h \leq S \right\} \tag{16}$$

where $\mathbf{y}_o^4 = \left\{ y_{k_{ji} o}, i = 1, \cdots I, j_i = 1, \cdots, J_i, k_{ji} = 1, \cdots, K_{ji} \right\}$, and $\boldsymbol{\mu}_{pqrs}^{\mathbf{4}}$ is the mean (centroid) of observations in cluster c_{pqrs}^4 at the indicator level.

Similar to the one-level K-Means algorithm, the Euclidean metric is used to calculate the distance between the cluster mean and the observation in the multi-level K-Means algorithm. Following Eqs. (13,14,15,16), the sum of squared Euclidean distance at each level is defined as Eqs. (17,18,19,20).

The sum of squared Euclidean distance between cluster mean $\boldsymbol{\mu}_p^1$ and the observation in cluster c_p^1 at the goal level is:

$$J\left(c_p^1\right) = \sum_{\mathbf{y}_o^1 \in c_p^1} \left\| \mathbf{y}_o^1 - \boldsymbol{\mu}_p^1 \right\|^2 \tag{17}$$

The sum of squared Euclidean distance between cluster mean $\boldsymbol{\mu}_{pq}^2$ and the observation in cluster c_{pq}^2 at the dimension level is:

$$J\left(c_{pq}^2\right) = \sum_{\mathbf{y}_o^2 \in c_{pq}^2} \left\| \mathbf{y}_o^2 - \boldsymbol{\mu}_{pq}^2 \right\|^2 \tag{18}$$

The sum of squared Euclidean distance between cluster mean $\boldsymbol{\mu}_{pqr}^3$ and the observation in cluster c_{pqr}^3 at the category level is:

$$J\left(c_{pqr}^3\right) = \sum_{\mathbf{y}_o^3 \in c_{pqr}^3} \left\| \mathbf{y}_o^3 - \boldsymbol{\mu}_{pqr}^3 \right\|^2 \tag{19}$$

The sum of squared Euclidean distance between cluster mean μ^4_{pqrs} and the observation in cluster c^4_{pqrs} at the indicator level is:

$$J\left(c^4_{pqrs}\right) = \sum_{\mathbf{y}^4_o \in c^4_{pqrs}} \left\| \mathbf{y}^4_o - \mu^4_{pqrs} \right\|^2 \tag{20}$$

The goal of the multi-level K-Means algorithm is to minimize the total sum of squared Euclidean distance over all clusters at the goal, dimension, category and indicator levels, aiming to obtain more homogeneous data points within the same cluster at each level, as expressed in Eqs. (21,22,23,24):

The minimum total sum squared Euclidean distance over all P clusters at the goal level, $C^P{}_1$ is defined as follows:

$$J\left(C^1_P\right) = \min \sum_{p=1}^{P} \sum_{\mathbf{y}^1_o \in c^1_p} \left\| \mathbf{y}^1_o - \mu^1_p \right\|^2 \tag{21}$$

The minimum total sum squared Euclidean distance over all P x Q clusters at the dimension level, $C^{PQ}{}_2$, is:

$$J\left(C^2_{PQ}\right) = \min \sum_{q=1}^{Q} \sum_{p=1}^{P} \sum_{\mathbf{y}^2_o \in c^2_{pq}} \left\| \mathbf{y}^2_o - \mu^2_{pq} \right\|^2 \tag{22}$$

The minimum total sum squared Euclidean distance over all P x Q x R clusters at the category level, $C^{PQR}{}_3$, is:

$$J\left(C^3_{PQR}\right) = \min \sum_{r=1}^{R} \sum_{q=1}^{Q} \sum_{p=1}^{P} \sum_{\mathbf{y}^3_o \in c^3_{pqr}} \left\| \mathbf{y}^3_o - \mu^3_{pqr} \right\|^2 \tag{23}$$

The minimum total sum squared Euclidean distance over all P x Q x R x S clusters at the indicator level, C^4_{PQRS}, is:

$$J\left(C^4_{PQRS}\right) = \min \sum_{s=1}^{S} \sum_{r=1}^{R} \sum_{q=1}^{Q} \sum_{p=1}^{P} \sum_{\mathbf{y}^4_o \in c^4_{pqrs}} \left\| \mathbf{y}^4_o - \mu^4_{pqrs} \right\|^2 \tag{24}$$

The following algorithm is based on our proposed multi-level K-Means algorithm, which is a combination of K-Means and hierarchical clustering.

Step 1: Define p = P and calculate the cluster mean μ^1_p of each group ($p = 1, \ldots, P$) at the goal level for all observations (i.e., **O**).

Step 2: Select the P point randomly as the P cluster's centroid and calculate the distance between each observation point and the P point for Eqs. (13,14,15,16) at the goal, dimension, category, and indicator levels, respectively, i.e., calculate the sum of the squared distance between data points and all centroids at the goal, dimension, category, and indicator levels.

Step 3: Evaluate the Euclidean metric of the observation with other clusters using Eqs. (21,22,23,24). For each p ($p = 1, \ldots, P$), if the total sum of the Euclidean metric of the observations in the p group is smaller than that for the other points, then that observation is clustered into the p group.

Step 4: Recalculate the p cluster mean using Eqs. (25,26,27,28) at the goal, dimension, category, and indicator levels by obtaining observations from Step 3.

$$\mu_p^1 = \frac{1}{\left|c_p^1\right|} \sum_{\mathbf{y}_o^1 \in c_p^1} \mathbf{y}_o^1, \forall p = 1, \ldots, P \tag{25}$$

$$\mu_{pq}^2 = \frac{1}{\left|c_{pq}^2\right|} \sum_{\mathbf{y}_o^2 \in c_{pq}^2} \mathbf{y}_o^2; \forall p = 1, \cdots, P, q = 1, \ldots, Q \tag{26}$$

$$\mu_{pqr}^3 = \frac{1}{\left|c_{pqr}^3\right|} \sum_{\mathbf{y}_o^3 \in c_p^3} \mathbf{y}_o^3; \forall p = 1, \cdots, P, q = 1, \ldots, Q, r = 1, \ldots, R \tag{27}$$

$$\mu_{pqrs}^4 = \frac{1}{\left|c_{pqrs}^4\right|} \sum_{\mathbf{y}_o^4 \in c_{pqrs}^4} \mathbf{y}_o^4; \forall p=1, \cdots, P, q=1, \ldots, Q, r=1, \ldots, R, s=1, \ldots, S \tag{28}$$

where $\left|c_p^l\right|, \left|c_{pq}^2\right|, \left|c_{pqr}^3\right|, \left|c_{pqrs}^4\right|$ represent the observations of $c_p^l, c_{pq}^2, c_{pqr}^3, c_{pqrs}^4$, respectively.

Step 5: If the level is the goal level, repeat Steps 3–5 until the final data set of the P groups at the goal level is determined, which is when the minimum sum of squared Euclidean distance over all P clusters at the goal level, $J\left(C_P^1\right)$, is reached. Repeat Step 2 to evaluate the dimension level.

Step 6: For all P x Q clusters at the dimension level, repeat Steps 3–5 until the minimum sum squared Euclidean distance over all clusters of P x Q at the dimension level $J\left(C_{PQ}^2\right)$ is reached. Repeat Step 2 to evaluate the category level.

Step 7: For all P x Q x R clusters at the category level, repeat Steps 3–5 until the final data set of the cluster in the category level is determined, which is when the minimum sum squared Euclidean distance over all clusters of P x Q x R at the category level $J\left(C_{PQR}^3\right)$ is reached. Repeat Step 2 to evaluate the indicator level.

Step 8: For all P x Q x R x S clusters at the category level, repeat Steps 3–5 until the minimum sum squared Euclidean distance over all clusters of P x Q x R x S at the indicator level $J\left(C^4_{PQRS}\right)$ is reached.

3 The Results from the H-DEA Model

One of the advantages of the DEA model is its ability to provide further information on how to improve the performance of countries. The H-DEA model results can benefit decision-makers in several ways. First, the model provides each country with the optimal weight in reaching the efficient frontier, thus minimizing the problem of bias weighting for international comparison. Second, the H-DEA model provides the optimal scores from different hierarchical levels in determining the key factors for improving national food security plans. The results and discussion of the GFSI using the H-DEA model are reproduced from Chen et al. (2019). Figure 2 presents the detailed distribution of overall scores for each country based on the H-DEA model. To determine whether differences exist in the means and rankings of the EIU and H-DEA scores, nonparametric statistical tests, such as the Wilcoxon signed-rank test and Spearman's rank correlation coefficient, are adopted in this study. As indicated by both statistical test results, the ranking of H-DEA is almost identical to that presented by the EIU expert panel if appropriate bounds are incorporated into the proposed H-DEA model.

For further discussion of the GFSI, we can organize and group the sample according to country characteristics. The country grouping can either use the standard classification based on individual preferences or the data science approach. In this section, the World Bank income classification and geographical location are used as in Chen et al. (2019). Overall, the sample is divided into four income groups. Table 1 shows that low-income countries are likely to underperform, in turn experiencing

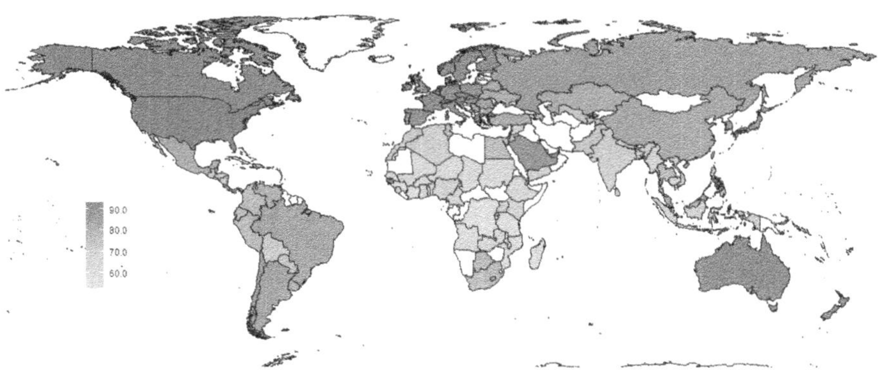

Fig. 2 Distribution of overall global food security scores based on the H-DEA model

Table 1 H-DEA scores and weight by income and geographical classification

	Score		Weight		
	Mean	Std Dev	Affordability	Availability	Quality & Safety
Entire sample	0.752	0.121	22%	30%	48%
By income level					
High income	0.894	0.027	32%	18%	50%
Upper middle income	0.774	0.058	18%	33%	49%
Lower middle income	0.680	0.058	18%	36%	46%
Low income	0.599	0.043	17%	39%	44%
By geographical region					
Asia & Pacific	0.760	0.097	22%	29%	49%
Central & South America	0.755	0.069	19%	32%	49%
Europe	0.876	0.048	29%	21%	50%
Middle East & North Africa	0.760	0.105	22%	31%	47%
North America	0.889	0.060	28%	22%	50%
Sub-Saharan Africa	0.608	0.062	18%	39%	43%

Source: Chen et al. (2019)

The weights given by EIU for affordability, availability, and quality & safety are 40%, 44% and 16%, respectively

EIU vs H-DEA model: Wilcoxon signed ranks test (P-value) is −0.643, and Spearman's rank correlation coefficient is 0.983

more threats to food security. Meanwhile, the result from the Kruskal–Wallis test confirms that performance differs significantly among the different income groups. The scores for the six geographical regions are presented in Table 1. Higher scores are recorded in the region of Europe and Northern America. Most selected countries obtained, on average, very low scores in the affordability category, and Sub-Saharan Africa was ranked as the worst performer. A detailed analysis of the scores for the level 2 indicators highlights the very low affordability score, which is due to low GDP per capita and limited access to financing. Insufficient public spending on agricultural R&D and high corruption are the leading causes of the second-lowest availability score.

In addressing the issue of the H-DEA weights, similar to H-DEA scores, the weights can also serve as a yardstick for future improvement. Based on the EIU weights, which are inspired by expert judgment, a higher weight suggests that a country must intensify its effort on the indicator that will improve its performance score, which is also known as a priori weights (Kao, 2010). However, the H-DEA model weights are more compelling than the weights from the EIU because the H-DEA models provide the relative weight of each indicator, which is derived from the data and known as the a posteriori weight (Kao, 2010). To identify which indicators can be substantially improved, the weights for income levels and geographical groups are compared in Table 1. The EIU weights suggest that priority should be granted to food availability and affordability. The relative weighting for the three categories has significantly changed in the H-DEA model. Greater emphasis is given to the quality and safety category due to the high power to discriminate between efficient and inefficient units. The model indicates that the appropriate weight within the allowable range is assigned to the quality and safety category in the sample countries under evaluation. Given that the majority of indicators under the quality and safety category are clearly linked to a country's nutritional status, the effort to reduce nutritional gaps by increasing food production and quality and safe food consumption is of primary importance.

Moreover, the H-DEA method provides the optimal weight for each country, thus allowing weights to be matched between different groups of countries. The H-DEA results in Table 1 show that as the income level decreases, the importance of food affordability decreases, while the importance of food availability increases. This finding indicates that in low-income countries, food availability is more important and has greater significance for the assessment process. Hence, national policies devoted to strengthening production systems to optimize agricultural production should be the top priority for low-income countries, while food affordability becomes increasingly important when income growth exists as a result of high purchasing power. For higher-income countries, food affordability is considered the most important category. To sustain the affordability of food, a country should prioritize macroeconomic policy implementation through balanced budget distribution via social programs. Analyzing the GFSI across six geographical regions, while food affordability is found to be more important in Europe and North America than in other regions, the availability of food is the most important category in Sub-Saharan African countries. The results suggest that the key drivers of food security in Sub-Saharan Africa are the eight indicators under the availability category. In contrast, the six indicators under the affordability category are more important in North America and Europe.

4 Country Grouping Using Multi-Level K-Means Cluster Analysis

As discussed in Section "The Results from the H-DEA Model", a sample of 110 countries is classified into several groups using the World Bank income classification and geographical regions. The country groupings are primarily based on the researcher's judgement or preference. To improve personal judgement, our proposed multi-level K-Means clustering approach, as presented in Section "Methodology", is used to cluster all observations into homogeneous and distinct groups. In this case, the performance scores or weightings of each country are treated as an observational data point to represent specific groups that share common characteristics. In other words, the observations within each group should have similar characteristics when compared to other groups. With the formation of such cluster groups, further recommendations and improvement plans can be considered for those countries within the same groups (e.g., introduce strategies to improve the performance scores of indicators).

Given the nature of observations in the GFSI, scoring clustering and weighting-clustering approaches are used to cluster these 110 countries into several groups.

To preserve space, the scoring clustering approach is used to demonstrate country grouping based on the multi-level K-Means clustering approach. The hierarchical structure of the GFSI is presented in Fig. 3. There is one goal level for each country, while the composite index score represents the overall food security performance of that country. The GFSI is later divided into three dimensions to verify food security, namely, affordability (A1), availability (A2), and quality and safety (A3). Different categories are then used to evaluate the affordability, availability, and quality and safety dimensions. There are six categories for affordability (AF1 ~ AF6), eight categories for availability (AV1 ~ AV8), and five categories for quality and safety (QU1 ~ QU5). From Fig. 3, we observe that a total of 28 indicators are used to construct the four-level GFSI. For more details on this structure, see Chen et al. (2019).

As described in the previous discussion, the scoring clustering approach can be seen as providing a performance cluster. Each value of the food security indicators has been standardized to between 0–100. The mean, standard deviation, maximum and minimum for the global food security performance score of all categories and indicators are reported in Table 2. From Table 2, we observe that the sampled countries obtained high scores in AF4 Agricultural import tariffs (i.e., 77.63), AV31 Existence of adequate crop storage facilities (i.e., 79.09) and QU22 National nutrition plan or strategy (i.e., 99.09) at the dimension level of affordability, availability, quality and safety, respectively. Apart from that, two quality and safety indicators also obtained high scores: QU51 Agency to ensure the safety and health of food (i.e., 95.45) and QU23 Nutrition monitoring and surveillance (i.e., 92.73). However, these sampled countries achieved lower scores in AF3 Gross domestic product per capita (i.e., 23.22), AV2 Public expenditure on agricultural R&D (i.e., 14.97) and QU32 Dietary availability of animal iron (i.e., 29.79) at each dimension level.

In this study, we used the score data set shown in Table 2 as our input data for our proposed cluster approach. A total of 110 countries have one composite index in the first level (i.e., goal), three variables in the second level (i.e., dimension level), 19 variables in the third level (i.e., category level), and 14 variables in the fourth level (i.e., indicator level).[1] The elbow approach is used to obtain the optimal number of clusters for each level.[2] On this basis, we partition all observations into P = 1, Q = 2, R = 2, and the number of clusters S is determined given on different clusters $c^3_{p,q,r}$ (e.g., number of clusters given on $c^3_{1,1,1}$ is 3 ($c^4_{1,1,1,1}$, $c^4_{1,1,1,2}$ and $c^4_{1,1,1,3}$), given on $c^3_{1,2,1}$ is 2 ($c^4_{1,1,2,1}$ and $c^4_{1,1,2,2}$) etc.). As a result, a total of 13 groups are

[1] We have assumed only one group at the goal level, and hence the goal level is removed from the data set. In this case, our suggested approach is used to cluster all observations at the dimension, category, and indicator levels into homogeneous and distinct groups.

[2] The elbow method is a popular approach for identifying the optimal number of clusters under the K-Means clustering approach. This study applied the variance as a clustering score and plotted it against the number of clusters. For more details on the elbow approach, please see Syajur et al. (2018) and Bholowalia and Kumar (2014).

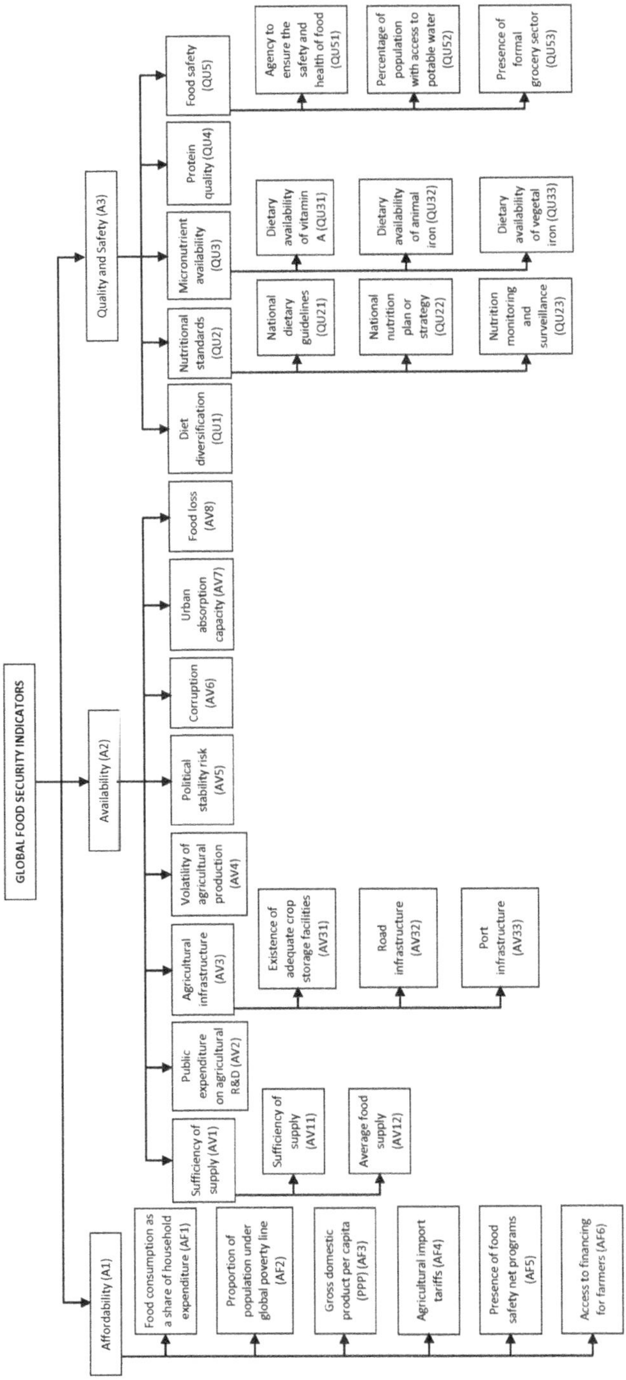

Fig. 3 Structure of the global food security index

Table 2 Global food security scores based on the H-DEA approach

Dimension	Code	Description	Average	Std Dev	Max	Min
Affordability	AF1	Food consumption as a share of household expenditure	57.57	27.12	100.00	0.00
	AF2	Proportion of population under global poverty line	71.19	33.24	100.00	0.00
	AF3	Gross domestic product per capita (PPP)	23.22	24.36	100.00	0.00
	AF4	Agricultural import tariffs	77.63	15.09	100.00	0.00
	AF5	Presence of food safety net programs	65.00	30.14	100.00	0.00
	AF6	Access to financing for farmers	59.09	30.99	100.00	0.00
Availability	AV11	Average food supply	55.95	23.47	100.00	0.00
	AV12	Dependency on chronic food aid	67.73	39.25	100.00	0.00
	AV2	Public expenditure on agricultural R&D	14.97	20.99	100.00	0.00
	AV31	Existence of adequate crop storage facilities	79.09	40.85	100.00	0.00
	AV32	Road infrastructure	42.50	27.63	100.00	0.00
	AV33	Port infrastructure	52.73	31.10	100.00	0.00
	AV4	Volatility of agricultural production	72.93	21.95	100.00	0.00
	AV5	Political stability risk	49.48	23.35	100.00	0.00
	AV6	Corruption	34.09	31.17	100.00	0.00
	AV7	Urban absorption capacity	63.19	10.18	100.00	0.00
	AV8	Food loss	75.90	16.81	100.00	0.00
Quality and safety	QU1	Diet diversification	55.98	26.60	100.00	0.00
	QU21	National dietary guidelines	74.55	43.76	100.00	0.00
	QU22	National nutrition plan or strategy	99.09	9.53	100.00	0.00
	QU23	Nutrition monitoring and surveillance	92.73	26.09	100.00	0.00
	QU31	Dietary availability of vitamin A	68.64	35.75	100.00	0.00
	QU32	Dietary availability of animal iron	29.79	22.38	100.00	0.00
	QU33	Dietary availability of vegetal iron	32.42	17.42	100.00	0.00
	QU4	Protein quality	44.15	26.78	100.00	0.00
	QU51	Agency to ensure the safety and health of food	95.45	20.93	100.00	0.00
	QU52	Percentage of population with access to potable water	72.88	28.67	100.00	0.00
	QU53	Presence of formal grocery sector	70.00	37.22	100.00	0.00

constructed at the indicator level, 4 groups at the category level and 2 groups at the dimension level.

To obtain a better understanding, the country grouping in global food security scores using multi-level K-Means clustering approach is illustrated in Fig. 4. At the dimension level, we used three dimensions for the evaluation in the 2 clusters, i.e., $c^2_{1,1}$ consists of 52 countries while $c^2_{1,2}$ consists of 58 countries. From Fig. 4, we find that $c^2_{1,2}$ has satisfactory performance at all three-dimension levels but not $c^2_{1,1}$. Ideally, $c^2_{1,1}$ should identify which category show unsatisfactory performance at the category level. Since two clusters ($c^2_{1,1,1}$, $c^2_{1,1,2}$) are formed given on $c^2_{1,1}$, $c^2_{1,1,1}$ is considered a better cluster group than $c^2_{1,1,2}$ at the category level. A similar evaluation to a group of clusters given on $c^3_{1,1,2}$ can be performed at the indicator levels, and we find that $c^2_{1,1,2,2}$ performed satisfactorily as compared to $c^2_{1,1,2,1}$.

It is interesting to note that country grouping formation offers greater insights. For instance, at the indicator level, most of the countries in $c^4_{1,2,2,1}$ (except for Singapore, Japan and South Korea) are considered to be members of the high-income group and located in the Western world, while the countries in $c^4_{1,2,2,2}$ are also classified within the high-income group but are located in the Middle East. The countries in both $c^4_{1,1,2,1}$ and $c^4_{1,1,2,2}$ (except for Cambodia, Bangladesh and Myanmar) are grouped in the low-income group and located in Africa. Therefore, the results of the country grouping are sensible, although income classification and geographical region are disregarded when we apply the multi-level K-Means clustering approach.

The centroids for the two groups at the dimension, category and indicator levels are presented in Tables 3, 4, and 5, while Euclidean distance between cluster centroids and the overall cluster centroid are calculated and summarized in Tables 6, 7, and 8. If the cluster centroid is greater than the overall cluster centroid, a positive value is shown in Tables 6, 7 and 8, implying that the cluster is above average in a particular level. With such information, the contributions of affordability, availability, and quality and safety can be identified, and such relevant information is useful in identifying the sources that affect the overall food security performance difference among the countries. For example, the United States, United Kingdom and Australia belong to $c^4_{1,2,2,1}$ which are characterized by high "affordability", "availability" and "quality and safety" scores. Most of the food security indicators (except for QU21 and QU32) are recognized as sources that make simultaneous high contributions in these three dimensions of global food security performance for $c^4_{1,2,2,1}$.

Table 3 Centroids of clusters at the dimension level

Dimension Level	Original	$c^2_{1,1}$	$c^2_{1,2}$
Size	110.00	52.00	58.00
A1	66.71	46.59	84.74
A2	66.15	55.36	75.83
A3	69.28	55.71	81.45

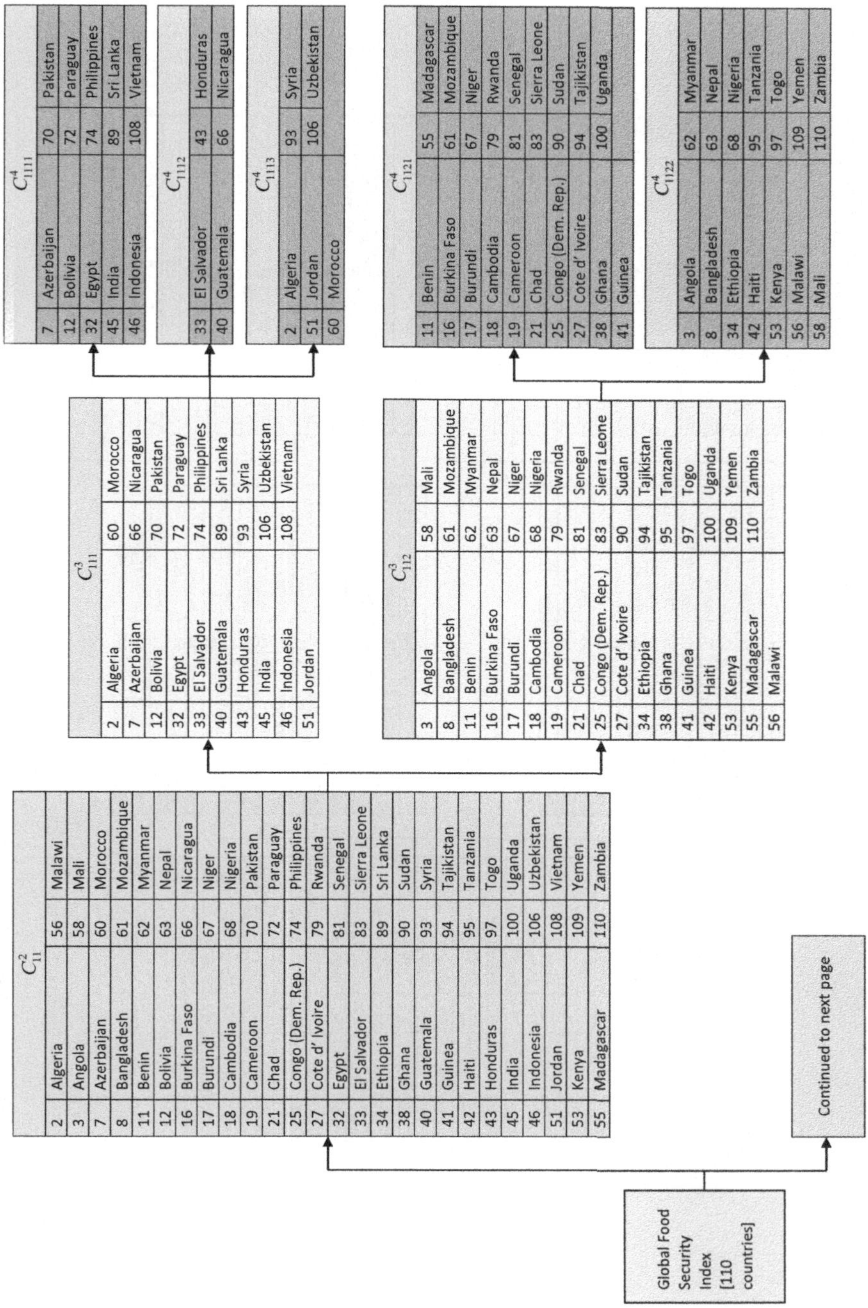

Fig. 4 Country grouping using multi-level K-means clustering approach

C^2_{12}

1	Taiwan	57	Malaysia
4	Argentina	59	Mexico
5	Australia	64	Netherlands
6	Austria	65	New Zealand
9	Belarus	69	Norway
10	Belgium	71	Panama
13	Botswana	73	Peru
14	Brazil	75	Poland
15	Bulgaria	76	Portugal
20	Canada	77	Romania
22	Chile	78	Russia
23	China	80	Saudi Arabia
24	Colombia	82	Serbia
26	Costa Rica	84	Singapore
28	Czech Republic	85	Slovakia
29	Denmark	86	South Africa
30	Dominican Republic	87	South Korea
31	Ecuador	88	Spain
35	Finland	91	Sweden
36	France	92	Switzerland
37	Germany	96	Thailand
39	Greece	98	Tunisia
44	Hungary	99	Turkey
47	Ireland	101	Ukraine
48	Israel	102	United Arab Emirates
49	Italy	103	United Kingdom
50	Japan	104	United States
52	Kazakhstan	105	Uruguay
54	Kuwait	107	Venezuela

C^3_{121}

4	Argentina	71	Panama
9	Belarus	73	Peru
13	Botswana	77	Romania
14	Brazil	78	Russia
15	Bulgaria	80	Saudi Arabia
23	China	82	Serbia
24	Colombia	86	South Africa
26	Costa Rica	96	Thailand
30	Dominican Republic	98	Tunisia
31	Ecuador	99	Turkey
52	Kazakhstan	101	Ukraine
57	Malaysia	105	Uruguay
59	Mexico	107	Venezuela

C^3_{122}

1	Taiwan	50	Japan
5	Australia	54	Kuwait
6	Austria	64	Netherlands
10	Belgium	65	New Zealand
20	Canada	69	Norway
22	Chile	75	Poland
28	Czech Republic	76	Portugal
29	Denmark	84	Singapore
35	Finland	85	Slovakia
36	France	87	South Korea
37	Germany	88	Spain
39	Greece	91	Sweden
44	Hungary	92	Switzerland
47	Ireland	102	United Arab Emirates
48	Israel	103	United Kingdom
49	Italy	104	United States

C^4_{1211}

24	Colombia	78	Russia
30	Dominican Republic		

C^4_{1212}

9	Belarus	101	Ukraine

C^4_{1213}

4	Argentina	80	Saudi Arabia
57	Malaysia	98	Tunisia
59	Mexico	99	Turkey
77	Romania	105	Uruguay

C^4_{1214}

13	Botswana	31	Ecuador
14	Brazil	71	Panama
15	Bulgaria	86	South Africa
23	China	96	Thailand
26	Costa Rica		

C^4_{1215}

52	Kazakhstan	107	Venezuela
73	Peru		

C^4_{1216}

82	Serbia		

C^4_{1221}

1	Taiwan	49	Italy
5	Australia	50	Japan
6	Austria	64	Netherlands
10	Belgium	65	New Zealand
20	Canada	69	Norway
22	Chile	75	Poland
28	Czech Republic	76	Portugal
29	Denmark	84	Singapore
35	Finland	85	Slovakia
36	France	87	South Korea
37	Germany	88	Spain
39	Greece	91	Sweden
44	Hungary	92	Switzerland
47	Ireland	103	United Kingdom
48	Israel	104	United States

C^4_{1222}

54	Kuwait	102	United Arab Emirates

Fig. 4 (continued)

Table 4 Centroids of clusters at the category level

Category Level	Original	$c^3_{1,1,1}$	$c^3_{1,1,2}$	Original	$c^3_{1,2,1}$	$c^3_{1,2,2}$
Size	52.00	19.00	33.00	58.00	26.00	32.00
AF1	35.83	47.66	29.02	77.06	63.48	88.10
AF2	43.38	68.95	28.65	96.13	91.57	99.84
AF3	4.22	7.94	2.08	40.25	20.83	56.03
AF4	75.21	71.53	77.33	79.80	79.62	79.95
AF5	38.94	50.00	32.58	88.36	75.00	99.22
AF6	34.13	50.00	25.00	81.47	60.58	98.44
AV5	34.40	37.12	33.00	63.01	48.08	75.14
AV6	13.94	15.79	29.02	52.16	29.81	70.31
AV7	64.45	63.28	32.83	62.05	66.21	58.68
AV8	67.16	77.94	12.88	83.74	78.90	87.67
QU1	34.27	44.55	65.13	75.44	64.07	84.69
QU2	88.81	92.98	60.96	97.98	96.78	98.96
AV1	41.04	55.84	28.35	89.93	83.95	94.78
AV2	6.25	6.58	86.41	22.80	17.31	27.26
AV3	46.69	63.29	32.53	83.82	75.76	90.37
AV4	75.85	74.13	6.06	70.31	68.33	71.91
QU3	36.44	39.33	37.13	58.12	50.51	64.30
QU4	20.67	31.12	76.85	65.21	51.41	76.41
QU5	45.00	57.44	34.78	81.00	68.50	91.16

Table 5 Centroids of clusters at the indicator level

Indicator Level	Original	$c^4_{1,1,1,1}$	$c^4_{1,1,1,2}$	$c^4_{1,1,1,3}$	Original	$c^4_{1,1,2,1}$	$c^4_{1,1,2,2}$	Original	$c^4_{1,2,1,1}$	$c^4_{1,2,1,2}$	$c^4_{1,2,1,3}$	$c^4_{1,2,1,4}$	$c^4_{1,2,1,5}$	$c^4_{1,2,1,6}$	Original	$c^4_{1,2,2,1}$	$c^4_{1,2,2,2}$
Size	19.00	10.00	4.00	5.00	33.00	19.00	14.00	26.00	3.00	2.00	8.00	9.00	3.00	1.00	32.00	30.00	2.00
AV11	50.47	46.25	41.13	66.40	31.98	33.15	30.40	61.05	54.17	72.30	71.05	52.44	61.47	55.50	79.75	79.43	84.65
AV12	50.00	50.00	25.00	70.00	28.79	31.58	25.00	90.38	33.33	100.00	100.00	100.00	83.33	100.00	100.00	100.00	100.00
AV31	78.95	100.00	0.00	100.00	42.42	0.00	100.00	100.00	100.00	100.00	100.00	100.00	100.00	100.00	100.00	100.00	100.00
AV32	27.63	25.00	31.25	30.00	18.94	21.05	16.07	45.19	33.33	25.00	53.13	52.78	25.00	50.00	73.44	73.33	75.00
AV33	48.68	52.50	43.75	45.00	24.24	25.00	23.21	54.81	58.33	37.50	62.50	55.56	41.67	50.00	82.81	81.67	100.00
QU21	73.68	100.00	100.00	0.00	39.39	21.05	64.29	96.15	100.00	100.00	100.00	100.00	100.00	0.00	93.75	100.00	0.00
QU22	100.00	100.00	100.00	100.00	100.00	100.00	100.00	96.15	100.00	50.00	100.00	100.00	100.00	100.00	100.00	100.00	100.00
QU23	89.47	90.00	100.00	80.00	87.88	89.47	85.71	92.31	100.00	0.00	100.00	100.00	100.00	100.00	100.00	100.00	100.00
QU31	57.89	60.00	37.50	70.00	36.36	50.00	17.86	78.85	66.67	100.00	100.00	50.00	100.00	100.00	100.00	100.00	100.00
QU32	16.65	18.35	13.53	15.76	9.49	10.05	8.74	35.52	32.13	38.80	36.05	35.29	37.63	30.60	53.86	69.55	47.05
QU33	30.45	25.31	40.75	32.48	41.14	42.52	39.26	26.55	16.80	26.30	34.25	25.00	25.43	12.10	29.37	28.06	49.15
QU51	100.00	100.00	100.00	100.00	84.85	84.21	85.71	100.00	100.00	100.00	100.00	100.00	100.00	100.00	100.00	100.00	100.00
QU52	74.49	77.32	71.35	71.36	38.37	38.31	38.46	84.28	74.33	98.00	89.75	83.52	68.03	98.50	98.24	98.17	99.30
QU53	68.42	65.00	87.50	60.00	25.76	21.05	32.14	90.38	100.00	100.00	93.75	94.44	50.00	100.00	100.00	100.00	100.00

Table 6 Euclidean distance between cluster centroids and the overall cluster centroid at the dimension level

Dimension Level	$c_{1,1}^{2}$	$c_{1,2}^{2}$
Size	52.00	58.00
A1	−20.12	18.04
A2	−10.79	9.67
A3	−13.57	12.17
Total	−44.48	39.88

Table 7 Euclidean distance between cluster centroids and the overall cluster centroid at the category level

Category Level	$c_{1,1,1}^{3}$	$c_{1,1,2}^{3}$	$c_{1,2,1}^{3}$	$c_{1,2,2}^{3}$
Size	19.00	33.00	26.00	32.00
AF1	−29.40	−48.04	−13.58	11.04
AF2	−27.18	−67.48	−4.56	3.71
AF3	−32.31	−38.17	−19.42	15.78
AF4	−8.28	−2.48	−0.18	0.15
AF5	−38.36	−55.79	−13.36	10.86
AF6	−31.47	−56.47	−20.89	16.97
AV5	−34.09	−57.40	−5.98	4.86
AV6	−16.22	−16.74	−5.49	4.46
AV7	−20.53	−46.69	−8.06	6.55
AV8	−1.73	0.99	−7.52	−3.94
QU1	−25.89	−30.18	−14.93	12.13
QU2	−36.37	−39.28	−22.35	18.16
AV1	−1.17	0.68	1.75	−5.78
AV2	−5.80	−22.78	−4.84	3.93
AV3	−30.90	−47.09	−11.38	9.24
AV4	−5.00	−11.57	−1.20	0.97
QU3	−18.79	−23.35	−7.61	6.18
QU4	−34.09	−50.55	−13.79	11.21
QU5	−23.57	−43.17	−12.50	10.16
Total	−421.13	−655.54	−185.89	136.63

Compared to Chen et al. (2019), our suggested approach also provides path information on how to manage and improve food security performance. For instance, a total of 10 countries, including India and Sri Lanka, are partitioned in $c_{1,1}^{2}$ at the dimension level, $c_{1,1,1}^{3}$ at the category level, and $c_{1,1,1,1}^{4}$ at the indicator level, as shown in Fig. 4. From Table 8, countries from $c_{1,1,1,1}^{4}$ performed satisfactorily in AV31, QU21, QU22 and QU51 but not satisfactorily in the other global food security indicators. Within the group of $c_{1,1,1,1}^{4}$, $c_{1,1,1,2}^{4}$ and $c_{1,1,1,3}^{4}$, each country could learn from peers that perform favorably in a particular indicator(s). Since the Euclidean distance of selected food security indicators in $c_{1,1,1,1}^{4}$ is larger than those in $c_{1,1,1,2}^{4}$ or $c_{1,1,1,3}^{4}$, as shown in Table 8, countries from $c_{1,1,2,1}^{4}$ should focus and

Table 8 Euclidean distance between cluster centroids and the overall cluster centroid at the indicator level

Indicator Level	$c^4_{1,1,1,1}$	$c^4_{1,1,1,2}$	$c^4_{1,1,1,3}$	$c^4_{1,1,2,1}$	$c^4_{1,1,2,2}$	$c^4_{1,2,1,1}$	$c^4_{1,2,1,2}$	$c^4_{1,2,1,3}$	$c^4_{1,2,1,4}$	$c^4_{1,2,1,5}$	$c^4_{1,2,1,6}$	$c^4_{1,2,2,1}$	$c^4_{1,2,2,2}$
Size	10.00	4.00	5.00	19.00	14.00	3.00	2.00	8.00	9.00	3.00	1.00	30.00	2.00
AV11	−33.50	−38.63	−13.35	−46.61	−49.35	−25.59	−7.45	−8.70	−27.31	−18.29	−24.25	−0.33	4.90
AV12	−50.00	−75.00	−30.00	−68.42	−75.00	−66.67	0.00	0.00	0.00	−16.67	0.00	0.00	0.00
AV31	0.00	−100.00	0.00	−100.00	0.00	0.00	0.00	0.00	0.00	0.00	0.00	0.00	0.00
AV32	−48.44	−42.19	−43.44	−52.38	−57.37	−40.10	−48.44	−20.31	−20.66	−48.44	−23.44	−0.10	1.56
AV33	−30.31	−39.06	−37.81	−57.81	−59.60	−24.48	−45.31	−20.31	−27.26	−41.15	−32.81	−1.15	17.19
QU21	3.85	3.85	−96.15	−75.10	−31.87	3.85	3.85	3.85	3.85	3.85	−96.15	3.85	−96.15
QU22	0.00	0.00	0.00	0.00	0.00	0.00	−50.00	0.00	0.00	0.00	0.00	0.00	0.00
QU23	−10.00	0.00	−20.00	−10.53	−14.29	0.00	−100.00	0.00	0.00	0.00	0.00	0.00	0.00
QU31	−40.00	−62.50	−30.00	−50.00	−82.14	−33.33	0.00	0.00	−50.00	0.00	0.00	0.00	0.00
QU32	−35.51	−40.34	−38.10	−43.82	−45.12	−21.73	−15.06	−17.81	−18.57	−16.23	−23.26	7.78	−6.81
QU33	−15.83	−0.39	−8.66	1.38	−1.88	−24.34	−14.84	−6.89	−16.14	−15.71	−29.04	−13.08	8.01
QU51	0.00	0.00	0.00	−15.79	−14.29	0.00	0.00	0.00	0.00	0.00	0.00	0.00	0.00
QU52	−20.92	−26.89	−26.88	−59.94	−59.78	−23.91	−0.24	−8.49	−14.72	−30.21	0.26	−0.07	1.06
QU53	−35.00	−12.50	−40.00	−78.95	−67.86	0.00	0.00	−6.25	−5.56	−50.00	0.00	0.00	0.00
Total	−315.68	−433.66	−384.41	−657.96	−558.53	−256.31	−277.51	−84.93	−176.37	−232.84	−228.71	−3.11	−70.26

learn about the global food security indicators AV32, QU23, QU33 and Q53 from $c^4_{1,1,1,2}$ and AV11, AV12, and QU51 from $c^4_{1,1,1,3}$.

Upon completion of their internal learning process within the same group of clusters, countries from $c^4_{1,1,1,1}$ should next consider the other group of clusters that showed better achievement in the global food security performance at the indicator level. For instance, $c^4_{1,1,1,1}$ should consider how to improve the food security indicators AV11, AV12, AV32, AV33, QU31, QU32, QU33, UQ52 and UQ53 from a group including $c^4_{1,2,1,1}$, $c^4_{1,2,1,2}$, $c^4_{1,2,1,3}$, $c^4_{1,2,1,3}$, $c^4_{1,2,1,4}$, $c^4_{1,2,1,5}$ and $c^4_{1,2,1,6}$. A similar learning process will continue as shown in Table 6, where $c^4_{1,1,1,1}$ could learn from the groups of $c^3_{1,2,2,1}$ and $c^3_{1,2,2,2}$ to improve their food security indicators, i.e., AV11, AV32, AV33, QU31, QU32, UQ52 and UQ53. Similar path information for other groups of clusters can be obtained from Fig. 4 and Tables 6, 7 and 8. Such path information cannot be obtained when country grouping is carried out using personal judgement.

5 Conclusions

This chapter provides an application of big data, data science, and data analytics methods in the H-DEA framework for the classification of high-dimensional data. The results and discussions of the GFSI using the H-DEA model are reproduced from Chen et al. (2019), in which the structure is classified into four levels, namely goal, dimension, category and indicator. Before making a decision, all the data were processed by H-DEA. From the H-DEA analysis, the aggregation of the sublevel efficiency indexes into an overall composite index efficiency measure for each country is obtained. The scoring results are developed as the inputs for K-Means clustering based on the data characteristics. Unlike in conventional approaches, we observe that our classification based on a hierarchical view with scoring clustering approaches is more appropriate for the government and policy makers in monitoring the food security standard.

In general, the H-DEA-based clustering approach provides several contributions to practical applications. Under theoretical reasoning and decision-making, policy makers cannot easily obtain all the data needed to make decisions due to data constraints. Under the scoring clustering approach, the results will help policy makers to understand the benchmarking process and the learning path. Overall, we made use of "significant" information to support better decisions based on our suggested clustering approach. With clustering information, policy makers within clusters can collaborate to make decisions and design policies to overcome food security issues in term of affordability, availability, quality and safety. Furthermore, each country can identify the most appropriate peers within the same cluster. H-DEA provides a benchmark for each observation and can be used to guide observations regarding how to learn and who to learn from in either internal or external groups. Consequently, a specific group can help a particular country change its weighting and performance gaps through similar recipes.

There are a few directions that can be suggested for future studies. The first direction is to assess the extension of the H-DEA model with a dynamic structure. Overall, composite indexes may be biased when unusual observations occur in a single period and are used to evaluate each level's weights and scores in a dynamic system. An extension to multiple periods may be more significant and provide useful insights. The second direction is to adopt our proposed model to evaluate other systems' performance to help managers identify improvement areas for each DMU. Third, the approach to determining the optimal number of groups for each level still has several gaps requiring improvement. One could take this issue into consideration in future studies.

References

Ball, G. H., & Hall, D. J. (1965). *ISODATA, a novel method of data analysis and pattern classification*. Technical report.

Bholowalia, P., & Kumar, A. (2014). EBK-means: A clustering technique based on elbow method and K-means in WN. *International Journal of Computer Application, 105*(9), 17–24.

Charnes, A., Cooper, W. W., & Rhodes, E. (1978). Measuring the efficiency of DMUs. *European Journal of Operational Research, 2*(6), 429–444. https://doi.org/10.1016/0377-2217(78)90138-8

Chen, M., Mao, S., & Liu, Y. (2014). Big data: A survey. *Mobile Networks and Applications, 19*(2), 171–209. https://doi.org/10.1007/s11036-013-0489-0

Chen, P. C., Yu, M. M., Shih, J. C., Chang, C. C., & Hsu, S. H. (2019). A reassessment of the global food security index by using a hierarchical data envelopment analysis approach. *European Journal of Operational Research, 272*(2), 687–698. https://doi.org/10.1016/j.ejor.2018.06.045

Cook, W. D., & Kress, M. (1990). A data envelopment model for aggregating preference rankings. *Management Science, 36*(11), 1302–1310. https://doi.org/10.1287/mnsc.36.11.1302

Demchenko, Y., Grosso, P., Laat, C., & Membrey, P. (2013). Addressing big data issues in scientific data infrastructure. In *IEEE 2013 international conference on collaboration technologies and systems* (pp. 48–55). http://dx.doi.org/10.1109/CTS.2013.6567203.

Emrouznejad, A., & Amin, G. R. (2009). DEA models for ratio data: Convexity consideration. *Applied Mathematical Modelling, 33*(1), 486–498. https://doi.org/10.1016/j.apm.2007.11.018

Fernandez-Castro, A., & Smith, P. (1994). Towards a general non-parametric model of corporate performance. *Omega, 22*(3), 237–249. https://doi.org/10.1016/0305-0483(94)90037-X

Kao, C. (2010). Weight determination for consistently ranking alternatives in multiple criteria decision analysis. *Applied Mathematical Modelling, 34*(7), 1779–1787. https://doi.org/10.1016/j.apm.2009.09.022

Kao, C., & Hung, H. T. (2008). Efficiency analysis of university departments: An empirical study. *Omega, 36*, 653–664. https://doi.org/10.1016/j.omega.2006.02.003

Kashima, H., Tsuda, K., & Inokuchi, A. (2003). Marginalized kernels between labeled graphs. In *proceedings of the twentieth international conference on machine learning* (pp. 321–328).

Krivonozhko, V. E., Førsund, F. R., & Lychev, A. V. (2015). Terminal units in DEA: Definition and determination. *Journal of Productivity Analysis, 43*(2), 151–164. https://doi.org/10.1007/s11123-013-0375-6

Lloyd, S. (1982). Least squares quantization in PCM. *IEEE Transactions on Information Theory, 28*(2), 129–137. https://doi.org/10.1109/TIT.1982.1056489

Loshin, D. (2013). *Big data analytics: From strategic planning to Enterprise integration with tools, techniques, NOSQL, and graph*. Elsevier.

MacQueen, J. (1967). Some methods for classification and analysis of multivariate observations. In L. Lecam & J. Neyman (Eds.), *Proceedings of the fifth Berkeley symposium on mathematical statistics and probability (pp. 281–297)*. University of California Press.

Olesen, O. B., Petersen, N. C., & Podinovski, V. V. (2015). Efficiency analysis with ratio measures. *European Journal of Operational Research, 245*(2), 446–462. https://doi.org/10.1016/j.ejor.2015.03.013

Popović, M., Savić, G., Kuzmanović, M., & Martić, M. (2020). Using data envelopment analysis and multi-criteria decision-making methods to evaluate teacher performance in higher education. *Symmetry, 12*, 563. https://doi.org/10.3390/sym12040563

Saaty, T. L. (1980). *The analytic hierarchy process*. McGraw-Hill.

Steinhaus, H. (1956). Sur la division des corp materiels en parties. *Bulletin de L'Academie Polonaise Des Sciences, IV*(12), 801–804.

Syajur, M. A., Khoutimah, B. K., Rochman, E. M. S., & Satoro, B. D. (2018). Integration K-means clustering method and elbow method for identification of the best customer profile cluster. *IOP Conference Series: Material, Science and Engineering, 336*, 012017.

Zhu, J. (2020). DEA under big data: Data enabled analytics and network data envelopment analysis. *Ann. Oper. Res.* https://doi.org/10.1007/s10479-020-03668-8

Dominance Network Analysis: Hybridizing Dea and Complex Networks for Data Analytics

L. Calzada-Infante and S. Lozano

1 Introduction

Data Envelopment Analysis (DEA) is a non-parametric technique aimed at benchmarking so-called Decision Making Units (DMUs). DEA has been mainly used in the production and service context so that DMUs are assumed to represent homogeneous operating units that consume inputs and produce outputs. However DEA can also be used to benchmark the performance of countries in the Summer Olympics (Lozano et al., 2002), human development index of countries (Lozano & Gutiérrez, 2008), eco-efficiency of products (Lozano et al., 2011), forest fire fighting (Gutiérrez & Lozano, 2013), F1 constructors (Gutiérrez & Lozano, 2014), NBA teams (e.g. Moreno & Lozano, 2014), team performance in a business game (Koltai et al., 2017), public finances (Moreno & Lozano, 2018), current account balance of payments of countries (Gutiérrez & Lozano, 2020a), F1 circuits (Gutiérrez & Lozano, 2020b), and many other types of entities not related to the classical production process context.

The fact is that, in the end, and as other researchers have also argued (most recently Zhu, 2020), DEA could as well be read as Data Envelopment Analytics. Thus, using a simple frontier analysis, DEA is able to benchmark multidimensional datasets (for convenience we may keep the term DMU for each observation). DEA is a quantitative technique and, hence, in principle, the attributes should be numeric, although methods for handling interval and ordinal data (Zhu, 2003, 2004) as well as fuzzy data (Hatami-Marbini et al., 2011; Emrouznejad et al., 2014) also exist.

L. Calzada-Infante
Facultad de Ciencias Económicas y Empresariales, Universidad de León, León, Spain
e-mail: calzadalaura@uniovi.es

S. Lozano (✉)
Department of Industrial Management, University of Seville, Seville, Spain
e-mail: slozano@us.es

J. Zhu, V. Charles (eds.), *Data-Enabled Analytics*, International Series in Operations Research & Management Science 312, https://doi.org/10.1007/978-3-030-75162-3_9

Another feature of DEA is that, although approaches with non-monotonic variables (Villa & Lozano, 2016) and dual-role factors (Toloo et al., 2018) exist, in principle the attributes must be either positive or negative. Positive attributes are those that the larger, the better, while for negative attributes it is the opposite. The negative attributes are can be called inputs while the positive attributes can be called outputs. As mentioned above, this terminology does not necessarily imply that the DMUs transform the former into the latter. Also, some inputs or outputs can be fixed (also called non-discretionary) so that the DMU could not possibly modify the observed value. This fact can be appropriately taken into account in DEA.

A key concept in DEA is dominance. A DMU A is said to strictly dominate another B if its positive attributes are higher and its negatives attributes are lower, that is, if A is better than B in all dimensions. If the attributes of A are as good as those of B and better in at least one dimension then A is said to dominate B (in a weak sense). A DMU is said to be inefficient if it is dominated (i.e. outperformed) by any other DMU. Conversely, a non-dominated DMU is said to be efficient.

When a DMU A dominates another DMU B the difference between the discretionary inputs and outputs of both gauges the difference in their performance. The efficient DMUs represent the best practice and as such are the benchmarks (reference units in DEA parlance) for the DMU they dominate. A useful way to model and visualize the dominance relationships between the DMUs is through a Dominance Network (DN). A DN is a directed weighted graph whose nodes are the different DMUs and there is an arc from a node B to a node A if A dominates B. The weight of that arc is related to the difference in the performance of the two nodes so that the larger that difference the larger the weight.

Since there cannot be cycles (i.e. it is directed acyclic graph, DAG), a DN has a special layered structure. Thus, the layer 0 is formed by efficient DMUs, which by definition are non-dominated and hence have no outgoing arcs. Layer 1 is formed by those DMUs that are dominated only by DMUs in layer 0. Similarly, DMUs in layer λ are dominated by DMUs in layer λ-1 and lower.

Due to the transitive nature of dominance relationships, (i.e. if C is dominated by B and B is dominated by A, then C is dominated by A), if a DMU belongs to a certain layer λ, then it has arcs towards one or more DMUs in layer λ-1 but also to all other layers up to layer 0. We can define the length of a path as the sum of the lengths of its arcs and we can use the distance from a node to the efficient frontier as a measure of its relative efficiency. However, as it is common in DEA, the distance from a node to the efficient frontier depends on the efficient DMU used as reference. Thus, a closest as well as a most-distant target can be identified.

Using this DN modelling all the tools that have been developed in the last decades for social and complex networks analysis (see, e.g. Newman, 2002, 2010) can be applied to process multidimensional datasets. This approach has been labelled DN analysis (Lozano & Calzada-Infante, 2018; Calzada-Infante & Lozano, 2020) and has been applied to Olympics (Calzada-Infante & Lozano, 2016) as well as to airlines performance (Lozano & Calzada-Infante, 2019). The strengths of this approach derive from the power of the network paradigm, which is common in data analytics, and has been used in DEA by many researches. A recent example of this is the use of Decision Trees, a well-known machine learning method, to estimate DEA production frontiers (Esteve et al., 2020).

Apart from DN analysis, different network-based approaches have already been used in DEA. This should not be confused with network DEA, in which it is the inner structure of the DMU that is modelled as a network of processes. Thus, for example, the context-dependent DEA approach of Seiford and Zhu (2003) classify the inefficient DMUs in different layers by a process called stratification that is similar to that carried out in DN. Simialrly, Ghahraman and Prior (2016) apply Dijkstra's shortest path algorithm over a network of DMUs to find the optimal stepwise improvement path from the inefficient DMUs to the Efficient Frontier (EF) reflecting the preferences of the decision maker in the weight of the edges. These weights consider the input similarities, an exponential penalty for long steps and the risk of failing to close efficiency gap between DMUs.

Liu et al. (2009) proposed a weighted network approach to rank the DMUs by measuring the importance of each reference unit by its eigenvector centrality. The weight of the edges is based on the optimal intensity variables computed by DEA for all possible input/output combinations. Liu and Lu (2010) apply this network-based approach to gauge the performance of Taiwan's government-supported research institutes and Ho et al. (2014) use it to assess the performance of technology transfer efficiencies in US universities. Other authors have applied this network-based approach with other centrality measures, such as Alpha-centrality (Liu & Lu, 2012), PageRank (Leem & Chun, 2015) and HITS (Simon de Blas et al., 2018). Hermoso et al. (2019) also applied the network-based approach with eigenvector centrality to rank European airlines from different points of view attending to their resource management, their position in the airlines network and in the social media. Lee and Choi (2019) applied closeness centrality and network weights based on the optimal intensity variables and the efficiency score of a super-efficiency DEA model. Finally, Kao et al. (2017) carry out an exploratory study about which complex networks metrics are correlated with the productive efficiency of firms and their corresponding level in a supply chain network, while Wang et al. (2019) use truncated-regression with bootstrap to assess the impact of interlock directorates, measured by centrality metrics and structural holes, and with the firm performance evaluated by the dynamic-slacks based measure DEA model.

From the above literature review, it can be concluded that it is possible to carry out multidimensional data benchmarking using DEA and complex networks analysis and that this marriage complies with the 3Vs (volume, variety and velocity) demanded by big data and discussed in Zhu (2020). Thus, as regards Volume, the network paradigm allows processing and modelling large-scale data sets. The size of the network depends on the number of observations (which is equal to the number of nodes) and on the number of input and output features and, in fact, the number of dominance relationships (and hence the density of the DN) increases as the number of observations grow. In other words, not only DN analysis can process large volumes of data but also its discriminating power actually increases as the number of observations increases. As regards Variety, the large number of DEA applications in many different sectors (education, transportation, banking, health care, energy, sports, all kinds of services, agriculture and farming, supply chains, etc.) attests to its capability to process a large variety of variables of different kinds. And, as regards velocity, the network paradigm allows a fast and efficient update of the information available as well as its quick processing. Finally, let us also say a word about the fourth V word, Value, mentioned in Zhu (2020). The frontier analysis carried out by DEA is different to the frequent statistics-based data analytics methods. Thus, it is based on identifying the best practices, not the central tendencies (as, for example, regression or neural networks do), and, instead of measuring deviations with respect to those central measures, it computes deviations (i.e. performance shortfalls) with respect to that best practice. In other words, DEA aims for excellence in performance and helps to identify and to promote it. It is in this sense that we claim that DEA is valuable data analytics tool.

Having made the above general reflections on the advantages and usefulness of hybridizing DEA and complex networks, in this chapter we specifically focus on DN analysis (DNA). Section 2 presents the basic concepts and the methodology while Sect. 3 applies it to analyse the data on the COVID-19 pandemic at the country level. The last Sect. 4 draws conclusions and discusses the limitations of the methodology as well as topics for further research.

2 DNA: Methodology

The aims of the DNA methodology are threefold. One is to model the whole dataset and its interrelationships. Thus, the DN represents the data as a whole, with its layer structure and distance distribution. Moreover, the transitivity of the network means that the DN has a skeleton, the subgraph formed only by the arcs between consecutive layers, that, without any information loss, is equivalent to the whole network. The second aim is to identify and characterize the role of the different observations. In particular, the best practice observations are distinguished from the inefficient ones. For the former, different indexes measuring how many inefficient observations can have it as reference and the total inefficiency of these observations are can be computed. For each inefficient observation, apart from the layer it belongs, the margin for improvement measured in terms of the distance to the EF can be gauged, among other measures. The third aim is to use the available network visualization algorithms to provide a picture of the dataset and its structure. As the saying goes, one picture is worth a thousand words.

Consider a dataset of n observations (DMUs), each one with its corresponding features. These features are classified into m "negative" attributes (inputs) and s "positive" ones (outputs). Let the input and output vectors of DMU j be $x_j = (x_{1j}, x_{2j}, \ldots x_{mj})$ and $y_j = (y_{1j}, y_{2j}, \ldots y_{sj})$ respectively. The *dominance relationships* among DMUs are represented in DNA as directed weighted edges, whose weights are the relative performance (i.e. the efficiency gap) between the two connected DMUs. Let D(r) be the set of DMUs that dominate a given DMU r, i.e. $D(r) = \{j \neq r : x_{ij} \leq x_{ir} \quad \forall\, i, \quad y_{kj} \geq y_{kr} \quad \forall\, k\} \subset D$. The dominance network

(D, E) is composed by a set of nodes D, which are the DMUs, and a set edges $E = \{(r, j) : r \in D \wedge j \in D(r)\}$, which correspond to the dominances relationships between the DMUs. Note that the edges are directed. Actually, these dominance relationships are anti-symmetric so that if there exists the edge (r, j) then there cannot exist the reciprocal edge (j, r). Mathematically, $(r, j) \in E \Rightarrow (r, j) \notin E$. Note also that the dominance relationship has the transitivity property. Thus, if DMU p dominates DMU r, and, in turn, DMU j dominates DMU p, then DMU j dominates DMU r, i.e. $p \in D(r) \wedge j \in D(p) \Rightarrow j \in D(r)$.

The weight e_{rj} associated to an edge (r, j) represents the relative inefficiency (i.e. underperformance) of DMU r with respect to DMU j. Different ways to define the edge weights can be used (Lozano & Calzada-Infante, 2017, 2018). In this chapter we consider the Normalized Additive Inefficiency (NAI) metric, defined by $e_{rj} = \begin{cases} 0 & \text{if } j \notin D(r) \\ \sum_i \frac{x_{ir} - x_{ij}}{g_i^x} + \sum_k \frac{y_{kj} - y_{kr}}{g_k^y} & \text{if } j \in D(r) \end{cases}$. This weight definition is units invariant as it is normalized by the vector $g = \left(g_i^x, g_k^y\right)$. There are multiple ways to choose the normalizing constants. One would be $g = \left(R_i, \hat{R}_k\right)$ where $R_i = x_i^{max} - x_i^{min} \quad \forall i$ and $\hat{R}_k = y_k^{max} - y_k^{min} \quad \forall k$ are the range of the input and output variables. Another option is to use the average of the input and output variables, $g = \left(x_i^{aver}, y_k^{aver}\right)$ where $x_i^{aver} = \frac{1}{n} \sum_{j \in D} x_{ij} \quad \forall i$ and $y_k^{aver} = \frac{1}{n} \sum_{j \in D} y_{kj} \quad \forall k$. None of the above alternatives, however, are recommended when we have data on different periods since these normalization coefficients would be different in each time period unless the range or the average of the observations are computed over the whole horizon. Lozano and Calzada-infante (2018) studied the influence of using different normalization vectors and how this affects the DN metrics. It concluded there is a high correlation among them, although the absolute magnitudes of some of the node indexes differ. Note that NAI can be adapted to the case in which some of the inputs or outputs are non-discretionary by excluding those dimension from the calculation of the edge weights e_{rj}. It can also reflect the preferences of the Decision Maker as regards the importance of the different input and output dimensions.

An interesting consequence of the use of the same normalization constants for the weights of all the edges when the NAI metric is used is the additive property. Thus, if $p \in D(r)$ and $j \in D(p)$, then $e_{rp} + e_{pj} = e_{rj}$. As a consequence of this property, the distance between any pair of nodes is independent of the path followed. The result is that all the paths are geodesics.

The DN is not always connected. As the DN is a directed acyclic network there are no strongly connected components. The DMUs can be grouped, however, into weakly connected components. If two DN nodes belong to the same weakly connected component, then there is a path between them in the corresponding undirected network. Each component *c* is a subgraph (D_c, E_c) composed by the set of nodes $D_c \subseteq D$ and the set of edges that connect them, $E_c = \{(r, j) \in E : r \in D_c \wedge j \in D_c\} \subseteq E$. Note that each edge and each DMU only can belong to one such component. The set of efficient DMUs is defined as $D^* = \{r : D(r) = \varnothing\}$ and it is formed by the DMUs that are not dominated by any other DMU. Given a DMU r the set of DMUs it dominates is defined as $D^{-1}(r) = \{j \neq r : x_{ij} \geq x_{ir} \ \ \forall i \ \ y_{kj} \leq y_{kr} \ \ \forall k\} \subset D$. In particular, for an efficient DMU j one can define $D^{--1}(j) = \{r : r \in D^{-1}(j) \wedge D(r) \cap D^* = \{j\}\} \subseteq D^{-1}(j)$, which is the set of (inefficient) DMUs that are dominated only by DMU j and thus have it as its only efficient reference unit.

Another important attribute of each node in the DN is the layer it belongs. These can be computed as in the stratification of Seiford and Zhu (2003). Thus, let NL_c be the number of layers of component c and $\lambda(r) = \begin{cases} 0 & \text{if } D(r) = \varnothing \\ 1 + \max\limits_{j \in D(r)} \lambda(j) & \text{otherwise} \end{cases}$ the layer to which DMU r belongs. By definition, the nodes in layer 0 are the efficient, i.e. non-dominated, ones.

Lozano and Calzada-infante (2018) has shown how many of the usual complex networks characterization measures can be directly applied to the DN scenario. Thus, at node level, for each DMU r, its In-degree, $d_r^{in} = \left|D^{-1}(r)\right|$, corresponds to the number of DMUs that it dominates while the Out-degree, $d_r^{out} = |D(r)|$, corresponds to the number of DMUs that dominate it. The latter is always lower than the *number of efficient benchmarks* (i.e. reference units) of DMU r, $|D^*(r)|$. For each DMU r, the length of the paths that reach the EF is in the range between the *maximum distance to EF* $e_r^{max} = \max\limits_{j \in D^*(r)} e_{rj}$ and the *minimum distance to EF*, $\tau_r^{min} = \min\limits_{j \in D^*(r)} e_{rj}$ Other more complex measures and indexes (like the clustering coefficient or the betweenness centrality) can also be computed.

For the efficient DMU $j \in D^*$ the corresponding In-degree, $|D^{-1}(j)|$, can be interpreted as a *benchmarking count*, i.e. for how many inefficient observations it can act as reference unit. Moreover, the *benchmark necessity* index, $|D^{--1}(j)|$, indicates the number of DMUs for which j is their only benchmark. The total strength of its dominance relationship, labelled *benchmarking potential index*, $\kappa_j = \sum\limits_{r \in D^{-1}(j)} e_{rj}$, measures the total inefficiency of the DMUs it dominates. If all these inefficient DMUs closed their efficiency gap with respect to DMU j this is the total efficiency improvement that would be achieved. Dividing this benchmarking potential index by the In-degree, we get the *average path length from the DMUs dominated by j*, χ_j, that can be compared with the average path length of the network. Another interesting measure that can be defined for efficient DMUs are the maximum distance from the dominated DMUs, called *Inefficiency radius*, $\sigma_j = \max\limits_{r \in D^{-1}(j)} e_{rj}$.

At the layer level, the percentage of DMUs that belongs to each layer is of interest as it provides information about the depth/shallowness of the DN. At the component level, for each component (D_c, E_c) both the percentage of nodes and edges that belong to the component as well as its share of efficient nodes, $\left|D_c^*\right| / |D^*|$, can be computed. This gives an idea about the relative size of each of these subgraphs with respect to the complete graph. The percentage of efficient nodes $\pi_c = \left|D_c^*\right| / |D_c|$

provides information cc subgraph. The diameter of the component, $\Delta_c = \max_{r,j \in D_c} e_{rj}$, measures the maximum inefficiency detected in the subgraph, while the *average distance of the component*, $\theta_c^{aver} = \frac{1}{|D_c|} \sum_{r \in D_c} e_r^{max}$, measures the average of the distances to the EF. Moreover, the degree distribution can be computed at the component level and the *average degree of the component*, $d_c^{aver} = \frac{1}{|D_c|} \sum_{r \in D_c} d_r^{in} = \frac{1}{|D_c|} \sum_{r \in D_c} d_r^{out}$ can be determined. The average degree allows assessing the density of the component, $\rho_c = \begin{cases} 1 & \text{if } |D_c| = 1 \\ \frac{d_c^{aver}}{|D_c|-1} & \text{otherwise} \end{cases}$, which is generally low (as there are no reciprocal arcs nor arcs between nodes in the same layer). Also, a measure of the connectivity among neighbours can be obtained computing the average clustering coefficient of each component, $CC_c^{aver} = \frac{1}{|D_c|} \sum_{r \in D_c} CC_r$.

At the network level, some metrics applied over the components such as diameter (Δ_c), the average distance to EF (θ_c^{aver}) and the percentage of efficient nodes (π_c) can be applied directly over the whole graph. Moreover, the number of components NC shows the large clusters of observations. Note that these clusters are defined based on dominance relationships, not on the levels of the observations attributes as it is generally done. It is important to highlight that each isolate node, that may be a maverick or an outlier, forms a single-node component.

As regards the graphical visualization of the DN that represents the multidimensional dataset, most complex network software packages provide tools and algorithms to visualize and work with networks. In the next section, some graphs generated with Gephi and with the R package igraph are shown. In general, a given network can be visualized in different ways. For instance, the DN can be visualized as a bipartite graph keeping only the dominance relationships between the inefficient DMUs and the efficient ones, so the nodes are split in two groups, the DMUs which are over the EF and those which are not. This is equivalent to pooling together (and ignoring the arcs between) all the nodes that do not belong to layer 0. A more informative visualization, however, is the one that shows all the different layers of the network and the arcs between them.

For large networks, although the overall density of the network is low, the graphical representation can still be cluttered. Multiple filters can be applied over the networks to visualize subgraphs that focus on specific sections of the DN (see Lozano & Calzada-Infante, 2018). Perhaps the most important filter is the *skeletonisation filter*, which removes all the transitive edges in the network. This does not lose any information because the removed arcs can be recovered at any time using the transitivity property. Another important filter is the *egonetwork filter*. Applied to a specific node it provides all the dominance relationships that involve that node and the nodes that dominate it or that it dominates.

These are, in summary, the basic elements of the DNA methodology. Extensions to the common Variable Returns to Scale (VRS) and Constant Returns to Scale (CRS) DEA technologies, which assume the convexity of the feasible region, the production possibility set in DEA parlance, have been studied in Calzada-Infante and Lozano (2020). Also, assuming that inputs and/or output prices are available, a multigraph DN approach can be used to evaluate two different types of dominance relationships: technical and economic, the latter based on the difference of cost, revenue or profit efficiency. Technical dominance implies economical dominance, but the reverse is not true. The difference between the economic inefficiency and the technical efficiency, called allocative inefficiency, is represented by a hyperedge that connects three nodes: the inefficient DMU, its technical efficient benchmark and its economic benchmark (see Lozano & Calzada-Infante, 2017).

3 DNA: Application to COVID-19 Data

This section aims at illustrating and showing the potential of DNA. The observations in this dataset correspond to 155 countries (labeled by their ISO 3166-1 alpha-2 code). The dataset can be obtained from the corresponding author upon request and involves four variables of which three are "negative" (i.e. inputs or undesirable outputs), namely Current health expenditure per capita in terms of PPP in 2017 (source WorldBank, 2020), total COVID-19 cases per million and total COVID-19 deaths per million (source Ritchie et al., 2020). There is also one "positive" variable, namely Life expectancy in 2020 (source World Population Review, 2020).

Note that the negative character of the health care expenditure variable should be interpreted in terms of the efficiency of that expenditure, i.e. the relative best practice is that of countries that even with limited expenditures in health care have been able to contain the spread of the pandemic and have small number of cases and deaths. Another way to put it is that a dominance relationship can only be established with a comparable DMU. Thus, to test whether your performance is good in relative terms you can compare it with that of a poorer country not the opposite.

The health care expenditure input, however, as well as the single output considered, is assumed non-discretionary and, hence, the potential reduction of health care expenditure per capita (same as the potential life expectancy increase) is not taken into account when computing the NAI edge lengths. Therefore, in the end, the inefficiency gap between any two countries, one of which dominates the other, measures only the corresponding differences in the number of COVID-19 cases and deaths. The normalization coefficients considered for the total COVID-19 cases and deaths are 100 and 1, respectively. This implicitly assumes that there is one death for every 100 cases, i.e. an approximate case fatality rate of 1%.

The aim is to assess and benchmark the effects of the pandemic in these 155 countries. Two different sets of observations are considered, in two different time periods, 06/01/2020 and 08/01/2020. The corresponding DNs are called DNJune and DNAugust, respectively.

As regards DNJune, the whole network can be visualized, using the skeleton filter and a layer layout, in Fig. 1. The network has 4 components: one giant component and three isolate DMUs (Central African Republic, Lesotho and Spain). These three

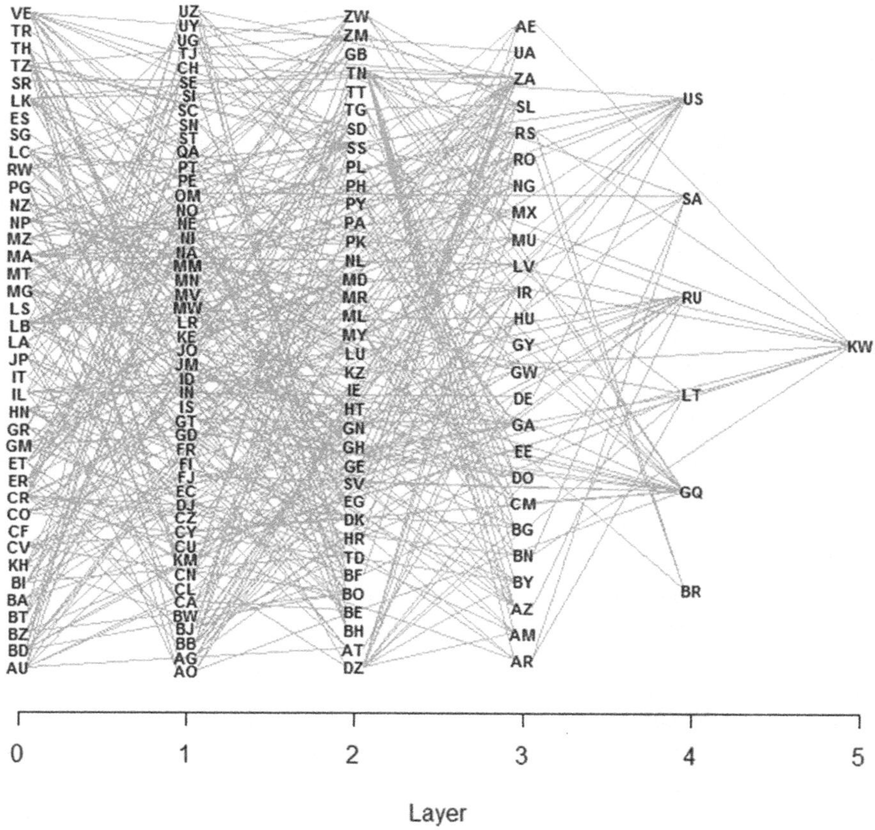

Fig. 1 Visualization of DNJune using a skeletonization filter and layer layout

countries do not dominate nor are dominated by any other country and hence belong to the EF. However, they are mavericks and their data cannot be benchmarked against those of the other countries. The giant component has 152 countries and the 829 edges and its average clustering coefficient is 0.35. The density of the giant component is 3.56%, which corresponds to an average degree 5.42. Countries are split in 6 layers as can be observe in Table 1. The number of efficient countries (i.e. in layer 0) of the giant component is 36 countries. In the whole network 25.2% of the countries are efficient, 31.0% belong to layer 1 and the 23.2% to layer 2. The average distance to EF within the giant component is 1.24 while the diameter is 15.89. The latter corresponds to the path from Qatar (layer 1) to Costa Rica (layer 0). This shows that the longest path does not necessarily start in the deepest layer and that the most inefficient node does not always belong to that deepest layer.

Table 1 shows the average In- and Out-Degree and the maximum and minimum distance to EF increase for each layer. It can be seen that the Out-degree increases

Table 1 Layer distribution of networks DNJune and DNAugust

DNJune		d_r^{in}			d_r^{out}			e_r^{max}			τ_r^{min}		
Layer	# DMUs	max	mean	min	max	mean	min	max	mean	min	max	mean	min
0	39	34	13.37	0	–	–	–	–	–	–	–	–	–
1	48	24	4.24	0	5	1.84	1	15.89	1.58	0.00	15.89	1.43	0.00
2	36	19	2.72	0	14	6.39	3	12.12	1.70	0.01	5.51	0.86	0.00
3	25	3	0.56	0	47	13.08	4	3.84	1.15	0.05	3.23	0.76	0.03
4	6	1	0.17	0	55	25.33	15	7.34	2.80	0.71	5.60	1.80	0.10
5	1	–	–	–	30	30.00	30	5.56	5.56	5.56	3.69	3.69	3.69
DNAugust		d_r^{in}			d_r^{out}			e_r^{max}			τ_r^{min}		
Layer	# DMUs	max	mean	min	max	mean	min	max	mean	min	Max	mean	min
0	42	36	12.19	0	–	–	–	–	–	–	–	–	–
1	43	27	5.09	0	7	1.95	1	10.03	1.33	0.00	10.00	1.23	0.00
2	37	21	2.78	0	20	6.32	2	13.16	2.39	0.06	10.16	1.50	0.00
3	21	5	1.19	0	20	10.71	5	8.73	2.13	0.11	7.36	1.40	0.01
4	10	1	0.30	0	54	23.60	12	9.13	3.58	0.15	5.91	2.14	0.05
5	2	–	–	–	57	41.50	26	4.14	3.50	2.86	2.25	1.62	1.00

while In-degree decreases as we move from layer 0 to layer 5. However, the mean of the minimum distance to the EF does not follow this pattern monotonously. Thus, it seems that, in average, the nodes in layer 3 are closer to the EF than those of layer 2. The maximum value 15.89 of e_r^{max} and τ_r^{min} for layer 1 are equal and refers to the same arc which has its origin in Qatar and ends in Costa Rica, Qatar's only benchmark. The layers with the fewer number of countries are layers 4 and 5. The countries in these layers have a minimum out-degree of 4 and 5, respectively. Kuwait is the only country in layer 5. It is dominated by 30 countries, it has 11 benchmarks in the EF (Bosnia Herzegovina, Colombia, Costa Rica, Greece, Israel, Lebanon, Morocco, New Zealand, Saint Lucia, Sri Lanka and Thailand) and the corresponding range of distances to the EF $\left[\tau_r^{min}, e_r^{max}\right] = [3.69, 5.56]$. The only reason why Kuwait does not belong to layer 4 is because it is dominated by Lithuania, which belongs to layer 4.

Tables 2 and 3 show the basic characterization measures of all the nodes and the additional measures for the efficient nodes, respectively. It can be seen, for example, that Botswana, Chile, Slovenia and Switzerland belong lo layer 1 and have a single efficient benchmark (Cambodia, Costa Rica, Greece and Japan, respectively). Botswana, Chile and Switzerland dominate one country each while Slovenia dominates two: Denmark (layer 2) and United States (layer 4).

Turning our attention to the maximum distance of the countries to the EF, shown in the left panel of Fig. 2, it can be seen that the median value increases as the layer increases but, except for the last layer, it remains relatively low, indicating that many of the nodes are rather close to the EF. Actually, it can be seen that some countries in layer 3 are closer to the EF than some nodes in layers 1 and 2. For instance, the distance to the EF of the following layer 2 countries are: Belgium [3.66, 12.12], United Kingdom [0.30, 8.70], Ireland [5.51, 7.25], Netherland [3.74, 5.49],

Table 2 Basic node characterization measures

Code	Country	DNJune						DNAugust					
		d_r^{in}	d_r^{out}	$e_r^{\max}$	$\tau_r^{\min}$	$\lambda(r)$	$\lvert D^*(r)\rvert$	d_r^{in}	d_r^{out}	$e_r^{\max}$	$\tau_r^{\min}$	$\lambda(r)$	$\lvert D^*(r)\rvert$
DZ	Algeria	13	3	0.28	0.26	2	2	16	3	0.48	0.46	2	2
AO	Angola	3	2	0.00	0.00	1	2	4	5	0.03	0.01	1	5
AG	Antigua & Barbuda	7	1	0.47	0.47	1	1	12	1	0.59	0.59	1	1
AR	Argentina	3	7	0.38	0.19	3	5	2	12	2.12	0.55	3	7
AM	Armenia	0	8	2.93	2.33	3	5	0	8	6.90	4.14	3	5
AU	Australia	12	0	–	–	0	0	12	0	–	–	0	0
AT	Austria	2	6	2.06	0.95	2	5	2	6	1.45	0.80	2	5
AZ	Azerbaijan	1	14	0.46	0.27	3	5	1	15	1.47	1.18	4	5
BH	Bahrain	0	5	5.34	5.33	2	2	0	10	8.73	7.36	3	5
BD	Bangladesh	12	0	–	–	0	0	10	0	–	–	0	0
BB	Barbados	14	1	0.32	0.32	1	1	20	0	–	–	0	0
BY	Belarus	0	9	3.84	3.23	3	6	0	8	2.98	2.68	3	4
BE	Belgium	0	11	12.12	3.66	2	7	0	11	10.38	3.33	2	6
BZ	Belize	12	0	–	–	0	0	15	0	–	–	0	0
BJ	Benin	10	2	0.01	0.01	1	2	9	2	0.05	0.03	1	2
BT	Bhutan	10	0	–	–	0	0	12	0	–	–	0	0
BO	Bolivia	0	10	0.91	0.26	2	7	0	11	4.72	1.97	3	7
BA	Bosnia-Herzegovina	4	0	–	–	0	0	3	0	–	–	0	0
BW	Botswana	1	1	0.01	0.01	1	1	1	7	0.12	0.07	2	6
BR	Brazil	0	17	3.29	1.23	4	9	0	18	8.53	4.62	4	10
BN	Brunei	2	8	0.26	0.12	3	5	2	6	0.15	0.13	3	3
BG	Bulgaria	0	10	0.45	0.27	3	6	0	12	1.09	0.74	4	6
BF	Burkina Faso	3	9	0.06	0.03	2	6	7	3	0.04	0.03	1	3
BI	Burundi	11	0	–	–	0	0	14	0	–	–	0	0
KH	Cambodia	8	0	–	–	0	0	9	0	–	–	0	0
CM	Cameroon	1	24	0.25	0.20	3	11	1	20	0.36	0.19	3	10
CA	Canada	4	4	3.70	1.96	1	4	4	3	3.23	2.62	1	3
CV	Cape Verde	5	0	–	–	0	0	4	0	–	–	0	0
CF	Central African Rep.	0	0	–	–	0	0	0	0	–	–	0	0
TD	Chad	0	8	0.07	0.05	2	6	0	2	0.05	0.02	1	2
CL	Chile	1	1	4.57	4.57	1	1	0	1	9.72	9.72	1	1
CN	China	24	1	0.04	0.04	1	1	27	1	0.03	0.03	1	1
CO	Colombia	8	0	–	–	0	0	1	0	–	–	0	0
KM	Comoros	5	5	0.12	0.10	1	5	5	5	0.22	0.06	1	5
CR	Costa Rica	28	0	–	–	0	0	15	0	–	–	0	0
HR	Croatia	3	3	0.51	0.51	2	2	5	2	0.46	0.40	1	2
CU	Cuba	2	1	0.03	0.03	1	1	2	0	–	–	0	0
CY	Cyprus	4	1	0.67	0.67	1	1	7	1	0.30	0.30	1	1

(continued)

Table 2 (continued)

		DNJune						DNAugust					
Code	Country	d_r^{in}	d_r^{out}	$e_r^{\max}$	$\tau_r^{\min}$	$\lambda(r)$	$\|D^*(r)\|$	d_r^{in}	d_r^{out}	$e_r^{\max}$	$\tau_r^{\min}$	$\lambda(r)$	$\|D^*(r)\|$
CZ	Czech Republic	2	2	0.81	0.60	1	2	3	2	0.53	0.53	2	1
DK	Denmark	1	9	2.44	0.70	2	6	1	8	1.73	1.08	2	5
DJ	Djibouti	0	2	2.96	2.71	1	2	1	2	2.14	1.64	1	2
DO	Dominican Republic	0	14	1.71	1.11	3	6	1	12	3.20	2.91	4	5
EC	Ecuador	0	1	3.65	3.65	1	1	1	1	4.83	4.83	1	1
EG	Egypt	1	6	0.24	0.06	2	4	1	6	0.73	0.46	2	4
SV	El Salvador	4	6	0.32	0.16	2	3	2	5	1.50	1.24	2	3
GQ	Equatorial Guinea	0	55	0.84	0.13	4	21	0	54	1.10	0.43	4	22
ER	Eritrea	23	0	–	–	0	0	22	0	–	–	0	0
EE	Estonia	0	4	1.47	1.46	3	2	3	3	0.73	0.67	2	2
ET	Ethiopia	25	0	–	–	0	0	21	0	–	–	0	0
FJ	Fiji	3	2	0.01	0.01	1	2	4	2	0.02	0.02	1	2
FI	Finland	6	2	1.34	1.18	1	2	6	2	0.91	0.70	1	2
FR	France	1	3	6.11	4.37	1	3	1	2	5.43	5.30	1	2
GA	Gabon	0	22	1.04	0.34	3	12	1	19	1.33	0.66	2	11
GM	Gambia	11	0	–	–	0	0	8	0	–	–	0	0
GE	Georgia	3	4	0.14	0.11	2	3	5	5	0.12	0.09	2	3
DE	Germany	1	8	2.59	0.85	3	6	1	7	1.80	1.15	3	5
GH	Ghana	2	8	0.22	0.17	2	7	2	7	0.44	0.17	1	7
GR	Greece	13	0	–	–	0	0	15	0	–	–	0	0
GD	Grenada	2	1	0.09	0.09	1	1	5	1	0.03	0.03	1	1
GT	Guatemala	10	1	0.06	0.06	1	1	3	1	1.69	1.69	1	1
GN	Guinea	2	6	0.24	0.23	2	5	3	5	0.21	0.15	2	4
GW	Guinea-Bissau	0	17	0.58	0.31	3	9	0	16	0.47	0.30	3	9
GY	Guyana	0	5	0.26	0.26	3	3	2	2	0.39	0.39	2	1
HT	Haiti	2	9	0.19	0.14	2	7	1	9	0.36	0.19	2	6
HN	Honduras	7	0	–	–	0	0	3	0	–	–	0	0
HU	Hungary	0	8	0.83	0.68	3	4	2	4	0.75	0.40	2	3
IS	Iceland	2	1	4.39	4.39	1	1	4	1	1.98	1.98	1	1
IN	India	3	3	0.15	0.10	1	3	2	4	0.67	0.40	2	3
ID	Indonesia	6	2	0.09	0.09	1	2	6	1	0.28	0.28	1	1
IR	Iran	0	12	2.34	1.30	3	7	0	12	3.19	1.04	3	8
IE	Ireland	0	6	7.25	5.51	2	5	1	5	5.21	4.56	2	4
IL	Israel	12	0	–	–	0	0	4	0	–	–	0	0
IT	Italy	2	0	–	–	0	0	2	0	–	–	0	0
JM	Jamaica	8	1	0.12	0.12	1	1	13	1	0.08	0.08	1	1
JP	Japan	14	0	–	–	0	0	14	0	–	–	0	0
JO	Jordan	9	1	0.02	0.02	1	1	11	1	0.03	0.03	1	1
KZ	Kazakhstan	1	4	0.46	0.43	2	3	2	9	2.01	1.72	3	5
KE	Kenya	8	3	0.03	0.02	1	3	5	2	0.14	0.12	1	2

(continued)

Table 2 (continued)

Code	Country	DNJune						DNAugust					
		d_r^{in}	d_r^{out}	$e_r^{\max}$	$\tau_r^{\min}$	$\lambda(r)$	$\|D^*(r)\|$	d_r^{in}	d_r^{out}	$e_r^{\max}$	$\tau_r^{\min}$	$\lambda(r)$	$\|D^*(r)\|$
KW	Kuwait	0	30	5.56	3.69	5	11	0	33	6.29	2.97	4	14
LA	Laos	12	0	–	–	0	0	13	0	–	–	0	0
LV	Latvia	1	9	0.54	0.36	3	6	2	6	0.36	0.34	3	3
LB	Lebanon	24	0	–	–	0	0	18	0	–	–	0	0
LS	Lesotho	0	0	–	–	0	0	0	20	0.15	0.05	4	11
LR	Liberia	4	5	0.10	0.08	1	5	2	4	0.22	0.15	1	4
LT	Lithuania	1	15	0.71	0.10	4	7	2	11	0.53	0.18	3	7
LU	Luxembourg	0	7	6.76	2.09	2	6	1	7	5.28	2.07	2	6
MG	Madagascar	21	0	–	–	0	0	11	0	–	–	0	0
MW	Malawi	11	3	0.01	0.00	1	3	8	4	0.13	0.06	1	4
MY	Malaysia	12	4	0.19	0.15	2	3	11	4	0.11	0.08	2	3
MV	Maldives	4	1	2.55	2.55	1	1	6	2	2.28	1.18	1	2
ML	Mali	4	6	0.08	0.08	2	5	4	4	0.09	0.06	2	3
MT	Malta	11	0	–	–	0	0	10	0	–	–	0	0
MR	Mauritania	3	8	0.13	0.09	2	7	2	10	0.79	0.52	3	7
MU	Mauritius	1	9	0.25	0.06	3	6	1	5	0.15	0.12	3	3
MX	Mexico	0	8	1.29	0.70	3	5	0	7	4.70	4.41	3	4
MD	Moldova	0	6	2.34	1.68	2	5	1	6	3.71	1.20	2	5
MN	Mongolia	2	1	0.00	0.00	1	1	4	0	–	–	0	0
MA	Morocco	26	0	–	–	0	0	26	0	–	–	0	0
MZ	Mozambique	8	0	–	–	0	0	8	1	0.01	0.01	1	1
MM	Myanmar	4	1	0.00	0.00	1	1	5	1	0.00	0.00	1	1
NA	Namibia	0	3	0.01	0.00	1	3	0	20	0.32	0.07	2	14
NP	Nepal	13	0	–	–	0	0	8	0	–	–	0	0
NL	Netherlands	0	6	5.49	3.74	2	5	1	5	4.52	3.87	2	4
NZ	New Zealand	11	0	–	–	0	0	11	0	–	–	0	0
NI	Nicaragua	8	2	0.08	0.06	1	2	10	2	0.31	0.27	1	2
NE	Niger	6	2	0.05	0.05	1	2	12	0	–	–	0	0
NG	Nigeria	0	16	0.05	0.03	3	10	1	16	0.11	0.01	3	10
NO	Norway	1	3	1.51	0.40	1	3	1	2	0.87	0.74	1	2
OM	Oman	3	2	1.72	1.71	1	2	2	4	5.71	4.56	2	3
PK	Pakistan	3	5	0.31	0.06	2	4	3	4	0.64	0.42	2	3
PA	Panama	0	3	3.11	3.11	2	2	0	4	8.06	6.90	2	3
PG	Papua New Guinea	18	0	–	–	0	0	18	0	–	–	0	0
PY	Paraguay	4	4	0.08	0.05	2	3	5	8	0.30	0.23	3	4
PE	Peru	0	2	5.35	5.33	1	2	0	2	10.03	10.00	1	2
PH	Philippines	2	3	0.18	0.17	2	2	3	2	0.43	0.19	1	2
PL	Poland	2	3	0.61	0.60	2	2	4	2	0.55	0.49	1	2
PT	Portugal	3	1	3.56	3.56	1	1	2	1	3.04	3.04	1	1
QA	Qatar	0	1	15.89	15.89	1	1	0	5	13.16	10.16	2	3

(continued)

Table 2 (continued)

		DNJune						DNAugust					
Code	Country	d_r^{in}	d_r^{out}	$e_r^{\max}$	$\tau_r^{\min}$	$\lambda(r)$	$\lvert D^*(r)\rvert$	d_r^{in}	d_r^{out}	$e_r^{\max}$	$\tau_r^{\min}$	$\lambda(r)$	$\lvert D^*(r)\rvert$
RO	Romania	1	14	1.42	0.38	3	8	1	11	2.08	1.73	3	6
RU	Russia	0	27	2.52	1.86	4	11	0	26	2.86	1.00	5	11
RW	Rwanda	15	0	–	–	0	0	18	0	–	–	0	0
LC	Saint Lucia	24	0	–	–	0	0	27	0	–	–	0	0
SA	Saudi Arabia	0	15	2.08	1.90	4	6	0	28	3.46	1.89	4	11
SN	Senegal	5	2	0.18	0.15	1	2	7	1	0.27	0.27	1	1
RS	Serbia	2	13	1.67	1.06	3	7	3	13	2.08	0.51	3	8
SC	Seychelles	0	1	0.01	0.01	1	1	0	1	0.34	0.34	1	1
SL	Sierra Leone	0	25	0.14	0.10	3	12	0	19	0.16	0.05	3	10
SG	Singapore	1	0	–	–	0	0	2	0	–	–	0	0
SI	Slovenia	2	1	0.70	0.70	1	1	4	1	0.57	0.57	1	1
ZA	South Africa	0	47	0.56	0.29	3	18	0	57	4.14	2.25	5	19
SS	South Sudan	1	14	0.08	0.03	2	11	3	13	0.11	0.00	2	9
ES	Spain	0	0	–	–	0	0	0	0	–	–	0	0
LK	Sri Lanka	34	0	–	–	0	0	35	0	–	–	0	0
SD	Sudan	2	12	0.16	0.11	2	10	2	8	0.26	0.18	2	7
SR	Suriname	1	0	–	–	0	0	1	18	1.40	0.75	4	8
SE	Sweden	0	3	7.17	5.43	1	3	0	3	8.19	8.06	2	2
CH	Switzerland	1	1	4.61	4.61	1	1	1	1	3.16	3.16	1	1
TJ	Tajikistan	3	2	0.32	0.10	1	2	4	1	0.05	0.05	1	1
TZ	Tanzania	15	0	–	–	0	0	20	0	–	–	0	0
TH	Thailand	33	0	–	–	0	0	36	0	–	–	0	0
TG	Togo	4	6	0.05	0.03	2	5	7	2	0.05	0.03	1	2
TT	Trinidad and Tobago	0	5	0.08	0.04	2	3	0	4	0.07	0.01	2	2
TN	Tunisia	19	3	0.07	0.05	2	2	21	2	0.06	0.06	2	1
TR	Turkey	1	0	–	–	0	0	8	0	–	–	0	0
UG	Uganda	8	1	0.01	0.01	1	1	10	0	–	–	0	0
UA	Ukraine	0	9	0.55	0.33	3	5	2	7	0.87	0.25	2	5
AE	United Arab Emirates	1	7	2.90	2.69	3	3	2	6	2.10	0.94	2	4
GB	United Kingdom	0	7	8.70	0.30	2	5	0	5	8.16	1.11	2	4
US	United States	0	23	7.34	5.60	4	8	0	26	9.13	5.91	4	10
UY	Uruguay	4	2	0.07	0.07	1	2	4	0	–	–	0	0
UZ	Uzbekistan	4	1	0.05	0.05	1	1	4	1	0.24	0.24	1	1
VE	Venezuela	22	0	–	–	0	0	15	0	–	–	0	0
ZM	Zambia	2	9	0.05	0.00	2	8	1	7	0.19	0.12	2	6
ZW	Zimbabwe	3	6	0.01	0.00	2	5	3	12	0.12	0.01	2	10

Table 3 Additional characterization measures for efficient nodes

Code	Country	DNJune				DNAugust			
		$\lvert D^{--1}(j)\rvert$	κ_j	σ_j	χ_j	$\lvert D^{--1}(j)\rvert$	κ_j	σ_j	χ_j
AU	Australia	–	63.47	12.03	5.29	–	55.36	10.21	4.61
BD	Bangladesh	–	12.29	2.71	1.02	–	18.14	4.09	1.81
BB	Barbados	–	–	–	–	–	66.14	8.89	3.31
BZ	Belize	–	8.71	2.48	0.73	–	25.06	4.67	1.67
BT	Bhutan	2	3.91	0.99	0.39	3	14.43	4.72	1.20
BA	Bosnia and Herzegovina	–	8.46	4.53	2.11	–	11.57	6.39	3.86
BI	Burundi	–	2.38	0.83	0.22	1	2.92	1.09	0.21
KH	Cambodia	1	2.74	1.03	0.34	–	8.08	4.14	0.90
CV	Cape Verde	–	4.27	1.86	0.85	–	8.20	2.87	2.05
CF	Central African Republic	–	–	–	–	–	–	–	–
CO	Colombia	–	16.50	4.96	2.06	–	4.62	4.62	4.62
CR	Costa Rica	4	67.46	15.89	2.41	1	69.02	12.03	4.60
CU	Cuba	–	–	–	–	–	9.49	6.16	4.74
ER	Eritrea	–	3.92	0.83	0.17	–	9.57	4.12	0.43
ET	Ethiopia	–	4.98	1.03	0.20	–	9.85	4.07	0.47
GM	Gambia	–	2.35	0.82	0.21	–	1.74	0.99	0.22
GR	Greece	3	48.64	11.91	3.74	4	60.87	13.16	4.06
HN	Honduras	–	11.38	3.25	1.63	–	7.31	4.14	2.44
IL	Israel	–	54.25	10.38	4.52	–	21.10	10.16	5.28
IT	Italy	–	3.96	3.66	1.98	–	4.43	3.33	2.22
JP	Japan	2	73.53	12.12	5.25	2	62.04	10.34	4.43
LA	Laos	1	2.86	1.04	0.24	1	7.92	4.14	0.61
LB	Lebanon	1	43.43	7.34	1.81	–	64.51	8.95	3.58
MG	Madagascar	–	7.48	2.96	0.36	–	9.99	3.97	0.91
MT	Malta	–	47.84	11.02	4.35	–	44.44	9.73	4.44
MN	Mongolia	–	–	–	–	–	5.56	4.11	1.39
MA	Morocco	1	33.63	5.38	1.29	1	67.64	8.24	2.60
MZ	Mozambique	–	2.06	0.83	0.26	–	–	–	–
NP	Nepal	–	5.78	2.28	0.44	–	9.09	3.89	1.14
NZ	New Zealand	–	60.97	12.06	5.54	–	54.77	10.38	4.98
NE	Niger	–	–	–	–	–	2.53	1.05	0.21
PG	Papua New Guinea	1	3.29	0.84	0.18	–	8.61	4.14	0.48
RW	Rwanda	–	3.95	1.02	0.26	1	10.15	4.08	0.56
LC	Saint Lucia	2	30.48	5.53	1.27	2	56.18	8.51	2.08
SG	Singapore	–	2.09	2.09	2.09	–	8.59	6.22	4.29
ES	Spain	–	–	–	–	–	–	–	–
LK	Sri Lanka	1	41.11	5.54	1.21	1	75.66	10.00	2.16
SR	Suriname	–	0.51	0.51	0.51	–	–	–	–
TZ	Tanzania	–	3.40	0.83	0.23	–	9.63	4.14	0.48
TH	Thailand	4	44.66	5.56	1.35	5	78.53	10.03	2.18
TR	Turkey	–	1.23	1.23	1.23	–	24.90	7.36	3.11
UG	Uganda	–	–	–	–	–	2.57	1.09	0.26
UY	Uruguay	–	–	–	–	–	18.40	8.73	4.60
VE	Venezuela	–	10.86	2.34	0.49	1	19.70	4.48	1.31

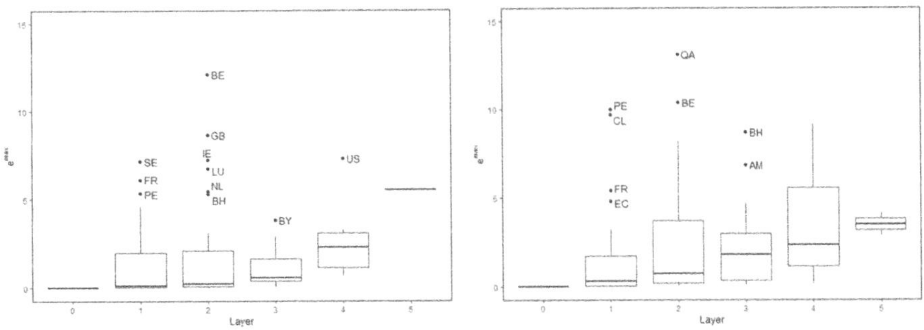

Fig. 2 Maximum distance to the EF by layer (Left panel DNJune, Right panel DNAugust)

Luxembourg [2.09, 6.76], Bahrain [5.33, 5.34], Panama [3.11, 3.11]. Similarly, in layer 1: Qatar [15.89, 15.89], Sweden [5.43, 7.17], Peru [5.33, 5.35], Ecuador [3.65, 3.65]. Interestingly, the In-degree of all these far-from-the EF nodes is equal to zero, i.e. they do not dominate any node. There are also two countries, like France [4.37, 6.11] and Switzerland [4.61, 4.61] that are also far from the EF but are intermediate benchmarks of other countries, i.e. they dominate Belgium [3.66, 12.12] and United States [5.60, 7.34], respectively, both in layer 4.

Focusing on the efficient nodes, let us consider the benchmarking necessity index. Costa Rica is the only benchmark for four inefficient countries (Qatar, Barbados, Chile and Maldives). The same happens to Thailand, which is the only reference unit for Antigua and Barbuda, China, Ecuador and Jordan. These two efficient nodes have a high In-degree. Actually, there is only one country that dominates more countries than them and it is Sri Lanka. They also have a high benchmarking potential index (73.53 and 44.66, respectively). The average path length from the DMUs dominated by each of these two DMUs, denoted by χ_j, is not high (2.41 for Costa Rica and 1.35 for Thailand) compared with the corresponding values for other efficient nodes such as New Zealand (5.54), Israel (4.52), Australia (5.29) or Japan (5.25). Most of the countries that these other DMUs dominate are far from the EF, i.e. there is a large efficiency gap between them and the countries they dominate.

The Inefficiency radius of an efficient node σ_j computes the maximum distance from the most inefficient DMU dominated by DMU j. The largest value for this index corresponds to Costa Rica. In Fig. 3 it can be seen that this metric is high correlated with the average path length of the dominated DMUs χ_j. In that figure, there are two well-differentiated groups. One group is formed by Japan, New Zealand, Australia, Israel, Malta, and Greece, with high values of both indexes. The second group have low values of both indexes and includes countries like Thailand, Sri Lanka and Italy. Apart from these two groups, there is one country (Costa Rica) with high value of σ_j and relative medium value of χ_j. For this country, although some of the countries it dominates are far from the EF, most of those dominated countries are close to the EF.

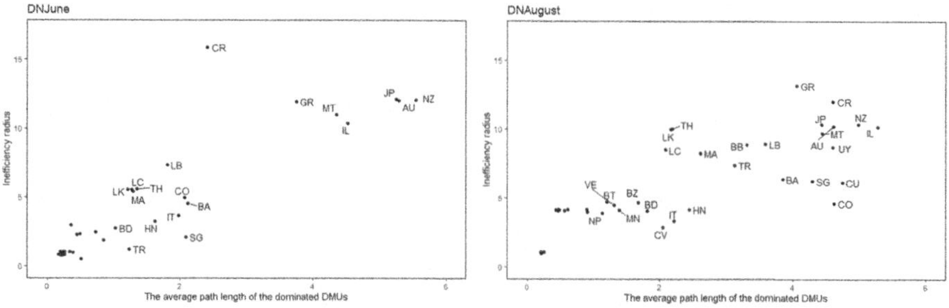

Fig. 3 Inefficiency radius (σ_j) versus average path length of DMUs dominated by j (χ_j)

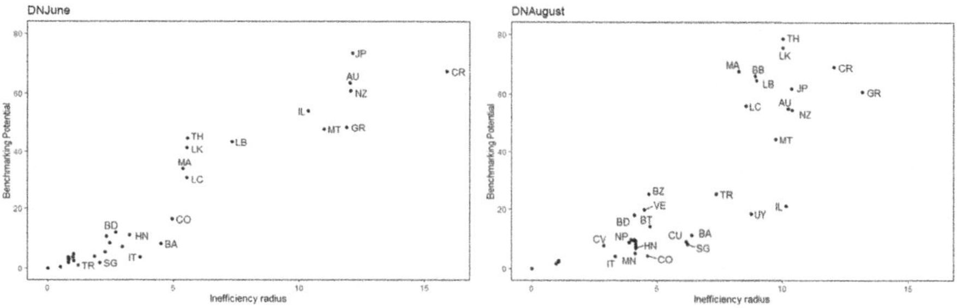

Fig. 4 Benchmarking potential (κ_j) versus Inefficiency radius (σ_j)

Figure 4 plots the Inefficiency radius versus the Benchmarking potential index. Again, the two measures are highly correlated. Again, two groups of countries can be distinguished. The one to the right has fewer nodes (namely, Costa Rica, Japan, New Zealand, Australia, Greece, Malta and Israel) but each of these countries dominates at least 11 other countries among which we can find the most inefficient ones. That explains why their inefficiency radius and benchmarking potential are high.

Apart from the 39 efficient nodes in DNJune, there are 23 layer 1 countries that are dominated by only one efficient country and of these 23, 16 have a maximum distance to the EF $e_r^{\max}$ that is lower than the average. These 16 nodes, although not efficient, are not far from it. Take, for example, Uganda (0.01), Myanmar (0.00), Jordan (0.02) or Guatemala (0.06). Some of these countries have also a high value of their In-degree, e.g. Barbados (14), China (24) or Guatemala (10).

As an example, Figs. 5 and 6 show the DNJune egonetwork of US and China, respectively. Interestingly, US does not dominate any node while is dominated by 23 countries, of which 8 (AU, CR, GR, IL, JP, LB, MT and NZ) are efficient. Conversely, China dominates 24 countries and it is dominated just by one (TH). It is apparent from the corresponding subgraphs that China is closer to the EF than USA. To be exact, the distance from China to the EF is rather low (0.04) while the range of distances from US to the EF is [5.60, 7.34].

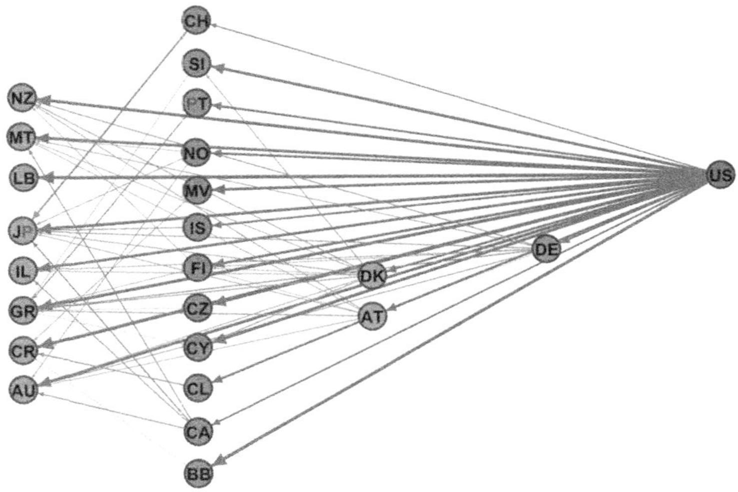

Fig. 5 US DNJune egonetwork

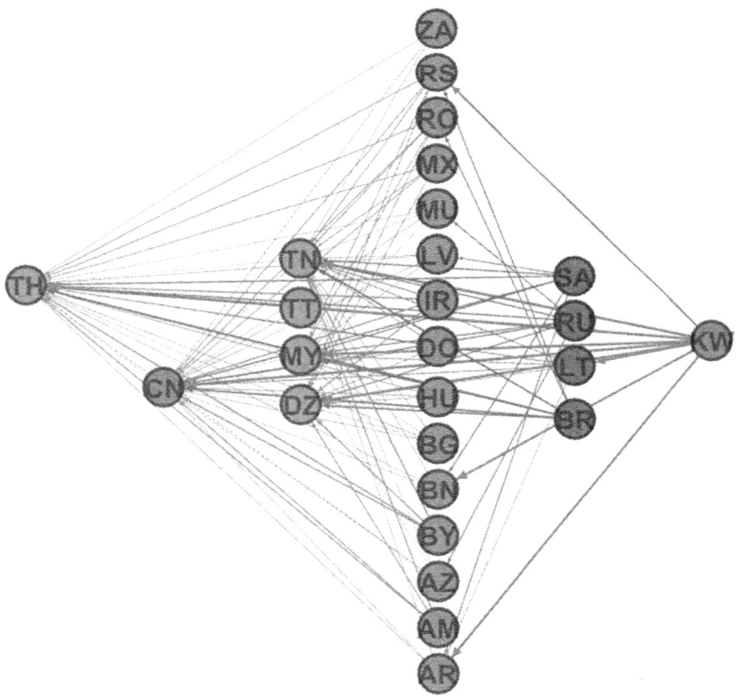

Fig. 6 China DNJune egonetwork

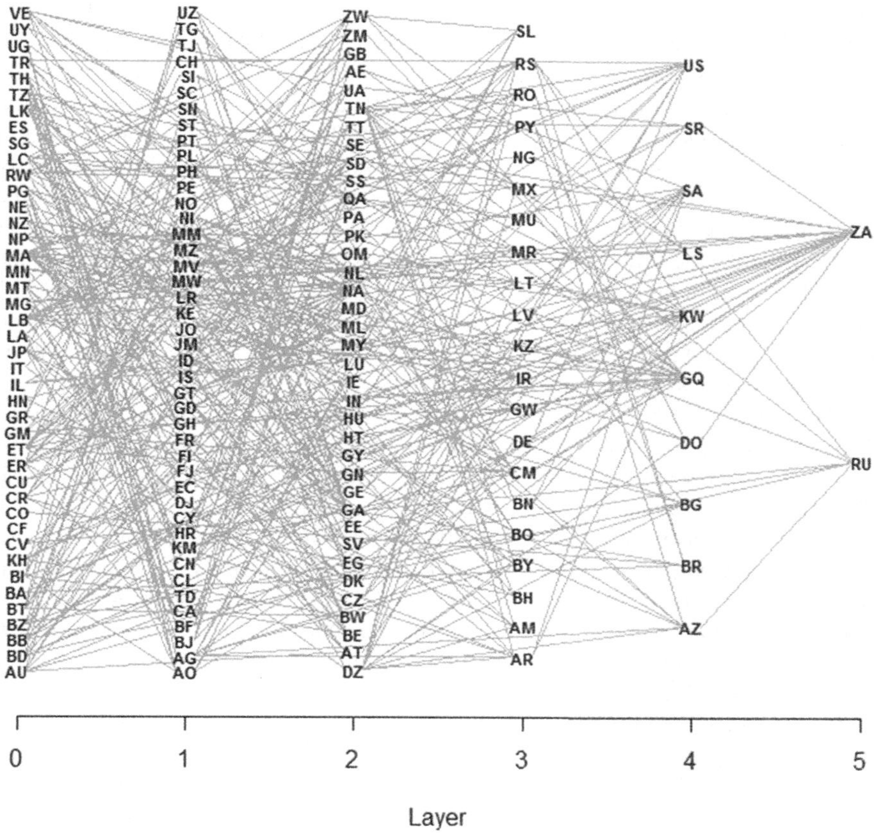

Fig. 7 Visualization of DNAugust using a skeletonization filter and layer layout

Now let us consider DNAugust, which corresponds to the situation of COVID-19 as of August 1st 2020 (see Fig. 7). The network has one giant component and two isolates (Spain and Central African Republic). That Spain keeps its maverick status is probably due to the fact that since the peak of the pandemic in early April, and for political reasons, the official number of deaths from COVID-19 acknowledged by the Spanish government grossly underestimates the true figure, probably by as much as 40%. This clearly distorts the corresponding data, making them unreliable and incomparable.

The giant component of DNAugust is composed of 153 countries and 862 edges and its average clustering coefficient is 0.38. Its density is equal to 3.71%. and its average degree is 5.63, very similar to DNJune. The countries of the network are structured in 6 layers with 27.10% in layer 0, 27.74% in layer 1 and 23.87% in layer 2. The deepest layer includes just two countries, Russia and South Africa, whose maximum distance to the EF is not too high, 2.86 and 4.14, respectively. The diameter of the giant component is 13.16, while its average path length is 1.53. The diameter corresponds to the path from Qatar (layer 2) to Greece (layer 0).

Table 1 shows how the In-degree decreases monotonously with the layer, while the Out-degree, and the minimum values of e_r^{max} and τ_r^{min} increase monotonously. The maximum and mean values of e_r^{max} and τ_r^{min} fluctuates, with local minima in layers 3 and 5. The median, however, as shown in Fig. 2, increases monotonously as the layer increases. The fluctuations of e_r^{max} and τ_r^{min} occur because some of the paths with the maximum values of e_r^{max} and τ_r^{min} have their origin in layers 1, 2 and 4. For instance, the following countries are the origin of some of the longest paths in the network: in layer 4, US $\left[\tau_r^{min}, e_r^{max}\right] = [5.91, 9.13]$ and Brazil [4.62, 8.53]; in layer 2, Qatar [10.16, 13.16], Belgium [3.33, 10.38], UK [1.11, 8.16], Sweden [8.06, 8.19]; in layer 1, Chile [9.72, 9.72] and Peru [10.00, 10.03]. These countries are among the most inefficient.

The range of the distance to EF of the two countries in layer 5 is narrow. Thus, Russia [1.00, 2.86] has 11 efficient benchmarks (Bangladesh, Barbados, Belize, Cape Verde, Costa Rica, Lebanon, Morocco, Saint Lucia, Sri Lanka, Thailand and Turkey) while South Africa [2.25, 4.14] has 19 efficient benchmarks (Bangladesh, Belize, Bhutan, Cambodia, Cape Verde, Eritrea, Ethiopia, Laos, Madagascar, Mongolia, Morocco, Nepal, Papua New Guinea, Rwanda, Saint Lucia, Sri Lanka, Tanzania, Thailand and Venezuela). These two countries have five efficient benchmarks in common. Russia belongs to layer 5 because it is dominated by Azerbaijan (layer 4) while South Africa is dominated by two countries from layer 4 (Suriname and Dominican Republic).

It can be seen, looking at the detailed results shown in Tables 2 and 3 that layer 1 has 43 countries, 20 of which have only one efficient benchmark. The layer 1 nodes with the highest e_r^{max} are Chile (9.72), Switzerland (3.16) and Portugal (3.04). Only two layer 1 countries (namely, Chile and Seychelles) do no dominate any other country ($d_r^{in} = 0$).

Regarding the efficient countries, Thailand and Greece have a benchmarking necessity equal to 4 and 5, respectively. In total, there are 23 inefficient countries with only one efficient benchmark, of which 20 countries belong to layer 1 and 3 countries (Czech Republic, Guyana and Tunisia) belong to layer 2. Thailand is not only the benchmark with the highest benchmarking necessity but also the highest In-degree and benchmarking potential. The countries with the highest benchmarking potential index are Thailand (78.53), Sri Lanka (75.66), Morocco (67.64) and Costa Rica (69.02) whose In-degree is 36, 35, 26 and 15, respectively. The average path length of their dominated DMUs χ_j are slightly over the average path length, except in the case of Costa Rica and Morocco whose χ_j are 4.60 and 2.60, respectively. However, the countries with the highest χ_j are Israel and New Zealand, with a value of 5.28 and 4.98.

As regards the inefficiency radius, Greece has the largest value, which corresponds, as it was mentioned above, to the edge from Qatar, which is the node with the largest e_r^{max} (13.16). In terms of inefficiency radius, Greece is closely followed by Costa Rica (12.03), New Zealand (10.38) and Japan (10.34). Plotting the inefficiency radius σ_j versus χ_j in Fig. 3, as we did with DNJune, we can distinguish two groups of countries and, in all cases, a positive correlation between the two measures.

Regarding the benchmarking potential index κ_j, the lowest values correspond to Gambia (1.74), Niger (2.53), Uganda (2.57) and Burundi (2.82), with In-degree 8, 12, 10 and 14, respectively. Therefore, although these countries are the inefficiency gap with those they dominate is small. As it can be seen in Fig. 4, comparing the benchmarking potential index with the inefficiency radius, the efficient nodes can be split in two groups: those whose benchmarking potential index is over 40 and those whose benchmarking potential index is under 26. Most of the countries of the first group dominate countries that are far from the EF. This group is composed by the following countries: Thailand, Sri Lanka, Morocco, Costa Rica, Greece, Lebanon, Barbados, Japan, New Zealand, Saint Lucia, Australia and Malta. All these countries dominate at least 10 countries each. The mean χ_j for this group is 3.59. The second group is composed by two thirds of the efficient countries. Except for Israel, whose inefficiency radius is 10.16, the rest of the countries have an inefficiency radius below 10, with an average χ_j of 4.18.

There are countries in the giant component that belong to layer 1 and are pretty close to the EF. In particular, the following countries have only one efficient reference which is rather close: Myanmar (0.00), Mozambique (0.01), Jordan (0.03), Grenada (0.03), China (0.03), Tajikistan (0.05), Jamaica (0.08). These countries can be labelled as almost efficient or minimal inefficient.

Comparing DNJune and DNAugust, the latter has 163 new dominance relationships while 130 dominance relationships that existed in DNJune disappeared. Hence, there are 3.98% more edges in DNAugust, so the network is slightly denser. Both networks have the same giant component and the two isolate countries are the same. The average path length is similar for both networks but the diameter of DNAugust is smaller, while the global clustering coefficient has slightly increased, i.e. the geodesic paths are shorter and there are more dominance relationships among the countries. The distribution of the countries in layers has changed. Table 1 shows that there are more efficient countries in DNAugust. Thus, in DNAugust 6 new countries become efficient (Barbados, Cuba, Mongolia, Niger, Uganda and Uruguay) while 2 efficient countries in DNJune have lost that status (Mozambique and Suriname). There have been similar changes in the other layers. In total, 41 countries have change their layer positions, and the countries are mow more equally distributed by layer. In DNAugust, Kuwait, which in DNJune was the only node in layer 5, has now advanced to layer 1 and is dominated by 33 countries instead of 30, while it does not dominate any country. Apart from this, e_r^{max} has increased while τ_r^{min} has decreased.

Comparing Fig. 2, the interquartile range of the maximum distance to EF by layer is slightly wider for DNAugust and the median value has also increased a little bit. This can also be seen in Fig. 3. The small increase in the geodesic distances (i.e. in the node inefficiencies) is also reflected in a small increase of the inefficiency radius of the efficient nodes and, to a less extent, of the benchmark potential, as it can be seen in Fig. 4.

In DNJune, China, dominated 24 countries while in DNAugust it dominates 27 and it has reduced both e_r^{max} and τ_r^{min}. Comparing the situation of the US in both networks, it still belongs to the same layer in both networks and still does not dominate any country. Its number of efficient benchmarks has risen from 8 to 10 (the new benchmarks are Barbados and Singapore) and its Out-degree from 23 to 26, i.e. it is dominated by three more countries. The distance from US to the EF, i.e. its relative inefficiency, has also increased, from [5.60, 7.34] in DNJune to [5.91, 9.13] in DNAugust.

Figure 8 shows Out- versus In-degree for both DNJune and DNAugust. The nodes with zero Out-degree (e.g. MA, CR, TH or LK) belong to the EF. On the contrary, the countries with high Our-degree (e.g. KW, RU, US) are dominated by many countries and hence tend to belong to the deepest layers. This does not mean they are necessarily highly inefficient. Thus, for example, Sierra Leone (SL) has Out-degree 25 and 19 in DNJune and DNAugust, respectively. However, as it can be seen in Table 2, its distance to the EF is small in both periods ([0.10, 0.14] in DNJune and [0.05, 0.16] in DNAugust). In fact, even though SL belonged to layer 3, its relative performance has been stable and high.

Figure 9 shows the range of distances to the EF in both periods. Since by definition $e_r^{max} \geq \tau_r^{min}$, the points always fall above the diagonal. The points on the diagonal generally corresponds to inefficient nodes that are dominated by just one efficient benchmark (and hence $e_r^{max} = \tau_r^{min}$). For a few countries, like United Kingdom (GB) or Belgium (BE), the range is wide as $e_r^{max} \gg \tau_r^{min}$. Both countries, however, have improved a little bit in DNAugust with respect to DNJune. For some countries, like Qatar (QA), both e_r^{max} and τ_r^{min} are high, indicating a large inefficiency. It can be seen in Fig. 8 and in Table 2 that for some countries (e.g. Peru, Colombia, Sweden, Bahrain, Brazil, US, Panama or Mexico) both its minimum as well as its maximum distance to the EF (in other words, its relative inefficiency) have increased from DNJune to DNAugust. Other countries, on the contrary, have more or less maintained their relative distance to the EF, e.g. France (from [4.37, 6.11] to [5.30, 5.43]), Germany (from [0.85, 2.59] to [1.15, 1.80]), Ghana (from [0.17, 0.22] to [0.17, 0.44]), Jordan (from[0.02, 0.02] to [0.03, 0.03]), Lithuania (from [0.10, 0.71] to [0.18, 0.53]), etc. while others have even improved it, e.g. Iceland (from [4.39, 4.39] to [1.98, 1.98]), Cyprus (from [0.67, 0.67] to [0.30, 0.30]), Czech Republic (from [0.60, 0.81] to [0.53, 0.53]), Estonia (from [1.46, 1.47] to [0.67, 0.73]), etc.

Figure 10 shows the change in the In- and Out-degrees between the two periods. There is high correlation in both cases, i.e. the nodes tend tnd to maintain their In- and Out-degrees. There are, however, some exceptions, like, for example. Costa Rica (CR), which went from an In-degree of 28 in DNJune to 15 in DNAugust. CR was efficient in both periods and this reduction in its In-degree only affected the number of countries for which it served as benchmark. Other countries, like Lesotho (LS), Namibia (NM) and Suriname (SR) have significantly increased the number of countries that dominate them. This is normally related to a worsening of their relative performance. In particular, LS and SR went from being efficient in DNJune to being inefficient in DNAugust (the corresponding distance to the EF were [0.05,

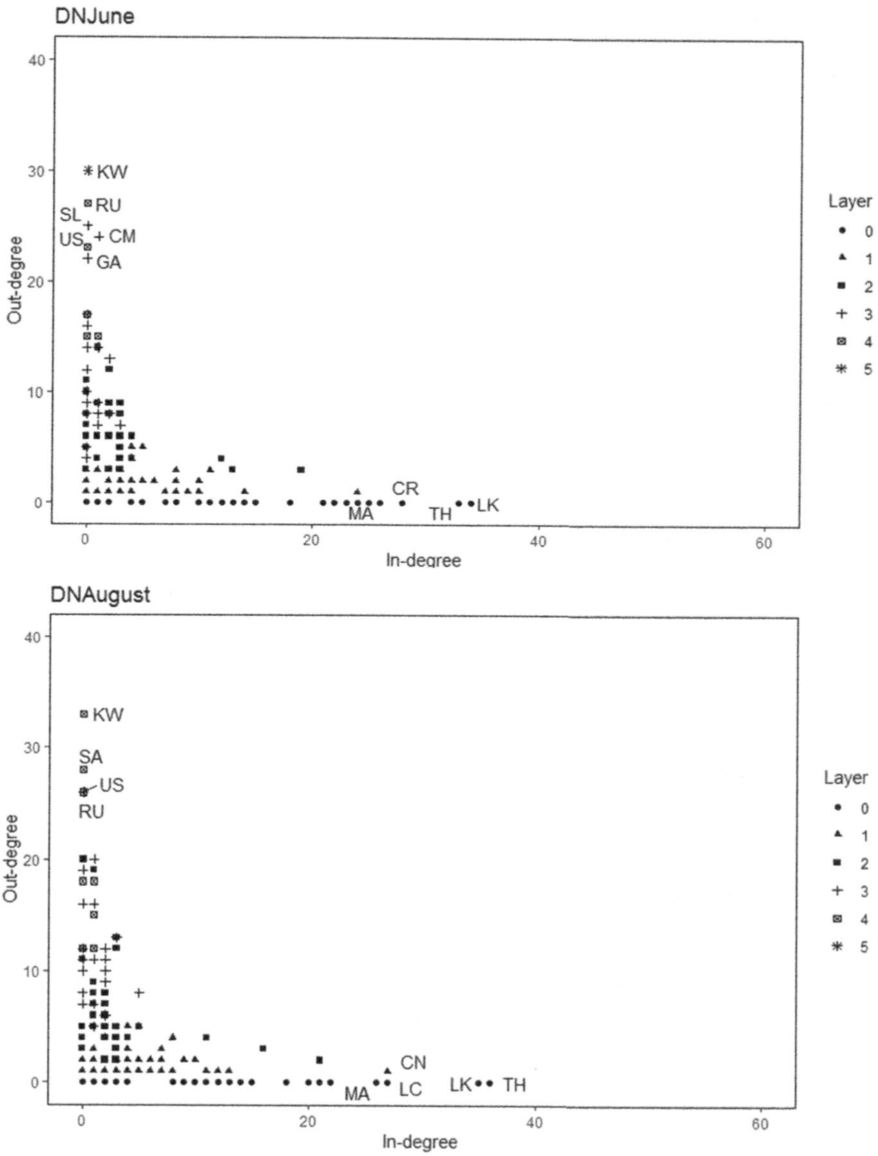

Fig. 8 Out-degree versus In-degree

0.15] and [0.75, 1.40] for LS and SR, respectively). Note also that the two countries with the largest In-degree, namely Thailand (TH) and Sri Lanka (SL) were the same in both periods. Those are the efficient countries that act as reference unit (i.e. as examples of best practice) for a higher number of countries. Other efficient countries

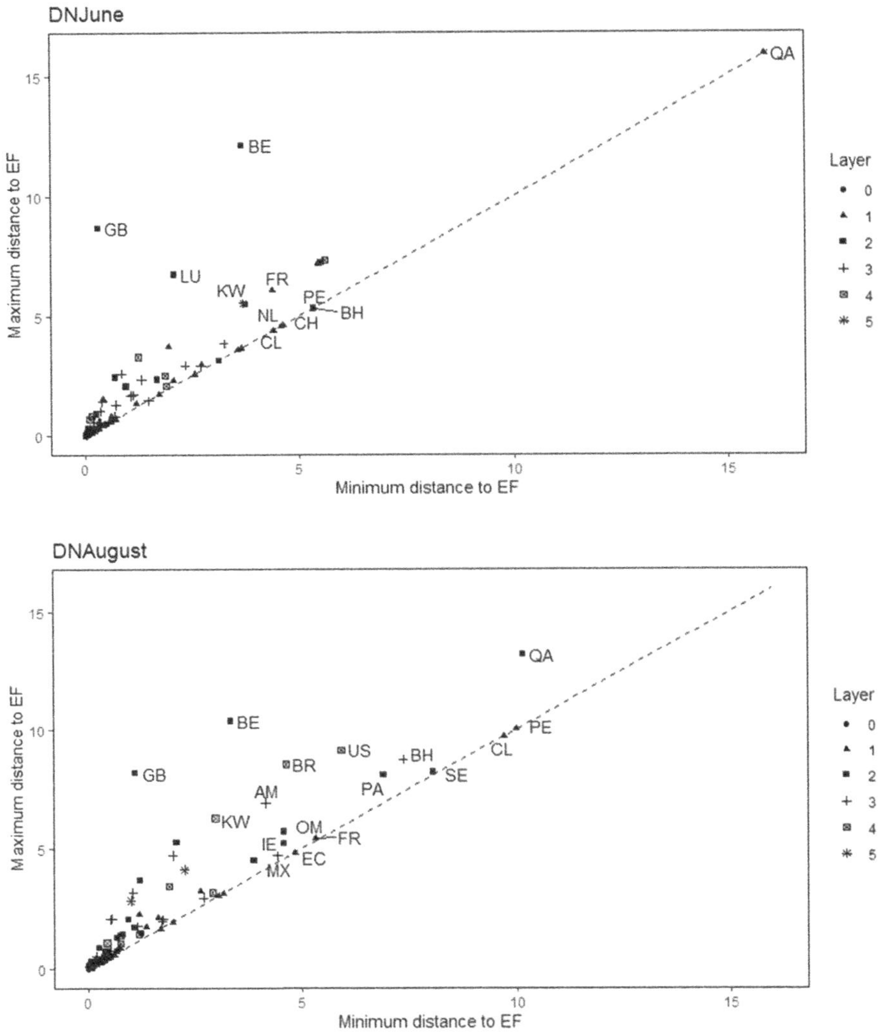

Fig. 9 Máximum versus Minimum distance to EF

that are also reference units for many other countries are Morocco (MA), Santa Lucia (SL) and Costa Rica (CR). Although not efficient, and hence is not a best practice reference unit, China (CN) has outperformed a large number of countries in both periods. Finally, note the two countries with the largest Out-degree, namely South Africa (ZA) and Equatorial Guinea (GQ). They are dominate by large number of countries, countries that even with smaller health expenditures per capita have managed to have a smaller number of COVID-19 cases and deaths. Although to a

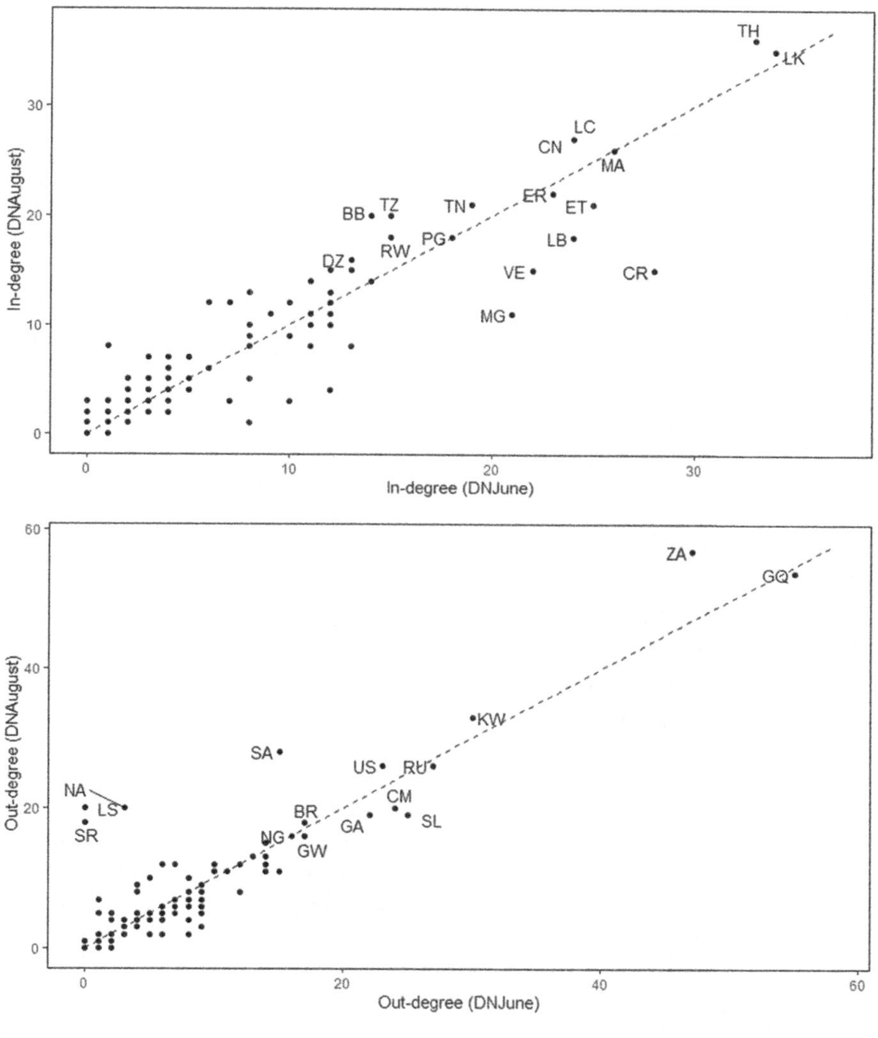

Fig. 10 DNAugust versus DNJune (In-degree and Out-degree)

lesser extent, other countries with a similar situation are Kuwait (KW), Saudi Arabia (SA), Russia (RU) and US.

Figure 11 shows the change in the maximum distance to EF between the two periods. Most countries, all that lay above the diagonal, have worsened their relative performance from DNJune to DNAugust. Most notably Peru (PE), Chile (CL), Brazil (BR), Panama (PN), Bahrain (BH), Armenia (AM) and Oman (OM) but also US and Sweden (SE). Although still rather inefficient in relative terms, Qatar (QA) and Belgium (BE), have improved somewhat from DNJune to DNAugust. United

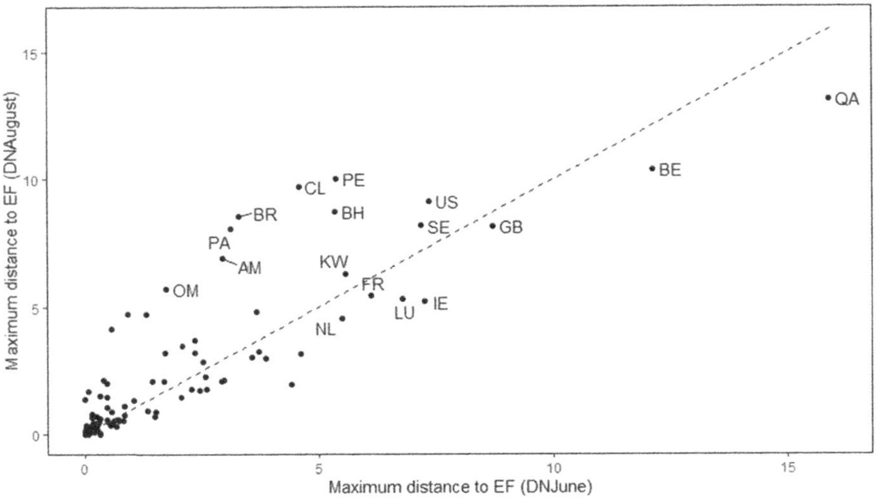

Fig. 11 DNAugust versus DNJune (Maximum distance to EF)

Kingdom (GB), although somewhat less inefficient, have not improved much in relative terms. That is because the two variables considered, number of COVID-19 cases and deaths, are cumulative. Thus, even if the corresponding rates of increase go down, past poor performances are not forgotten.

Figure 12 shows the change in the benchmarking potential between the two periods. Note that this measure only applies to efficient countries, which means that the points plotted correspond to countries that have been labelled efficient in both periods. These are the best performers in relative terms. Some of these countries, namely those on the top half, are those that have the largest benchmarking potential, i.e. they seems to have the cumulative largest inefficiency differences with the countries they dominate. It is not surprising to see in that group countries like Japan (JP), Costa Rica (CR), Australia (AU), New Zealand (NZ), Thailand (TH), Morocco (MA) or Greece (GR). Other countries, like Israel (IL), Turkey (TR), Venezuela (VE) or Bangladesh (BD) have also performed well so far with the pandemic. As we said before, all the countries plotted in the figure have shown the best practice so far.

Finally, Figs. 13 and 14 show the geographical distribution of the efficiency in fighting the COVID-19 pandemic measured through the maximum distance of each country to the EF. Both figures have the same scale to allow their comparison. The darker the fill, the higher the maximum distance to the EF. It can be seen the efficiency of some countries, like Brazil, Chile, Peru, Bolivia, Mexico or South Africa, worsened in DNAugust with respect to DNJune. Other countries however,

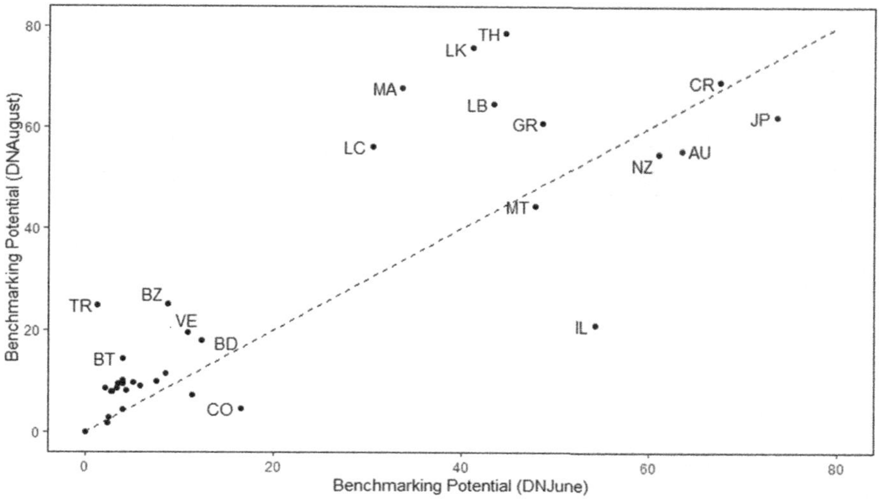

Fig. 12 DNAugust versus DNJune (Benchmarking Potential)

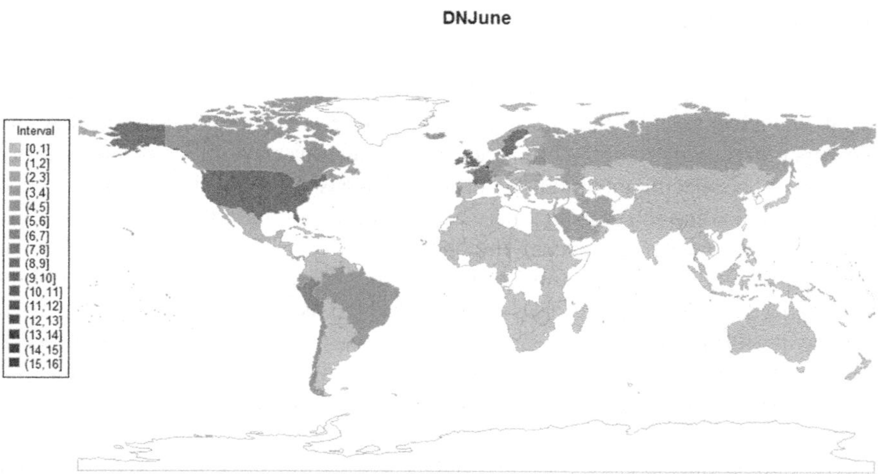

Fig. 13 Maximum distance to EF of the different countries benchmarked (DNJune)

like Iceland, Ireland, Sweden, Russia, UK or France, improved their efficiency in DNAugust with respect to DNJune. Finally, many countries, in Africa, India, South East Asia but also countries like the US, Iran or Saudi Arabia maintain similar efficiency levels.

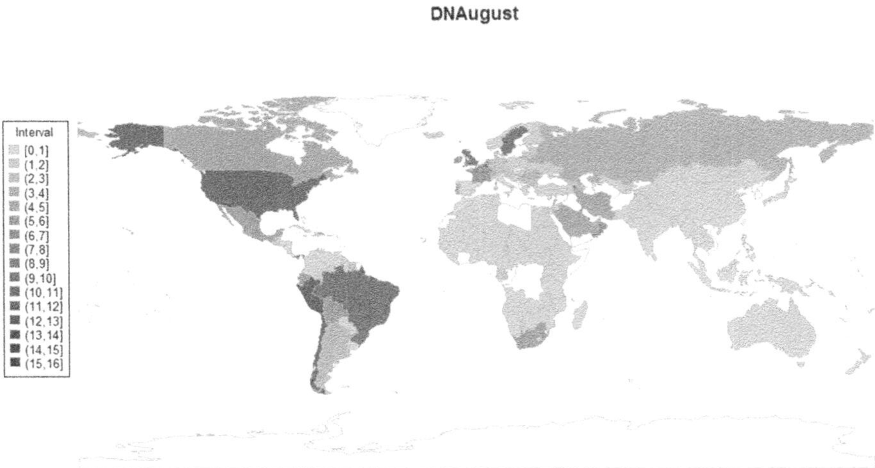

Fig. 14 Maximum distance to EF of the different countries benchmarked (DNAugust)

4 Conclusions

In this chapter the hybridization of DEA and complex networks is advocated. It brings the advantages of the multidimensional benchmarking prowess of DEA and the versatility, computational efficiency and modelling capabilities of the network paradigm. The specific methodology presented is based on DN analysis. It is argued that the dominance relationships between the observations reflect the information on their respective performances. The observations can refer to different entities or to a given entity in different time periods. In any case, their benchmarking allows identifying the best practice cases and, using them as references, the inefficiency, i.e. the performance shortfall, associated to every other observation. For each inefficient observation, the overall margin for improvement and the specific amount to be improved in each dimension can be easily deduced from these dominance relationships.

The simple, layered structure of the DN and its transitive property allows for an easy interpretation of the characterization measures commonly used in complex network analysis (in- and out-degree, geodesic distances, diameter, components, etc.). These indexes can be computed at the node level, at the component level and the whole network level and provide a detailed quantitative picture of the topology and structure of the DN and of the distribution and relative efficiency of the different observations. The graphical capabilities of complex network tools together with appropriate visualization filters (skeleton, bipartite, egonetwork, etc.) enhance the usefulness of this methodology.

As the COVID-19 application presented above has shown, among the strengths of the DNA methodology and its advantages over traditional DEA we can highlight:

- It provides a complete picture of the dataset, providing information not only on the relative inefficiency of an observation with respect to the best practice, but also on the relative inefficiency gap between an inefficient observation and other inefficient observations it dominates.
- While DEA focuses on the EF, DNA determines the whole layered structure of the dataset, identifying the relative position of each observation within the sample.
- While DEA computes a single target for each DMU, normally either the furthest or the closest one, DNA determines all the efficient observations that dominate each DMU, computing the distance to each of them.
- Except for low dimensional datasets, DEA does not provide any visualization capabilities while DNA can plot, for low as well as for high dimensional datasets, the whole set of relationships between the observations and its implicit network structure.

Among the drawbacks and limitations of this approach we can mention the large number of characterization measures that are computed. Paradoxically, the wealth of information provided by the methodology is itself difficult to summarize and communicate. Simplifying the quantitative analysis, focusing on the specific features that may be of interest in each application would be desirable. Another limitation is that the observations that contain the maximum value of a positive variable or the minimum value of a negative variable tend to be (unless there are ties) non-dominated just because of this fact and irrespective of the values of the other variables. This effect, however, is limited unless the number of variables is large with respect to the number of observations, something which do not generally occur in practice. Although knowing which nodes are isolates provides information about those observations but, because they are outliers or mavericks, they cannot be related (in dominance terms) to the rest of the dataset and hence their performance cannot be properly assessed.

There are many other topics for further research in this exciting area of DEA-based data analytics in general and of hybridizing DEA and complex networks in particular. One is integrating the temporal dimension more tightly in the DN analysis. In Sect. 3 we have considered multiple time periods but we have carried a contemporaneous analysis, i.e. we have benchmarked the observations within each period. In some cases, it may make more sense to carry out an intertemporal analysis, pooling all observations of all the periods together. This allows a more direct comparison of the observations in different periods. The issue of missing data (i.e. some observations lack some input or output values) should also be studied. Extending the analysis to interval, ordinal and fuzzy data is also a worthy endeavour. But, most of all, applying the methodology to real-world large multidimensional datasets and proving its usefulness is the true challenge ahead.

Acknowledgements This research was carried out with the financial support of the Spanish Ministry of Economy, Industry and Competitiveness, and the European Regional Development Fund (ERDF), grant DPI2017-85343-P.

References

Calzada-Infante, L., & Lozano, S. (2016). Analyzing Olympic Games through dominance networks. *Physica A, 462*, 1215–1230. https://doi.org/10.1016/j.physa.2016.07.001

Calzada-Infante, L., & Lozano, S. (2020). Some variants and extensions of dominance network analysis. *Computers and Industrial Engineering, 141*, 106275. https://doi.org/10.1016/j.cie.2020.106275

Emrouznejad, A., Tavana, M., & Hatami-Marbini, A. (2014). The state of the art in fuzzy data envelopment analysis. *Studies in Fuzziness and Soft Computing, 309*, 1–45. https://doi.org/10.1007/978-3-642-41372-8_1

Esteve, M., Aparicio, J., Rabasa, A., & Rodriguez-Sala, J. J. (2020). Efficiency analysis trees: A new methodology for estimating production frontiers through decision trees. *Expert Systems with Applications*. https://doi.org/10.1016/j.eswa.2020.113783

Ghahraman, A., & Prior, D. (2016). A learning ladder toward efficiency: Proposing network-based stepwise benchmark selection. *Omega, 63*, 83–93. https://doi.org/10.1016/j.omega.2015.10.004

Gutiérrez, E., & Lozano, S. (2013). Avoidable damage assessment of forest fires in European countries: An efficient frontier approach. *European Journal of Forest Research, 132*, 9–21. https://doi.org/10.1007/s10342-012-0650-5

Gutiérrez, E., & Lozano, S. (2014). A DEA approach to performance-based budgeting of formula one constructors. *Journal of Sports Economics, 15*(2), 180–200. https://doi.org/10.1177/1527002512447629

Gutiérrez, E., & Lozano, S. (2020a). Efficiency performance of current account-BoP flows in advanced world economies considering GHG emissions. *Journal of Cleaner Production, 254*, 120139. https://doi.org/10.1016/j.jclepro.2020.120139

Gutiérrez, E., & Lozano, S. (2020b). Benchmarking formula one auto racing circuits: A two stage DEA approach. *Operational Research*. https://doi.org/10.1007/s12351-018-0416-z

Hatami-Marbini, A., Emrouznejad, A., & Tavana, M. (2011). A taxonomy and review of the fuzzy data envelopment analysis literature: Two decades in the making. *European Journal of Operational Research, 214*, 457–472. https://doi.org/10.1016/j.ejor.2011.02.001

Hermoso, R., Latorre, M. P., & Martinez-Nuñez, M. (2019). Multivariate data envelopment analysis to measure airline efficiency in European airspace: A network-based approach. *Applied Sciences, 9*, 5312. https://doi.org/10.3390/app9245312

Ho, M. H. C., Liu, J. S., Lu, W. M., & Huang, C. C. (2014). A new perspective to explore the technology transfer efficiencies in US universities. *Journal of Technology Transfer, 39*(2), 247–275. https://doi.org/10.1007/s10961-013-9298-7

Kao, T. W. D., Simpson, N. C., Shao, B. B. M., & Lin, W. T. (2017). Relating supply network structure to productive efficiency: A multi-stage empirical investigation. *European Journal of Operational Research, 259*(2), 469–485. https://doi.org/10.1016/j.ejor.2016.11.008

Koltai, T., Lozano, S., Uzonyi-Kecskés, J., & Moreno, P. (2017). Evaluation of the results of a production simulation game using a dynamic DEA approach. *Computers and Industrial Engineering, 105*, 1–11. https://doi.org/10.1016/j.cie.2016.12.048

Lee, J., & Choi, G. (2019). A dominance-based network method for ranking efficient decision-making units in data envelopment analysis. *Sustainability, 11*(2059), 1–20. https://doi.org/10.3390/su11072059

Leem, B. H., & Chun, H. (2015). Measuring the influence of efficient ports using social network metrics. *International Journal of Engineering Business Management, 7*(1), 1–8. https://doi.org/10.5772/60040

Liu, J. S., & Lu, W. M. (2010). DEA and ranking with the network-based approach: A case of R&D performance. *Omega, 38*, 453–464. https://doi.org/10.1016/j.omega.2009.12.002

Liu, J. S., & Lu, W. M. (2012). Network-based method for ranking of efficient units in two-stage DEA models. *Journal of the Operational Research Society, 63*, 1153–1164. https://doi.org/10.1057/jors.2011.132

Liu, J. S., Lu, W. M., Yang, C., & Chuang, M. (2009). A network-based approach for increasing discrimination in data envelopment analysis. *The Journal of the Operational Research Society, 60*(11), 1502–1510. https://doi.org/10.1057/jors.2009.35

Lozano, S., & Calzada-Infante, L. (2017). Dominance network analysis of economic efficiency. *Expert Systems with Applications, 82*, 53–66. https://doi.org/10.1016/j.eswa.2017.04.004

Lozano, S., & Calzada-infante, L. (2018). Efficiency assessment using network analysis tools. *Journal of the Operational Research Society, 69*(11), 1803–1818. https://doi.org/10.1080/01605682.2017.1409866

Lozano, S., & Calzada-Infante, L. (2019). Efficiency ranking using dominance network and multiobjective optimization indexes. *Expert Systems with Applications, 126*, 83–91. https://doi.org/10.1016/j.eswa.2019.02.016

Lozano, S., & Gutiérrez, E. (2008). Data envelopment analysis of the human development index. *International Journal of Society Systems Science, 1*(2), 132–150. https://doi.org/10.1504/IJSSS.2008.021916

Lozano, S., Villa, G., Guerrero, F., & Cortés, P. (2002). Measuring the performance of nations at the Summer Olympics using data envelopment analysis. *Journal of the Operational Research Society, 53*(5), 501–511. https://doi.org/10.1057/palgrave/jors/2601327

Lozano, S., Adenso-Díaz, B., & Barba-Gutiérrez, Y. (2011). Russell non-radial eco-efficiency measure and scale elasticity of a sample of electric/electronic products. *Journal of the Franklin Institute, 348*, 1605–1614. https://doi.org/10.1016/j.jfranklin.2011.02.005

Moreno, P., & Lozano, S. (2014). A network DEA assessment of team efficiency in the NBA. *Annals of Operations Research, 214*, 99–124. https://doi.org/10.1007/s10479-012-1074-9

Moreno, P., & Lozano, S. (2018). Super SBI Dynamic Network DEA approach to measuring efficiency in the provision of public services. *International Transactions in Operational Research, 23*, 715–735. https://doi.org/10.1111/itor.12257

Newman, M. E. J. (2002). Assortative mixing in networks. *Physical Review Letters, 89*(20), 208701. https://doi.org/10.1103/PhysRevLett.89.208701

Newman, M. E. J. (2010). *Networks of networks – An introduction*. Oxford University Press.

Ritchie, H., Ortiz-Ospina, E., Beltekian, D., Mathieu, E., Hasell, J., Macdonald, B., Giattino, C., & Rose, M. (2020). *Coronavirus Pandemic (COVID-19) – The data*. https://ourworldindata.org/coronavirus-data. Accessed 2 July 2020.

Seiford, L. M., & Zhu, J. (2003). Context-dependent data envelopment analysis – Measuring attractiveness and progress. *Omega, 31*, 397–408. https://doi.org/10.1016/S0305-0483(03)00080-X

Simon de Blas, C., Simon Martin, J., & Gomez Gonzalez, D. (2018). Combined social networks and data envelopment analysis for ranking. *European Journal of Operational Research, 266*(3), 990–999. https://doi.org/10.1016/j.ejor.2017.10.025

Toloo, M., Keshavarz, E., & Hatami-Marbini, A. (2018). Dual-role factors for imprecise data envelopment analysis. *Omega, 77*, 15–31. https://doi.org/10.1016/j.omega.2017.05.005

Villa, G., & Lozano, S. (2016). DEA with non-monotonic variables. Application to EU governments' macroeconomic efficiency. *Journal of the Operational Research Society, 67*, 1510–1523. https://doi.org/10.1057/jors.2016.36

Wang, W.-K., Lu, W.-M., Kweh, Q. L., Nourani, M., & Hong, R.-S. (2019). Interlocking directorates and dynamic corporate performance: The roles of centrality, structural holes and number of connections in social networks. *Review of Managerial Science*, 1–21. https://doi.org/10.1007/s11846-019-00347-2

World Population Review. (2020). *Life expectancy by country 2020*. https://worldpopulationreview.com/countries/life-expectancy. Accessed 15 June 2020.

WorldBank. (2020). *Current health expenditure per capita, PPP (current international $) | Data*. https://data.worldbank.org/indicator/SH.XPD.CHEX.PP.CD. Accessed 6 July 2020.

Zhu, J. (2003). Imprecise data envelopment analysis (IDEA): A review and improvement with an application. *European Journal of Operational Research, 144*, 513–529. https://doi.org/10.1016/S0377-2217(01)00392-7

Zhu, J. (2004). Imprecise DEA via standard linear DEA models with a revisit to a Korean mobile telecommunication company. *Operations Research, 52*(2), 323–332. https://doi.org/10.1007/s10479-020-03668-8

Zhu, J. (2020). DEA under big data: Data enabled analytics and network data envelopment analysis. *Annals of Operations Research*. https://doi.org/10.1007/s10479-020-03668-8

Value Extracting in Relative Performance Appraisal with Network DEA: An Application to U.S. Equity Mutual Funds

Hirofumi Fukuyama and Don U. A. Galagedera

1 Introduction

Data analytics practitioners aim to assist businesses acquire data-driven insights so that their implementation may help businesses gain competitive edge over their competitors. Two ways in which businesses may gain competitive edge over their competitors are cost reduction and best practices adoption. An analytical procedure designed to assess performance of an entity relative its peers is data envelopment analysis (DEA). DEA application enable ranking of entities based on relative performance where a relatively poor performing entity is able to identify its competitors that adopt best practices under the technology that they operate. Hence, analysts recognise DEA as an empirically useful relative performance appraisal methodology. Conventional DEA models can accommodate multiple inputs and multiple outputs in performance appraisal. This is a clear advantage in DEA application as in performance appraisal many measures of performance may be included. As we shall see later, when applying DEA to assess performance of financial entities, studies consider many different factors of performance along with risk and return. Risk and return measures are the commonly used factors in financial entity performance appraisal. Several DEA studies assess financial entity performance with cost measures along with risk and return. When assessing performance with DEA, it is possible to incorporate multiple factors, thereby assess relative performance from a multidimensional framework, and at the same time

H. Fukuyama (✉)
Fukuoka University, Fukuoka, Japan
e-mail: fukuyama@fukuoka-u.ac.jp

D. U. A. Galagedera
Department of Econometrics and Business Statistics, Monash Business School, Monash University, Caulfield East, VIC, Australia
e-mail: Tissa.Galagedera@monash.edu

J. Zhu, V. Charles (eds.), *Data-Enabled Analytics*, International Series in Operations Research & Management Science 312, https://doi.org/10.1007/978-3-030-75162-3_10

identify role models of poor performers. Mahajan (1991) recognising diversity of information that DEA application extracts, brand DEA as *data envelopment analytics*.

Now, there are other challengers in relative performance appraisal. Most notably, the advent of big data. Recent DEA literature reveal attempts of DEA researchers and practitioners addressing this issue (Badiezadeh et al., 2018; Zhu et al., 2018; Khezrimotlagh et al., 2019; Zhu, 2020). Citing production efficiency evaluation and productivity measurement as traditional uses of DEA, Zhu (2020) points out that DEA is evolving into *data enabled analytics* with studies relying on DEA as a data-oriented scientific analytical tool. Zhu (2020) highlights that network DEA link closely to certain desirable outcomes in big data analytics. Specifically, Zhu (2020) emphasises value of insights obtained through network DEA as a significant contribution that network DEA could make in the big data field. In this chapter, we provide empirical evidence to support this claim by applying network DEA to managed fund performance appraisal.

A managed fund is a financial entity where investors deposit money with the aim of receiving financial rewards sometime in the future. There are many types of managed funds. Among the relatively heavily resourced managed fund types is mutual funds. Generally, mutual funds are favoured as a retirement plan option. The country with the largest share of the global mutual fund (MF) industry by far is U.S. with $21.3 trillion in total net assets and approximately 8000 MFs at year-end 2019.[1] There are different types of MFs: equity funds, fixed-income funds, money market funds and balanced funds. The investment objectives of these types differ by level of risk and reward potential. MF portfolio managers invest funds in accordance with the stated objectives of the fund. This is not an easy task as investment landscape is widening, investors continue to demand positive rewards, global stock prices experience downward pressure, and equity market volatility is unstable. Operation of a MF is a task of the investment firm. Investment firms raise money by selling shares to the investors. MF manager invests money to accomplish the stated objectives of the fund. Therefore, when assessing MF performance, it is not prudent to ignore the role of the investment firm.

Early MF performance measures are two-dimensional. Two-dimensional performance measures combine return and risk of a MF into a single measure. Sharpe (1966) ratio and Treynor (1965) ratio belong to this category. Sharpe ratio considers total risk measured as the standard deviation of the returns whereas Treynor ratio considers systematic risk referred to as beta. These measures use historical information (time series of returns) in their computation. MF industry consider them as important indicators of MF performance. However, it is questionable whether two-dimensional risk-adjusted return measures really represent investment firm performance. It is true that portfolio management performance is an influential factor of investment firm performance. However, processes such as administration management and cost management are also important factors of investment firm

[1] https://www.icifactbook.org/ch3/20_fb_ch3

performance. When appraising performance with multiple factors of performance in addition to risk and return that becomes an appraisal under a multi-dimensional framework. Such an undertaking requires data on many factors and linking them to a composite performance measure of MF management process. DEA is a methodology that can address all these issues. DEA is a data driven performance appraisal methodology. The ability of DEA to accommodate management processes comprising of multiple sub processes demonstrates versatility in its application. Network DEA is the strand of literature that covers this aspect of DEA application. Network in this case is the layout (structure) of an overall process conceptualised as a linked set of sub processes. Zhu (2020) points out that conceptualisation of an overall management process as a network representation in performance appraisal of decision-making units (DMUs) using DEA is consistent with the value dimension of big data.

Discussion on how to conceptualise overall MF management process is ongoing. MF performance appraisal using DEA started in the late 1990s (Murthi et al., 1997; McMullen & Strong, 1998). Murthi et al. (1997) develop a performance index incorporating transaction costs such as expense ratio, turnover and loads under the DEA framework. They highlight that their DEA based performance index is consistent with measures of efficiency encountered in production economics. Murthi et al. (1997) interpret ratio of output to input as an indicator of how well the MF manager utilises input (resources) in generating output. McMullen and Strong (1998) demonstrate that DEA is a useful performance appraisal methodology in MF selection when an investor's expected utility function contains more attributes than the traditional mean and variance terms. They highlight that a single composite score that DEA estimates is appealing because interpretation of a single score is much simpler than interpreting several statistics. Application of DEA for MF performance appraisal advanced from then on.

Recent studies of MF performance appraisal using DEA focus on extracting value through conceptualization of the overall fund management process as a network of sub processes. For example, Premachandra et al. (2012) and Galagedera et al. (2016) consider MF management process as comprising of two serially linked sub processes as shown in Panel (a) of Fig. 1. These studies label the sub processes as operational management and portfolio management. They differ in the selection of input, intermediate and output measures. The two-dimensional performance measures commonly used in the finance industry such as Sharpe ratio and Treynor ratio are concentrated on the returns of the fund portfolio and hence aligns with assessment of the portfolio management aspect of the overall fund management process. Premachandra et al. (2012) argue that when assessing overall performance of a MF, one should not ignore the role played by the investment firm that operates the MF. Their suggestion is instead of considering MF management process as a black box operation, should expand (look inside) the black box to gain information on MF performance from different aspects of fund management. The network DEA model that Premachandra et al. (2012) propose is not restrictive in orientation (input orientation and output orientation) and returns to scale (constant and variable)

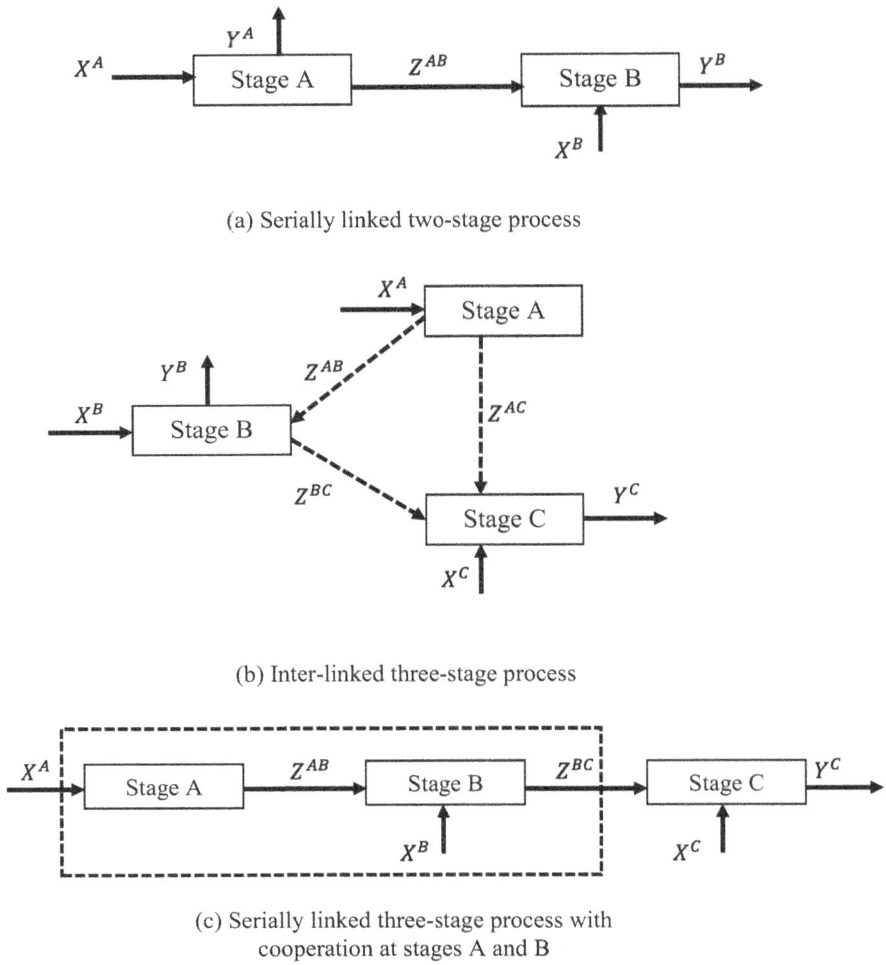

Fig. 1 Some network configurations adopted in mutual fund performance appraisal

and enable decomposition of the overall performance at the sub process level (Premachandra et al., 2016).

Sánchez-González et al. (2017) consider a variant of the governance structure proposed by Berkowitz and Qiu (2003) as MF management process. Their network structure shown in Panel (b) of Fig. 1 comprise of three sub processes; portfolio management, marketing and selling, and operations management. In Premachandra et al. (2012) network (two serially linked sub processes) representation of the MF management process, use one link (intermediate measure) which is net asset value. In Sánchez-González et al. (2017) representation of MF overall management process, three links are established with four intermediate measures. Number of funds and fund types link portfolio management, and marketing and selling sub processes. Gross returns links portfolio management and operational management sub processes. Money flows links marketing and selling, and operational management sub processes. In Premachandra et al. (2012) and Sánchez-González et al. (2017) studies operational management is one of the sub processes of the network representation of the MF overall management process. These two studies conceptualise operational management function differently. Premachandra et al. (2012) view operational management function as a sub process aiming to secure highest net asset value whereas Sánchez-González et al. (2017) consider operational management function as a sub process aiming to obtain highest possible profit. These applications clearly demonstrate versatility of network representation of MF management process in particular and production processes in general.

Galagedera et al. (2018) conceptualize MF management process as a serially linked three-stage process shown in Panel (c) of Fig. 1. They label the sub processes as operational management, resource management, and portfolio management. They extend the two-stage representation of MF management process in Premachandra et al. (2012) to a three-stage representation. Galagedera et al. (2018) consider net assets of a MF as its total assets transformed through a management process different from operational and portfolio management processes. Galagedera et al. (2018) refer to the MF management process that transforms total assets to net assets as resource management process. Moreover, they argue that the individual stage-level management processes operate under different levels of risk exposure. Galagedera et al. (2018) argue that MF portfolio management is a high-risk undertaking compared to operational and resource management and refer to them collectively as an allied process. In the network structure in Panel (c) of Fig. 1, the dashed-lined rectangle indicates alliance between operational and resource management sub processes.

In essence, the studies that appraise MF management performance using DEA where a network of sub-processes represents the overall MF management process provide insightful information on performance of the MF management process from different aspects of management. See Kao (2014) for a review of network structures used in DEA studies, and Basso and Funari (2016) for a survey of MF performance appraisal studies that use DEA. Table 10 in Appendix A gives details including input, intermediate and output measures used in network DEA studies of MF performance appraisal.

As illustrated in Fig. 1, network representation of MF management process adopted in DEA-based performance appraisal studies have different structures. Each network structure represents multiple functional areas of the MF management process as linked sub-processes. Basso and Funari (2016) point out that the number of papers published yearly on MF performance appraisal using DEA increased rapidly from 2003. They tabulate 77 DEA-based MF performance appraisal papers published during the period 2003 to 2014. Out of these 77 papers only one study (Premachandra et al., 2012) conceptualises MF overall management process as a network process. The other 76 papers differ in stylised facts such as input-output measures, returns to scale assumption and orientation of DEA application.

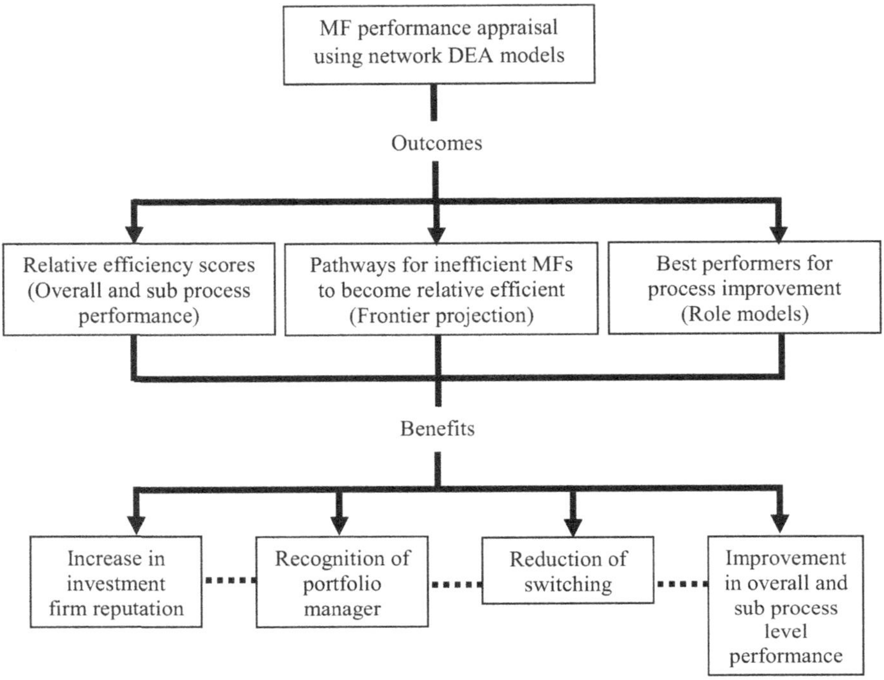

Fig. 2 Value extraction in performance appraisal with network DEA

All studies that consider network representation of MF management process for performance appraisal highlight that network representation uncover information that black box representation cannot. A clear advantage is that overall performance assessed through network structures may be decomposed to assess sub process or stage-level performance. This way, investment firms can determine MF performance from different aspects of management. Figure 2 presents a flow chart depicting how network DEA application for MF performance appraisal may add value to different stakeholders.

Figure 2 portrays value extraction through network DEA application of MF performance appraisal as a two-tier progression. The first tier indicates the direct outcomes (results) of network DEA model application. We classify direct outcomes broadly into three categories; relative efficiency scores, frontier projections and role models. DEA efficiency scores are relative and therefore could rank MFs based on overall performance as well as on sub process performance. This information is valuable to both portfolio managers and investors. Portfolio manager will be able to gauge her investment management performance compared to competitors. Investors on the other hand will be able to make informed decisions in fund selection. Frontier projections provide pathways for inefficient performers to reach frontier of best performance and become efficient. This information is valuable to fund managers to manage resources efficiently. In addition to pathways, DEA identifies from among best performers who would serve as role models to inefficient performers. Inefficient MFs may adopt the best practises of their role models or at least consider the best practices for guidance towards performance improvement. The second tier shows benefits of adoption of using first tier information favourably. Clearly, an outcome of implementation of best practices is improved MF overall and sub process level performance. When investors realise their investment is well managed they are more likely to be loyal to the investment firm resulting in less redemption and brand switching. Portfolio managers who implement best practices will gain recognition in the fund management industry resulting in increased demand for their services. When fund managers of the investment firm perform well with good outcomes to investors, investment firm reputation will increase. Interestingly, the aforementioned benefits match standards expected in performance evaluation criteria such as risk-adjusted return, performance relative to markets, performance relative to peers, and portfolio quality. In this, we see that network DEA application for MF performance appraisal may bring benefits to all stakeholders.

The remainder of this chapter is organized as follows. Section 2 provides a discussion of MF performance appraisal from disbursement management perspective. Section 3 develops network DEA models for technical and disbursement efficiency evaluation. An empirical application of the models proposed in this chapter is in Sect. 4 followed by a discussion on extending MF performance appraisal from multiple input management perspectives. The chapter concludes in Sect. 6 with some remarks after a discussion of robustness of the results in Sect. 5.

2 Mutual Fund Performance Appraisal from Disbursement Management Perspective

MF fees is a concern to many investors. This is evident in the U.S. with number of investment firms being subject of litigation for charging excessive fees (Galagedera et al., 2020). U.S. Securities and Exchange Commission (SEC) point out the need for investors to assess a fund's costs because they can have an enormous impact on returns (U.S. Securities and Investment Commission, 2000). During the period 2014–2018, management fees that makes up a fund's total expense increased by 5%. Change in total expense ratios vary by fund type. For example, total expense and management fee ratios for equity, bond, and mixed asset mutual funds decreased by at least three percent since 2014 (McCune, 2019). In 2018, active-fund investors in U.S. paid 4.5 times than passive-fund investors (Morningstar, 2019).

Management fees charged by managed funds in relatively smaller economies is no exception. Canstar is a company that rates thousands of products including banking, insurance and superannuation in Australia.[2] According to the managed

[2]https://www.canstar.com.au/

funds rated in 2018 by Canstar, management fees generally ranges between 0.19% and 2.50%, with the range varying across fund types.[3] An additional ongoing fee that some managed funds charge is administration fee. Canstar reports that the maximum administration fee charged in 2018 was 0.15%. Another type of fee charged by some managed funds is performance fee. The highest performance fee of managed funds reported in 2018 is 35%. Performance fees are generally paid when the managed fund outperforms a benchmark. In this, we see that fees charged by managed funds is highly variable. This makes MF selection from among many a difficult task. Hence, it is important to assess MF performance focussing exclusively on cost and expense management. One such study is Galagedera et al. (2020). They refer to costs and expenses as disbursements and propose a network DEA model to determine how efficiently MFs manage disbursements in generating output (return) and refer to the DEA-based efficiency measure as disbursement efficiency. SEC's mutual fund tips suggest fund size, tax consequences, risks, and volatility are factors that investors should consider in addition to past performance (returns) when considering investments in MFs (U.S. Securities and Investment Commission, 2000). Therefore, for a comprehensive assessment of MF performance, it is prudent to consider multiple factors of MF performance and MF management structure. Network DEA is ideally suited for such a task.

In this chapter, we discuss a methodology to determine MF management efficiency from different management perspectives conceptualising MF overall management process as a network process. The idea is to develop a framework to assess management performance from different perspectives of management within an overall performance-appraisal framework. We determine overall performance (technical efficiency) by considering all measures depicted in Fig. 3 as factors of performance.[4]

When determining disbursement management performance, we consider a subset of the input measures considered in technical efficiency evaluation and retain the intermediate measures (total net assets), and output measures (1-year, 3-year and 5-year return) as observed.[5] The subset of input measures include measures associated with disbursement only. Turnover ratio, total risk, downside risk and systematic risk are risk measures. Disbursement measures are marketing and distribution expenses, operational costs and management fees. The input, intermediate and output measures shown inside the dashed-lined rectangle of Fig. 3 are the measures

[3] https://www.canstar.com.au/star-rating-reports/managed-funds-2018/

[4] The measures shown in Figure 3 are from Galagedera et al. (2020). When selecting measures for MF performance appraisal, Galagedera et al. (2020) refer to previous DEA based MF performance appraisal studies (Murthi et al., 1997; Morey & Morey, 1999; Choi & Murthy, 2001; Basso & Funari, 2001; Premachandra et al., 2012; Basso & Funari, 2017). Choi and Murthy (2001) consider total risk and transaction costs as inputs. They assume that expense ratios, loads and turnover ratios measure transaction costs involved in constructing a portfolio, which is the stage 2 sub-process that we refer to as the portfolio management stage. Use of returns as output is common in all studies.

[5] Including 1-year, 3-year and 5-year return together is not a problem because it is not necessary to aggregate outputs prior to DEA application.

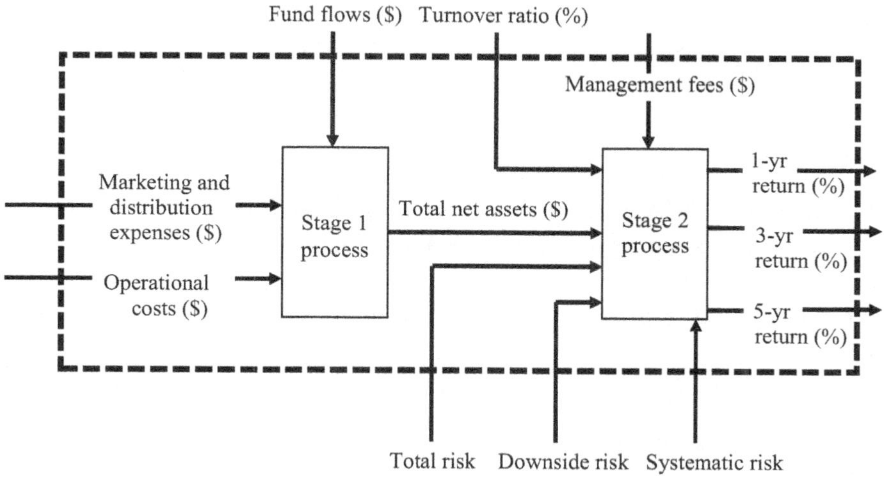

Fig. 3 Configuration of measures in disbursement and overall efficiency evaluation

considered in disbursement-management performance appraisal. Disbursement efficiency score reflects how efficiently a given MF manages its costs and expenses in return generation. The measures shown outside the rectangle (fund flows and risk) are the additional measures considered in overall performance appraisal (technical efficiency computation).

As shown in Fig. 3, intermediate measures and output measures are common in technical efficiency, risk and disbursement efficiency, and disbursement efficiency evaluation. The difference is only in the choice of input measures. The proposed variable configurations nests within an overall variable structure and hence could compare separately assessed technical efficiency and disbursement efficiency meaningfully. According to our modelling framework, disbursement efficiency score of a given MF is less than or equal to its technical efficiency score. This we state in Sect. 3 under Remark 2. When developing network DEA models for MF disbursement management performance, we are guided by the concept underlying 'capacity' discussed in Johansen (1968). See Färe (1984) for necessary and sufficient conditions for applicability of the capacity concept in production. Färe et al. (1989) develop linear programming models to determine production capacity from observed input-output data. Färe et al. (1989) developed linear programming models do not require input prices. Hence, capacity under the notion of production management can be determined by formulating models under the DEA framework. When appraising disbursement management performance we follow this lead.

Fig. 4 Variable configuration for technical efficiency evaluation of DMU_j

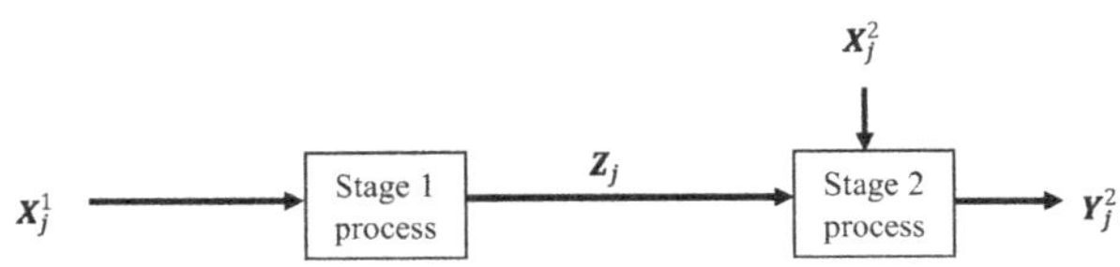

3 Network DEA Models for Technical Efficiency and Disbursement Efficiency Evaluation

3.1 Technical Efficiency Evaluation (Fig. 4)

Let $\boldsymbol{X}_j^1 = \left(x_{1j}^1, x_{2j}^1, \ldots, x_{N_{X1}j}^1\right)$ denote the inputs and $\boldsymbol{Z}_j = \left(Z_{1j}, Z_{2j}, \ldots, Z_{N_Z j}\right)$ denote the outputs at stage 1, $\boldsymbol{X}_j^2 = \left(x_{1j}^2, x_{2j}^2, \ldots, x_{N_{X2}j}^2\right)$ denote the inputs and $\boldsymbol{Y}_j^2 = \left(Y_{1j}^2, Y_{2j}^2, \ldots, Y_{N_{Y2}j}^2\right)$ denote the outputs at stage 2 of DMU_j. $\boldsymbol{Z}_j$ links stage 1 and stage 2 and therefore they serve as intermediate measures. Further, let $\boldsymbol{\lambda}^{(1)} = \left(\lambda_1^{(1)}, \lambda_2^{(1)}, \ldots, \lambda_J^{(1)}\right)$ and $\boldsymbol{\lambda}^{(2)} = \left(\lambda_1^{(2)}, \lambda_2^{(2)}, \ldots, \lambda_J^{(2)}\right)$ such that $\sum_{j=1}^{J} \lambda_j^{(1)} = 1$ and $\sum_{j=1}^{J} \lambda_j^{(2)} = 1$ be vectors of intensity measures applicable to stage 1 and stage 2 respectively where J is the number of DMUs investigated. The conditions imposed on the intensity variables implies that DMUs operate under variable returns to scale (VRS) technology.[6] Then, an output-oriented network DEA model that computes the overall (technical) efficiency score of DMU_j under the VRS assumption[7] is

$$\max \beta_k^T \tag{1}$$

$$\text{Subject to} \sum_{j=1}^{J} \lambda_j^{(1)} \boldsymbol{X}_j^1 \le \boldsymbol{X}_k^1$$

$$\sum_{j=1}^{J} \lambda_j^{(1)} \boldsymbol{Z}_j \ge \boldsymbol{Z}_k$$

$$\sum_{j=1}^{J} \lambda_j^{(2)} \boldsymbol{Z}_j \le \boldsymbol{Z}_k$$

[6]The production possibility sets of the individual stages together forms the production possibility set of the two-stage system (Färe & Grosskopf, 2000).

[7]Choi and Murthy (2001) point out that MFs generally operate at different scales (increasing, constant, or decreasing), which may affect portfolio performance regardless of managerial skill. Therefore, it is important to control for scale effect in evaluating performance.

$$\sum_{j=1}^{J} \lambda_j^{(2)} \boldsymbol{X}_j^2 \leq \boldsymbol{X}_k^2$$

$$\sum_{j=1}^{J} \lambda_j^{(2)} \boldsymbol{Y}_j^2 \geq \beta_k^T \boldsymbol{Y}_k^2$$

$$\sum_{j=1}^{J} \lambda_j^{(1)} = 1$$

$$\sum_{j=1}^{J} \lambda_j^{(2)} = 1$$

$$\boldsymbol{\lambda}^{(1)} \geq \boldsymbol{0};\ \boldsymbol{\lambda}^{(2)} \geq \boldsymbol{0};\ \beta_k^T \text{unrestricted}$$

Let β_k^{T*} denote the optimal objective function value of model (1). Then, technical efficiency score of DMU_k (TE_k^*) is given by 1/ β_k^{T*}. β_k^{T*}reveals whether or not DMU_k operates at its full potential with current levels of inputs. $\beta_k^{T*} = 1$ indicates that DMU_k is technical efficient implying that all inputs are utilized to its full potential within the production process and $\beta_k^{T*} > 1$ indicates that DMU_k is technical inefficient suggesting that output augmentation is possible with current levels of inputs. $\beta_k^{T*} > 1$ may be interpreted as the proportional increase in the outputs that renders DMU_k technical efficient. In other words, technical inefficient DMU_k will get projected on to the technical efficient frontier when the outputs $\boldsymbol{Y}_k^2$ increase to $\beta_k^{T*}\boldsymbol{Y}_k^2$. Hence, $\beta_k^{T*}\boldsymbol{Y}_k^2$ provide a set of output targets to inefficient DMU_k to become technical efficient.

3.2 *Disbursement Efficiency Evaluation (Fig. 5)*

Now let, $\boldsymbol{X}_j^{1D} = \left(X_{1j}^{1D}, X_{2j}^{1D}, \ldots, X_{N_{D1}j}^{1D}\right) \subset \boldsymbol{X}_j^1$ and $\boldsymbol{X}_j^{2D} = \left(X_{1j}^{2D}, X_{2j}^{2D}, \ldots, X_{N_{D2}j}^{2D}\right) \subset \boldsymbol{X}_j^2$ denote subsets of disbursement measures as stage 1 inputs and stage 2 inputs respectively and $\boldsymbol{X}_j^{1O} = \left(X_{1j}^{1O}, X_{2j}^{1O}, \ldots, X_{N_{O1}j}^{1O}\right) = \boldsymbol{X}_j^1 - \boldsymbol{X}_j^{1D}$ and $\boldsymbol{X}_j^{2O} = \left(X_{1j}^{2O}, X_{2j}^{2O}, \ldots, X_{N_{O2}j}^{2O}\right) = \boldsymbol{X}_j^2 - \boldsymbol{X}_j^{2D}$ denote the subsets of 'other inputs' at stage 1 and stage 2 respectively of DMU_j.[8] Other inputs are the inputs

[8] $\boldsymbol{X}_j^1 = \boldsymbol{X}_j^{1D} \cup \boldsymbol{X}_j^{1O}$ and $\boldsymbol{X}_j^2 = \boldsymbol{X}_j^{2D} \cup \boldsymbol{X}_j^{2O}$. $N_{D1} + N_{O1} = N_{X1}$ and $N_{D2} + N_{O2} = N_{X2}$.

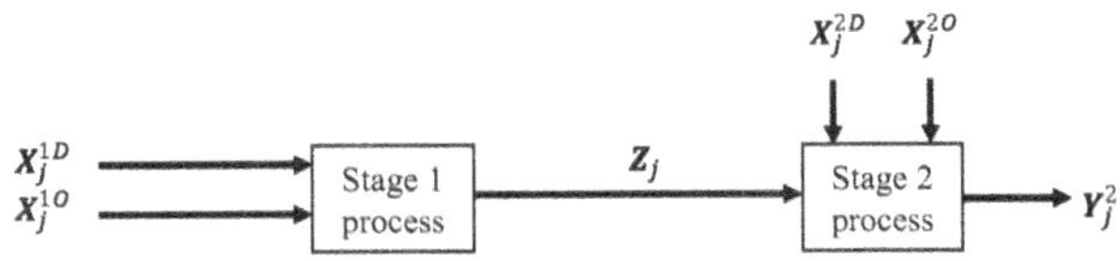

Fig. 5 Variable configuration for disbursement efficiency evaluation of DMU_j

not classified as disbursements. Then, an output-oriented network DEA model that computes disbursement efficiency of DMU_j under the VRS assumption is

$$\max \beta_k^D \tag{2}$$

$$\text{Subject to } \sum_{j=1}^{J} \mu_j^{(1)} \boldsymbol{X}_j^{1D} \leq \boldsymbol{X}_k^{1D}$$

$$\sum_{j=1}^{J} \mu_j^{(1)} \boldsymbol{X}_j^{1O} \leq \boldsymbol{\delta}_{X_{1O}} \boldsymbol{X}_k^{1O}$$

$$\sum_{j=1}^{J} \mu_j^{(1)} \mathbf{Z}_j \geq \widetilde{\mathbf{Z}}_k$$

$$\sum_{j=1}^{J} \mu_j^{(2)} \mathbf{Z}_j \leq \widetilde{\mathbf{Z}}_k$$

$$\sum_{j=1}^{J} \mu_j^{(2)} \boldsymbol{X}_j^{2D} \leq \boldsymbol{X}_k^{2D}$$

$$\sum_{j=1}^{J} \mu_j^{(2)} \boldsymbol{X}_j^{2O} \leq \boldsymbol{\delta}_{X_{2O}} \boldsymbol{X}_k^{2O}$$

$$\sum_{j=1}^{J} \mu_j^{(2)} \boldsymbol{Y}_j^2 \geq \beta_k^D \boldsymbol{Y}_k^2$$

$$\sum_{j=1}^{J} \mu_j^{(1)} = 1$$

$$\sum_{j=1}^{J} \mu_j^{(2)} = 1$$

$$\boldsymbol{\mu}^{(1)}, \boldsymbol{\mu}^{(2)}, \boldsymbol{\delta}_{X_{1O}}, \boldsymbol{\delta}_{X_{2O}} \geq \mathbf{0};\ \beta_k^D, \widetilde{\mathbf{Z}}_k \text{ unrestricted}$$

$\boldsymbol{\delta}_{X_{1O}}, \boldsymbol{\delta}_{X_{2O}}$ are introduced in model (2) to relax the bound on other inputs. The third and fourth sets of constraints ensure that intermediate measures are not restricted. In the context of MFs, these relaxations imply when assessing MF disbursement efficiency under the empirical set up shown in in Fig. 3, net assets, risk and fund flows do not impede return generation.

Let β_k^{D*} denote the optimal objective function value of model (2). Then, disbursement efficiency score of DMU_k (DE_k^*) is given by 1/ β_k^{D*}. β_k^{D*}= 1 implies DMU_k is *disbursement efficient*. Disbursement efficiency implies all disbursements are utilized efficiently. That is, at the current level of disbursements, DMU_k produces maximum possible outputs and therefore is disbursement efficient. On the other hand, $\beta_k^{D*} > 1$ implies that DMU_k is disbursement inefficient. Disbursement efficiency is essentially a measure of overall efficiency of DMU_k in the use of its disbursements. The coefficients associated with other inputs, $\boldsymbol{X}_k^{1O}$ and $\boldsymbol{X}_k^{2O}$ have no influence on β_k^{D*} calculation and hence, in this case, they are effectively unrestricted inputs. If DMU_k is disbursement inefficient, then it can become disbursement efficient by increasing its outputs to $\beta_k^{D*}\boldsymbol{Y}_k^2$. We refer to $\beta_k^{D*}\boldsymbol{Y}_k^2$ with which disbursement inefficient DMU_k may become efficient by utilizing its disbursements fully as disbursement output. Disbursement output $\beta_k^{D*}\boldsymbol{Y}_k^2$ and target output $\beta_k^{T*}\boldsymbol{Y}_k^2$ are output levels desired in the short-term. Yang and Fukuyama (2018) formulate generalised versions of models (1) and (2) for a single-stage production process adopting directional distance function approach. They consider good outputs and bad outputs and accommodate direct links between bad outputs and some inputs. Yang and Fukuyama (2018) apply their models to measure regional production potential of Chinese provinces. Models (1) and (2) can accommodate Yang and Fukuyama (2018) extensions with suitable modification.

Remark 1 According to our modelling framework, if DMU_k is disbursement efficient $\left(1/\beta_k^{D*} = 1\right)$, then it is technical efficient $\left(1/\beta_k^{T*} = 1\right)$ as well. The converse may not hold.

This follows, as the constraint set of model (2) is a subset of the constraint set of model (1) and the two models have similar objective functions. Therefore, the optimal value of model (2) must be at least as that of model (1). That is $\beta_k^{D*} \geq \beta_k^{T*}$. Now since β_k^{D*} and β_k^{T*} are $\geq$1, $\beta_k^{D*} = 1$ (DMU_k is disbursement efficient) implies $\beta_k^{T*} = 1$ (DMU_k is technical efficient). When $\beta_k^{T*} = 1$, we have that $\beta_k^{D*} \geq 1$ and hence the converse may not hold.

Remark 2 According to our modelling framework, disbursement efficiency score of DMU_k is less than or equal to its technical efficiency score.

This follows from $\beta_k^{D*} \geq \beta_k^{T*}$ as disbursement efficiency score and technical efficiency score of DMU_k is computed as $1/\beta_k^{D*}$ and $1/\beta_k^{T*}$ respectively.

3.3 Disbursement Utilisation

For DMU_k, we define *disbursement utilisation* (DU) labelled DU_k as the ratio of β_k^{T*} to β_k^{D*} as

$$DU_k = \frac{\beta_k^{T*}}{\beta_k^{D*}} = \frac{DE_k^*}{TE_k^*} \tag{3}$$

Then we have $DU_k \leq 1$. According to (3), for DMU_k, $DU_k = 1$ may result when $\beta_k^{T*} = \beta_k^{D*} = 1$ and when $\beta_k^{T*} = \beta_k^{D*} > 1$. This means, disbursement utilisation of DMU_k equals 1 when DMU_k is disbursement efficient and when DMU_k is both technical inefficient and disbursement inefficient with equal inefficiency score. $DU_k = 1$ with $\beta_k^{D*} = \beta_k^{T*} = 1$ indicates that disbursements $\boldsymbol{X}_j^{1D}$ and other inputs $\left(\boldsymbol{X}_j^1 - \boldsymbol{X}_j^{1D}\right)$ of DMU_k are utilized to its full potential to produce the observed outputs. When $DU_k = 1$ with $\beta_k^{D*} = \beta_k^{T*} > 1$, disbursements of DMU_k are not utilized to its full potential and therefore DMU_k is not disbursement efficient. By increasing the outputs of DMU_k by factor β_k^{D*} while maintaining disbursements $\boldsymbol{X}_j^{1D}$ at the observed (current) levels renders DMU_k disbursement efficient. Then, according to Remark 1, DMU_k will be technical efficient as well. This is a situation where attaining disbursement efficiency through output augmentation resulting in efficient disbursement utilisation.

$DU_k < 1$ indicates that $\beta_k^{D*} > \beta_k^{T*}$. Hence, $DU_k < 1$ implies that DMU_k is disbursement inefficient. When DU_k is greater than 1, it is possible that DMU_k is technical efficient $\left(\beta_k^{T*} = 1\right)$ or technical inefficient $\left(\beta_k^{T*} > 1\right)$. According to Remark 1, output augmentation for technical efficiency does not guarantee disbursement efficiency. Therefore, in this case, augmenting the output by factor β_k^{T*} does not guarantee efficient disbursement utilisation. In this case, efficient disbursement utilisation is guaranteed when output is augmented by factor β_k^{D*}. This is possible whether or not DMU_k is technical efficient.

We formulate models (1) and (2) assuming that production technology operates under variable returns to scale. If production technology exhibits non-increasing returns to scale, replace $\sum_{j=1}^{J} \lambda_j^{(1)} = 1$ and $\sum_{j=1}^{J} \lambda_j^{(2)} = 1$ in model (1) with $\sum_{j=1}^{J} \lambda_j^{(1)} \leq 1$ and $\sum_{j=1}^{J} \lambda_j^{(2)} \leq 1$. If production technology exhibits constant returns to scale, remove $\sum_{j=1}^{J} \lambda_j^{(1)} = 1$ and $\sum_{j=1}^{J} \lambda_j^{(2)} = 1$ in model (1). If production technology exhibits non-increasing or constant returns to scale technology, constraints on the intensity variables of model (2) require similar modification.

3.4 *Effect of Random Variation on Disbursement Utilisation*

Random variation in the data does not substantially distort disbursement utilisation given in (3) because its effect on β_k^D and β_k^T estimates are almost the same (Holland & Lee, 2002). Hence, an estimate of disbursement output derived using observed output and an unbiased estimate of disbursement utilisation is reliable.

3.5 *Extension to Input-Orientation Case*

So far, we discussed output-oriented models. We could modify these models to accommodate input orientation case as well. When applying input-orientation on MFs, that would be cost minimisation. The aim would then be to determine whether a given MF operates at minimum cost given the observed output. An application of this type is available in Kang and Kim (2015). They study cost minimization and associated utilization rate of 34 regional public hospitals in South Korea. Kerstens et al. (2019) discuss input-oriented technical and economic capacity notions under convex and nonconvex technologies. They compare technical and capacity notions empirically in a panel data framework using a sample of French fruit producers. Under the output-orientation case discussed in this chapter, the objective of DEA models is output (returns) maximization. Input-orientation requires formulation of DEA model for MF performance appraisal from cost-minimisation objective. Such application is straightforward in the black box or single-stage production processes. Extending the input-orientation concept under DEA to a two-stage process to assess disbursement efficiency and associated utilization assessment require an input, intermediate and output measures configuration amenable to such an assessment.

4 Application to U.S. Equity Mutual Funds

To illustrate the application of disbursement and technical efficiency measures and disbursement-utilisation measure, we use cross-sectional data, as at 31 December 2017 on 1706 U.S. equity MFs.[9] Empirical framework is the same as the one

[9]In this application, we focus on the overall management process and one other aspect of management, namely costs and expenses management. Following this approach, we can assess MF performance from other perspectives such as risk as well. Appendix B gives an example.

Table 1 Summary statistics of technical and disbursement efficiency scores and disbursement utilisation rate

Panel A: Size category (No of MFs)	Technical efficiency (Average)	No of technical efficient MFs (Percentage)	Disbursement efficiency (Average)	No of disbursement efficient MFs (percentage)	Disbursement utilisation (Average)
Large (512)	0.3455	33 (6.45)	0.1739	5 (0.98)	0.6206
Mid (682)	0.2767	30 (4.40)	0.1598	3 (0.44)	0.7045
Small (512)	0.2717	31 (6.05)	0.1585	6 (1.17)	0.7216
All (1706)	0.2959	94 (5.51)	0.1637	14 (0.82)	0.6844
Panel B: Style category (No of MFs)					
Blend (587)	0.3095	33 (5.62)	0.1451	0 (0.00)	0.6442
Growth (709)	0.3258	49 (6.91)	0.1997	14 (1.97)	0.7158
Value (410)	0.2244	12 (2.93)	0.1279	0 (0.00)	0.6879
All (1706)	0.2959	94 (5.51)	0.1637	14 (0.82)	0.6844

Notes: Disbursement utilisation is computed as disbursement efficiency/technical efficiency

depicted in Fig. 1. That is, to assess MF performance, we adopt the two-stage structure as its overall management process and consider the input, intermediate and output measure set shown in Fig. 3. Table 1 presents the technical efficiency, disbursement efficiency and disbursement utilisation computed in model (1) and model (2) and in (3) by MF size and Morningstar MF style category. Panel A of Table 1 reports a summary by large funds, mid-size funds, and small funds categories. The small category includes the bottom 30%, the large category includes the top 30% of the funds and mid-size category includes the mid 40% of the funds by assets under management. Panel B of Table 1 reports a summary by three Morningstar style categories; blend, growth and value. Value funds follow investment strategies that seek to invest in undervalued stocks. Growth funds comprise of diversified portfolios of stocks with primary goal of capital appreciation with little or no dividend payment. Blend funds are equity funds comprising of a mix of both value and growth stocks.

Results reported in Table 1 reveal that technical efficiency of large funds, on average, is higher than mid-size and small funds. However, according to the distribution of technical efficient funds reported in the third column of Table 1, there is no discernible difference across fund size. In the case of MF style, growth funds, on average, do better in technical efficiency than value and blend funds. Majority of MFs that makes up the technical efficient frontier are also growth funds. Pattern of the distribution of disbursement efficiency of MFs is consistent with that of observed with technical efficiency. A noticeable difference is in the composition of the disbursement efficient frontier. Fourteen MFs are disbursement efficient and all of them belongs to growth type. Here we uncover that, over the investigation period, the number of disbursement efficient funds (14) is small compared to the number of technical efficient funds (94). Numerically, on average, disbursement utilisation of large funds is smaller than disbursement utilisation of mid-size funds and small

Table 2 Distribution of mutual funds by technical efficiency, disbursement efficiency and disbursement utilisation

Mutual fund efficiency type	Technical efficiency (TE) $1/\beta_k^{T*}$	Disbursement efficiency (DE) $1/\beta_k^{D*}$	Disbursement utilisation (DE/TE) $\beta_k^{T*}/\beta_k^{D*}$	Number of mutual funds belonging to efficiency type
Type 1	1	1	1	14 (0.82%)
Type 2	<1	<1	1	48 (2.81%)
Type 3	1	<1	<1	80 (4.69%)
Type 4	<1	<1	<1	1564 (91.68%)
	Technical efficient MFs	Disbursement efficient MFs	Disbursement utilisation efficient MFs	
	94 (5.51%)	14 (0.82%)	62 (3.63%)	

Notes: Technical and disbursement efficient MFs belongs to Type 1. Technical and disbursement inefficient MFs with equal technical and disbursement inefficient scores belongs to Type 2. Technical efficient and disbursement inefficient MFs belongs to Type 3. Technical and disbursement inefficient MFs with unequal technical and disbursement inefficient scores belongs to Type 4

funds record the highest average. When analysed by fund style, the results reveal that disbursement utilisation of blend funds is smaller, on average, than disbursement utilisation of value funds. Growth funds record the highest average disbursement utilisation.

Table 2 presents the number of MFs that belongs to four possible efficiency types. (i) Disbursement efficient and technical efficient (type 1), (ii) disbursement inefficient and technical efficient (type 2), (iii) disbursement inefficient and technical inefficient with equal inefficiency scores (type 3) and (iv) disbursement inefficient and technical inefficient with unequal inefficiency scores (type 4).

As expected, the technical efficiency score of all MFs are greater than or equal to the corresponding disbursement efficiency score. According to Table 1 entries, 14 MFs (0.82%) are disbursement efficient and approximately 5.5% of the MFs (94 MFs) are technical efficient. The 14 disbursement efficient MFs are technical efficient as well. This is an indication that these 14 MFs are operating at the maximum potential output levels with the given levels of disbursements. The results reveal that 80 of the 94 technical efficient MFs are disbursement inefficient. This is an important message to MF managers who are interested in costs and expenses management performance as well as overall performance of their MFs. Under our modelling framework, we can identify technical efficient MFs that are disbursement inefficient. It is not possible to do that with technical efficiency evaluation alone. In technical efficiency evaluation, all input, intermediate and output measures deemed relevant for performance appraisal are considered together. In our modelling framework, we evaluate disbursement efficiency with a subset of the input measures considered in technical efficiency evaluation. The difference in the two sets is that the input measures not classified as disbursements are unbounded. In both sets, intermediate measures are unrestricted and output measures remain unchanged.

Table 3 Difference in mutual fund ranking based on TE and DE scores

Difference in TE rank and DE rank (x)	Number of mutual funds (%)
$x < -10$	648 (37.98)
$-10 \leq x < -5$	10 (0.59)
$-5 \leq x < 0$	21 (1.23)
$x = 0$	20 (1.17)
$0 < x \leq 5$	28 (1.64)
$5 < x \leq 10$	26 (1.52)
$x > 10$	953 (55.87)
Total	1706 (100)

Notes: TE is technical efficiency and DE is disbursement efficiency. The lower the rank, the better the performance

Then, a MF deemed disbursement efficient is technical efficient as well. See Remark 1. Table 2 reveal that a very high percentage (99.18%) of MFs are disbursement inefficient. This is an indication that majority of MFs are not generating return efficiently with given levels of disbursements.

4.1 Comparison of MF Rankings Based on Technical Efficiency and Disbursement Efficiency Scores

Here we discuss the association of MF rankings based on technical efficiency and disbursement efficiency scores. Spearman rank correlation between the two sets of rankings is 0.79. Generally, the association between the rankings based on technical efficiency and disbursement efficiency scores is positive and strong. Table 3 gives the distribution of MFs by ranking difference. Difference in the rankings vary considerably across MFs. For example, in 679 of the 1706 (39.8%) MFs, disbursement efficiency score based ranking is worse than technical efficiency score based ranking.[10] In majority of these MFs (648 out of 679) the difference in the ranking exceeds 10. In approximately 60% of the MFs, the ranking based on the disbursement efficiency score is better than the ranking based on the technical efficiency score. In majority of these MFs (953 out of 1007), the difference in the ranking exceeds 10. Nevertheless, Spearman rank correlation between the two sets of rankings is 0.7893. See Table 9 for correlation with other performance measures.

Through this empirical study, we highlight that generally disbursement management performance of a given MF may not be consistent with overall management performance. This is useful information to MF managers. Commonly used managed fund performance measures refer to return and risk as factors of performance. Even when DEA studies consider other factors of performance including risk and

[10]The lower the rank, the better the performance.

return in MF performance appraisal, they are generally put together in performance appraisal. In this paper, we demonstrate performance assessment from two different management perspectives. That is, from cost and expense management perspective and overall management perspective. The difference in the two approaches is that in the one case, input measures are purely representative of cost and expenses in managing the MF and in the other case, all input measures representative of many different factors including costs and expenses are considered. The outputs in both cases are the same. Because we consider the same set of outputs in both cases of assessment, comparison of the effect of different input sets on the same output set is meaningful.

4.2 Practical Use of Disbursement Efficiency Score

To describe practical use of disbursement efficiency score, we select one MF that belongs to type 2, 3 and 4. Their names, observed input, intermediate and output measures are in Table 4.

The technical and disbursement efficiency scores and disbursement utilisation of the three MFs chosen for discussion are in the bottom three rows of Table 4. The selected MF under type 2 is Oberweis Small-Cap Opportunities. It has the same technical efficiency score and disbursement efficiency score and hence its disbursement utilisation is 1. However, Oberweis Small-Cap Opportunities is both technical inefficient and disbursement inefficient. To obtain the maximum potential output, we multiply the observed output by reciprocal of efficiency score.[11] Since in this case, technical efficiency score is equal to the disbursement efficiency score, the maximum potential outputs with observed disbursements and unbounded other inputs are the same as the maximum potential outputs with all observed inputs. In both cases, intermediate measures are free or unrestricted which is the common practice in envelopment type formulation for performance appraisal of DMUs in serially linked two stage processes.

Plumb Equity that belongs to type 3 type is technical efficient and disbursement inefficient. In this case, Plumb Equity operates at its maximum potential outputs with all observed inputs. However, Plumb Equity does not operate at its maximum potential output when only the observed disbursements are considered. This is an example where output levels required for disbursement efficiency is more than the output levels required for technical efficiency. This is possible owing to our modelling framework where outputs remain the same in technical efficiency and disbursement efficiency evaluation and the difference in the two types is that the

[11] When computing relative efficiency scores using the proposed models, we transformed the input-output data to make them positive. Hence, the Pareto efficient output reported in Table 5 are the retransformed values.

Table 4 Input-output measures of type 2, 3 and 4 mutual funds selected for discussion

	Oberweis Small-Cap Opportunities (Type 2 example)	Plumb Equity (Type 3 example)	BlackRock Advantage Large Cap Gr Inv A (Type 4 example)
Stage 1 inputs			
Fund flows ($\$10^6$)	−2.9968	1.1823	302.0526
Marketing and distribution fees ($\$10^6$)	0.0188	0.0647	1.6716
Operational costs ($\$10^6$)	0.0733	0.1091	0.3302
Intermediate measure			
Total net assets ($\$10^6$)	0.0008	0.0026	0.0669
Stage 2 inputs			
Total risk	14.91	11.50	10.62
Turnover ratio (%)	150.00	18.00	130.00
Systematic risk	1.07	0.99	1.01
Downside risk	6.29	3.52	1.93
Management fees ($\$10^6$)	0.0726	0.1407	1.8687
Stage 2 outputs			
1-year return (%)	25.54	39.70	23.13
3-year return (%)	12.31	13.67	11.08
5-year return (%)	14.84	16.67	11.31
Performance			
Technical efficiency	0.2883	1	0.2154
Disbursement efficiency	0.2883	0.4071	0.1093
Disbursement utilisation	1	0.4071	0.5074

Notes: Disbursements comprise of marketing and distribution fees and operational costs (stage 1 inputs), and management fees (stage 2 input). Disbursement utilisation (≤ 1) is the ratio of disbursement efficiency score and technical efficiency score

inputs in the disbursement efficiency evaluation is a sub set of the inputs in the technical (overall) efficiency evaluation.

Table 5 gives the output targets for disbursement Pareto efficient frontier projection of Oberweis Small-Cap Opportunities, Plumb Equity and BlackRock Advantage Large Cap Gr Inv A. BlackRock Advantage Large Cap Gr Inv A that belongs to case 4 type is technical inefficient and disbursement inefficient with disbursement utilisation less than 1. BlackRock Advantage Large Cap Gr Inv A could become disbursement efficient by achieving return of 36.99% in the short-

Table 5 Output targets for disbursement efficient (Pareto) frontier projection

	Oberweis Small-Cap Opportunities (Type 2 example)	Plumb Equity (Type 3 example)	BlackRock Advantage Large Cap Gr Inv A (Type 4 example)
1-year return (%)	40.48	42.33	36.99
3-year return (%)	15.22	15.51	20.04
5-year return (%)	17.64	18.53	20.22

Notes: Technical and disbursement inefficient MFs with equal technical and disbursement inefficient scores belongs to Type 2. Technical efficient and disbursement inefficient MFs belongs to Type 3. Technical and disbursement inefficient MFs with unequal technical and disbursement inefficient scores belongs to Type 4

term, return of 20.04% in the medium-term and return of 20.22% in the long-term. The output targets for disbursement efficiency in this case is higher than the output targets for technical efficiency. See Remark 2. From Remark 1, we know that achieving disbursement efficiency renders the MF technical efficient as well. Hence, in this case, we may select the set of output for disbursement efficiency as the targets for implementation in practice. Plumb Equity is technical efficient but disbursement inefficient. Plumb Equity could achieve disbursement efficiency with 42.33% return in the short-term, 15.51% return in the medium term and 18.53% return in the long-term. Oberweis Small-Cap Opportunities could become disbursement efficient by achieving return of 40.48% in the short-term, return of 15.22% in the medium-term and return of 17.64% in the long-term. The aforementioned output targets are valid in the short-term. Hence, output targets facilitate management decisions that could have implications in the short-term. As with forecasting (is based on past patterns), output targets are based on past performance and hence should be used for guidance in the short-term as the business world operates in climate of uncertainty.

Färe et al. (1989), Färe et al. (1994) and Kirkley et al. (2002) interpret $\boldsymbol{\delta}^*_{\boldsymbol{X10}}$ and $\boldsymbol{\delta}^*_{\boldsymbol{X20}}$ as measures of other input utilisation rates where $\boldsymbol{\delta}^*_{\boldsymbol{X10}}$ and $\boldsymbol{\delta}^*_{\boldsymbol{X20}}$ are a set of optimal values obtained in the solution to model (2). For example, $\delta_{10} \in \boldsymbol{\delta}_{\boldsymbol{X10}}$ associated with inequality $\sum_{j=1}^{J} \mu_j^{(1)*} X_j^1 \leq \delta_{10}^* x_{1k}^1$ measures degree of utilisation of input x_{1k}^1. We refer to δ_{10}^* as utilisation index of input x_{1k}^1. The utilisation index $\delta_{10}^* = 1$ and $\sum_{j=1}^{J} \mu_j^{(1)*} X_j^1 = x_{1k}^1$, indicates that input x_{1k}^1 is utilized in full in generating maximum potential output with the current levels of disbursements. Thus if the index is greater than unity, the corresponding input is underutilised and if the index is less than unity, that input is over utilized.

Table 6 gives input utilisation index of 'other' inputs of the three MFs selected for the discussion. The results reported in Table 6 reveal that Oberweis Small-Cap Opportunities manage fund flows, total risk and systematic risk well with utilisation index close to 1. On the other hand, turn over utilisation index of Oberweis Small-Cap Opportunities is low at 0.37 and its downside risk utilisation index is 0.84. An indication here is that Oberweis Small-Cap Opportunities should pay attention to asset management in return generation. Plumb Equity and BlackRock Advantage Large Cap Gr Inv A manage fund flows well. However, they have

Table 6 Other inputs' utilisation

Mutual fund	Fund flows δ^*_{x1O}	Total risk δ^*_{x2O}	Turnover ratio	Systematic risk	Downside risk
Oberweis Small-Cap Opportunities	1	0.99	0.37	0.99	0.84
Plumb Equity	1	1.29	3.18	1.08	1.61
BlackRock Advantage Large Cap Gr Inv A	1.05	1.17	0.01	0.95	2.25

Notes: Other inputs are the inputs considered in the analysis and not belonging to disbursements category

high and low asset management performance in return generation with turnover utilisation index at 3.18 and 0.01 respectively. The low utilisation index of 0.01 for turnover ratio indicate that BlackRock Advantage Large Cap Gr Inv A should reduce turnover ratio considerably and Plumb Equity with high utilisation index of 3.18 for turnover should increase turnover ratio considerably for better performance in asset management.

5 Extending to Assess Performance from Different Input Management Perspectives

We may extend the discussion in Sect. 4 to assess performance from several different input management perspectives. Table 7 lists input and output measures used in DEA studies of MF performance appraisal. We classify the input measures broadly under five types. They are risk, fund manager characteristics, fund size characteristics, financial ratios, and disbursements. Out of the five types clearly two types stand out. They are risk and disbursements.

If we are to extend application to different input management perspectives, we could assess performance from disbursement management perspective, risk and disbursement management perspective and overall management perspective, for example. When assessing performance under each of these three perspectives three input sets may be used. One set with input measures relevant to assess disbursement management performance and another set to assess risk and disbursement management performance by augmenting the first input measures set with measures of risk. The third input measures set, may include more input measures to assess performance from an overall management perspective. Appendix B gives a set of DEA models applicable to assess disbursement management performance, risk and disbursement management performance and overall management performance under the two-stage framework shown in Fig. 4.

Table 7 Input-output measures used in single-stage DEA studies of MF performance appraisal

Input measures	Output measures
Risk	*Returns*
Asset allocation in risk holdings, asset allocation in risk free assets, cash	Annual return
Higher partial moments, lower partial moments	1-year, 3-year, 5-year, 10-year return
Jensen's alpha, Sharpe index	Cumulative return
Systematic risk, value at risk, downside risk	Stochastic monthly rate of return
Expense ratio	*Others*
Standard deviation, skewness, kurtosis	Total market value at end of period
Turnover, drawdown	Ethical indicator
Percentage of stocks	Information service quality
Fund manager characteristics	Product innovation speed
Fund manager average tenure	Total shares
Fund size characteristics	Upper partial moments
Number of fund types, number of funds	Maximum period of consecutive gain
Total assets, fund size, net asset value	
Financial ratios	
Price to book ratio, price to earnings ratio	
5-year standard deviation/earnings per share	
Disbursements (cost, expenses and fees)	
Sales charge, transaction cost rate	
Minimum initial investment	
Subscription costs, redemption fees	
Marketing and distribution fees, 12b-1 fees	
Front-end load, back-end load, differed load	

In this study, we do not discuss efficiency at the individual stage level. To obtain stage-level efficiency scores multiplier models are needed (Zhu, 2020). Stage-level or sub-process inefficiency may vary across MFs. Hence, their contribution towards overall inefficiency may reveal strengths and weaknesses of the overall MF management process from different aspects of management compared to their peers. Such managerial information is obtainable when performance appraisal adopts network DEA. This is a case of value extraction or value mining of data through a technique or method different from those discussed in big data analytics literature. Put simply, big data analytics techniques process data that comes from different sources and formats. Choi et al. (2017) refer to statistics, machine learning, data mining, optimisation and others such as social network analysis, clustering network analysis and visualisation analysis as big data analytic techniques. We suggest formulation of multiplier models under network structures to assess performance from different management perspectives and decomposition of performance at the stage level as direction for future work.

Table 8 Summary statistics of two-stage model and black box model estimated efficiency

	Average	Standard deviation	Minimum	Maximum	No of technical efficient MFs (percentage)
Panel A: Technical efficiency					
Single-stage model	0.3750	0.3033	0.0160	1.0000	237 (13.89)
Two-stage model	0.2959	0.2084	0.0140	1.0000	94 (5.51)
Panel B: Disbursement efficiency					
Single-stage model	0.1826	0.1240	0.0059	1.0000	28 (1.64)
Two-stage model	0.1637	0.0926	0.0059	1.0000	14 (0.82)

6 Robustness Check

6.1 Comparison of the Results with Black Box Model

In this sub-section, we demonstrate usefulness of representing MF management process as a network (two-stage process) over the standard representation (single-stage process) in performance appraisal. In the single-stage process 1-year, 3-year and 5-year returns are the outputs and the other measures including the intermediate measure shown in Fig. 3 are the inputs. Panel A of Table 8 presents summary statistics of technical efficiency and Panel B of Table 8 presents summary statistics of disbursement efficiency estimated in the single-stage and two-stage models. When MF performance is assessed using the single-stage model, the results reveal that 237 MFs are technical efficient. This is 143 more MFs than in the two-stage case. With respect to disbursement management performance, the number of efficient MFs under the single-stage case is twice as that of under the two-stage case. It is therefore not surprising that average technical and disbursement efficiency in the single-stage case is higher than in the two-stage case. Findings here suggest that network representation may increase discriminatory power of performance and thereby demonstrate network DEA-based performance appraisal add value.

6.2 Comparison of DEA-Based Performance and Risk-Adjusted Return Performance

In this sub-section, to empirically validate our choice of DEA models for MF performance appraisal, we investigate the association between MF rankings obtained under DEA-based performance measures technical efficiency, disbursement efficiency and disbursement utilisation and two risk-adjusted return measures (total risk-adjusted 1-year return and downside risk-adjusted 1-year return). Panel A of Table 9 reports Spearman's rank correlation between technical efficiency, disbursement efficiency and disbursement utilisation estimated in the single-stage model and risk-adjusted return. Panel B of Table 9 reports Spearman's rank correlation between technical efficiency, disbursement efficiency and disbursement utilisation estimated in the two-stage model and risk-adjusted return. The results in both panels reveal that rank correlation between MF rankings based on risk-adjusted return measures and the rankings based on technical efficiency and disbursement efficiency scores are positive. Moreover, correlation between the rankings based on risk-adjusted return measures and technical efficiency scores is stronger than that with disbursement efficiency scores. This may be because risk measures are not included as factors of performance in disbursement efficiency calculation. Previous studies provide evidence in support of strong positive association between DEA-based performance and risk-adjusted return performance. For example, Choi and Murthy (2001) in a DEA study of 761 U.S. MFs report positive association between technical efficiency and Sharpe index based performance. Further, when technical efficiency is estimated with one input (total risk) and with multiple inputs (risk and three proxies for costs and expenses) with gross return as output, Choi and Murthy (2001) report that the correlation between the Sharpe ratio and DEA measure increases substantially in the multiple input case. Network DEA-based performance appraisal generally use more measures than corresponding conventional DEA-based appraisal. We find that, MF rankings based on traditional risk-adjusted return measures is stronger with rankings based on network DEA-based performance than with that of single-stage DEA-based performance when both types of analysis consider the same set of output measures.

The results in both panels of Table 9 reveal that association between MF rankings based on disbursement utilisation and the two risk-adjusted return measures are negative and strong. Again, the evidence is stronger in the case with two-stage model estimated disbursement utilisation compared to the case with the single-stage model estimated disbursement utilisation. The negative correlation between disbursement utilisation and risk-adjusted return efficiency based rankings suggest that disbursement utilisation of high risk-adjusted return MFs are lower than with low risk-adjusted return MFs. Under our variable selection for technical and disbursement efficiency estimation, this is expected. In disbursement-management performance appraisal, risk measures are not considered. Lower disbursement utilisation implies technical efficiency is relatively higher than disbursement efficiency. We advance our findings as empirical validation of our network-DEA model based measures, disbursement efficiency and disbursement utilisation.

7 Concluding Remarks

Businesses can gain competitive edge over its competitors by improving operational efficiency of its processes. Operational efficiency in this context include better utilisation of limited resources among others. Through extraction of information of value, businesses can improve key performance indicators of business processes. Big data analytics can aid in this process. Big data comes in volume at high speed

Table 9 Rank correlation

	Technical efficiency (TE)	Disbursement efficiency (DE)	Disbursement utilisation (DU)	Total risk-adjusted 1-year return	Downside risk-adjusted 1-year return
Panel A: TE, DE and DU estimated in the single-stage model					
Technical efficiency (TE)	1				
Disbursement efficiency (DE)	0.7766	1			
Disbursement utilisation (DU)	−0.7060	−0.2358	1		
Total risk-adjusted 1-year return	0.5853	0.4993	−0.4268	1	
Downside risk-adjusted 1-year return	0.5961	0.4339	−0.5778	0.7883	1
Panel B: TE, DE and DU estimated in the two-stage model					
Technical efficiency (TE)	1				
Disbursement efficiency (DE)	0.7893	1			
Disbursement utilisation (DU)	−0.7083	−0.2432	1		
Total risk-adjusted 1-year return	0.6399	0.5226	−0.4819	1	
Downside risk-adjusted 1-year return	0.6796	0.4447	−0.7125	0.7883	1

(velocity) and in different forms (variety). According to big data literature, veracity (quality) and value (data transformed into business value) added to these three aspects (volume, velocity and variety) make up the five big-Vs of big data. This chapter demonstrates how network DEA may contribute positively towards value aspect of big data analytics. In this chapter, we highlight that network DEA model applications in mutual fund performance appraisal extract information consistent with value aspect of big data analytics. Further, we extend new measures of mutual fund performance appraisal discussed in the recent literature to demonstrate value extraction in network DEA application of MF performance appraisal. We demonstrate this through an empirical application to U.S. equity MFs.

The models proposed in this chapter are applicable in situations where different sets of input measures capture overall management performance from different aspects of management. We develop models to assess performance from different management perspectives with progressively increasing sets of input measures. We start with a set of input measures that measures performance from a specific management perspective. Thereafter, we augment the already chosen input measure set with another set of input measures so that the new input measure set is applicable for performance appraisal from a combination of two management perspectives of which one is the management perspective identified with the input measure set chosen first. We continue this process until all input measures selected for performance appraisal is exhausted. At each step of input measure set augmentation, a new management perspective is included. When analysing performance with each input measure, the output measures set remains the same. Retaining output measures in performance appraisal under different input management perspectives makes comparison of rankings based on relative performance valid. This approach is applicable in performance appraisal of other types of financial entities such as pension funds and superannuation funds and may not be limited to applications in the finance sector.

A.1 Appendices

A.1.1 Appendix A (Table 10)

A.1.2 Appendix B

Network DEA Models for Performance Appraisal from Three Different Management Perspectives

Here we present network DEA models to appraise performance from three management perspectives. They are overall management, risk and disbursement management and disbursement management perspectives. In the three network representations shown in Figs. 6, 7 and 8, the difference is in the input measures. In all three cases, the intermediate measures and the outputs remain the same.

Technical Efficiency Evaluation

Let $\boldsymbol{X}_j^{1T} = \left(x_{1j}^{1T}, x_{2j}^{1T}, \ldots, x_{N_{X1}j}^{1T}\right)$ denote the inputs and $\boldsymbol{Z}_j^T = \left(Z_{1j}^T, Z_{2j}^T, \ldots, Z_{N_Z j}^T\right)$ denote the outputs at stage 1, $\boldsymbol{X}_j^{2T} = \left(x_{1j}^{2T}, x_{2j}^{2T}, \ldots, x_{N_{X2}j}^{2T}\right)$ and $\boldsymbol{Z}_j^T = \left(Z_{1j}^T, Z_{2j}^T, \ldots, Z_{N_Z j}^T\right)$ denote the inputs and $\boldsymbol{Y}_j^{2T} = \left(Y_{1j}^{2T}, Y_{2j}^{2T}, \ldots, Y_{N_{Y2}j}^{2T}\right)$ denote the outputs at stage 2 of DMU_j. $\boldsymbol{Z}_j^T$ links stage 1 and stage 2 and therefore

Table 10 Input, intermediate and output measures used in network DEA representation of overall MF management process

Reference	Sub processes	Input measures	Intermediate measures	Output measures
Premachandra et al. (2012)	Operational management Portfolio management	Management fees Marketing and distribution fees Net asset value Fund size Net expense ratio Turnover Standard deviation	Net asset value	Net asset value Average return
Galagedera et al. (2016)	Operational management Portfolio management	Management fees Marketing and distribution fees Net asset value Fund size Net expense ratio Turnover Standard deviation	Net asset value	Total cash flow to investors Net asset value Average return
Sánchez-González et al. (2017)	Portfolio management Marketing and selling Operational management	Labour Capital Risk Expense ratio Turnover Number of funds Fund size Net returns Money flows Gross returns Total assets	Number of funds Fund size Money flows Gross returns	Number of funds Fund size Gross returns Unitholders flows Money flows Profits
Galagedera et al. (2018)	Operational management Resource management Portfolio management	Management fees Marketing and distribution fees Fund size Expense ratio Turnover Net asset value Total risk Downside risk Systematic risk	Fund size Net asset value	Fund size Net asset value Returns
Galagedera et al. (2020)	Operational management Portfolio management	Fund flows Marketing and distribution fees Operational costs Total net assets Management fees Total risk Systematic risk Downside risk	Total net assets	Total net assets 1-year return 3-year return 5-year return

Notes: Details on the above references are available in the reference section

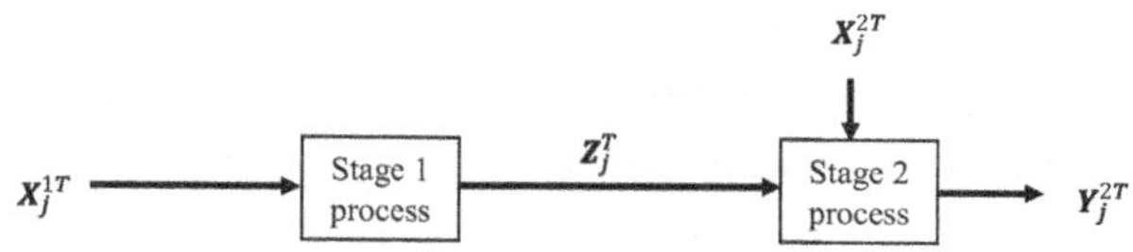

Fig. 6 Variable configuration for technical efficiency evaluation of DMU_j

they serve as intermediate measures. Further, let $\boldsymbol{\lambda}^{(1)} = \left(\lambda_1^{(1)}, \lambda_2^{(1)}, \ldots, \lambda_J^{(1)}\right)$ and $\boldsymbol{\lambda}^{(2)} = \left(\lambda_1^{(2)}, \lambda_2^{(2)}, \ldots, \lambda_J^{(2)}\right)$ such that $\sum_{j=1}^{J}\lambda_j^{(1)} = 1$ and $\sum_{j=1}^{J}\lambda_j^{(2)} = 1$ be vectors of intensity variables applicable to stage 1 and stage 2 respectively. The conditions imposed on the intensity variables implies that technology operates under variable returns to scale (VRS) assumption. Then, an output-oriented network DEA model that computes the technical efficiency score of DMU_j under the VRS assumption is

$$\max \beta_k^T \tag{4}$$

$$\text{Subject to} \sum_{j=1}^{J} \lambda_j^{(1T)} \boldsymbol{X}_j^{1T} \leq \boldsymbol{X}_k^{1T}$$

$$\sum_{j=1}^{J} \lambda_j^{(1T)} \boldsymbol{Z}_j^T \geq \boldsymbol{Z}_k^T$$

$$\sum_{j=1}^{J} \lambda_j^{(2T)} \boldsymbol{Z}_j^T \leq \boldsymbol{Z}_k^T$$

$$\sum_{j=1}^{J} \lambda_j^{(2T)} \boldsymbol{X}_j^{2T} \leq \boldsymbol{X}_k^{2T}$$

$$\sum_{j=1}^{J} \lambda_j^{(2T)} \boldsymbol{Y}_j^{2T} \geq \beta_k^T \boldsymbol{Y}_k^{2T}$$

$$\sum_{j=1}^{J} \lambda_j^{(1T)} = 1$$

$$\sum_{j=1}^{J} \lambda_j^{(2T)} = 1$$

$$\boldsymbol{\lambda}^{(1T)} \geq \boldsymbol{0};\ \boldsymbol{\lambda}^{(2T)} \geq \boldsymbol{0};\ \beta_k^T \text{ unrestricted}$$

Let β_k^{T*} denote the optimal objective function value of model (1). Then, technical efficiency score of DMU_k (TE_k^*) is given by $1/\beta_k^{T*}$. β_k^{T*}reveals whether or not DMU_k operates at its full potential with current levels of inputs. $\beta_k^* = 1$ indicates that DMU_k is technical efficient and $\beta_k^{T*} > 1$ indicates that DMU_k is inefficient suggesting that output augmentation is possible with current levels of inputs. $\beta_k^{T*} > 1$ may be interpreted as the proportional increase in the outputs that renders DMU_k technical efficient. In other words, inefficient DMU_k will get projected on to the technical efficient frontier when the outputs $\boldsymbol{Y}_k^2$ increase to $\beta_k^{T*}\boldsymbol{Y}_k^2$. Hence, $\beta_k^{T*}\boldsymbol{Y}_k^2$ provide a set of output targets to inefficient DMU_k to become technical efficient.

Risk and Disbursement Efficiency Evaluation

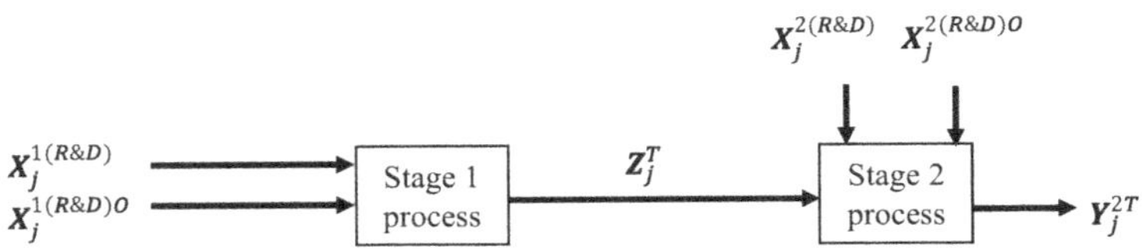

Fig. 7 Variable configuration for risk and disbursement efficiency evaluation of DMU_j

Now let, $\boldsymbol{X}_j^{1(R\&D)} = \left(X_{1j}^{1(R\&D)}, X_{2j}^{1(R\&D)}, \ldots, X_{N(R\&D)1j}^{1(R\&D)}\right) \subset \boldsymbol{X}_j^{1O}$ and $\boldsymbol{X}_j^{2(R\&D)} = \left(X_{1j}^{2(R\&D)}, X_{2j}^{2(R\&D)}, \ldots, X_{N(R\&D)2j}^{2(R\&D)}\right) \subset \boldsymbol{X}_j^{2O}$ denote the subsets of risk and disbursement measures as stage 1 inputs and stage 2 inputs respectively and $\boldsymbol{X}_j^{1(R\&D)O} = \left(X_{1j}^{1(R\&D)O}, X_{2j}^{1(R\&D)O}, \ldots, X_{N(R\&D)O1j}^{1(R\&D)O}\right) = \boldsymbol{X}_j^{1O} - \boldsymbol{X}_j^{1(R\&D)}$ and $\boldsymbol{X}_j^{2(R\&D)O} = \left(X_{1j}^{2(R\&D)O}, X_{2j}^{2(R\&D)O}, \ldots, X_{N(R\&D)O2j}^{2(R\&D)O}\right) = \boldsymbol{X}_j^{2O} - \boldsymbol{X}_j^{2(R\&D)O}$ denote the subsets of other inputs at stage 1 and stage 2 respectively of DMU_j.[12] Then, an output-oriented network DEA model that computes the risk and disbursement efficiency score $\beta_k^{R\&D}$ of DMU_j under the VRS assumption is

$$\max \beta_k^{R\&D} \tag{5}$$

$$\text{Subject to} \sum_{j=1}^{J} \mu_j^{(1R\&D)} \boldsymbol{X}_j^{1(R\&D)} \leq \boldsymbol{X}_k^{1(R\&D)}$$

[12] $\boldsymbol{X}_j^1 = \boldsymbol{X}_j^{1(R\&D)D} \cup \boldsymbol{X}_j^{1(R\&D)O}$ and $\boldsymbol{X}_j^2 = \boldsymbol{X}_j^{2(R\&D)D} \cup \boldsymbol{X}_j^{2(R\&D)O}$. $N_{(R\&D)1} + N_{(R\&D)O1} = N_{X1}$ and $N_{(R\&D)2} + N_{(R\&D)O2} = N_{X2}$.

$$\sum_{j=1}^{J} \mu_j^{(1R\&D)} \boldsymbol{X}_j^{1(R\&D)O} \leq \boldsymbol{\delta}_{X_{1O}} \boldsymbol{X}_k^{1(R\&D)O}$$

$$\sum_{j=1}^{J} \mu_j^{(1R\&D)} \boldsymbol{Z}_j^{T} \geq \overset{\sim(R\&D)}{\boldsymbol{Z}_k}$$

$$\sum_{j=1}^{J} \mu_j^{(2R\&D)} \boldsymbol{Z}_j^{T} \leq \overset{\sim(R\&D)}{\boldsymbol{Z}_k}$$

$$\sum_{j=1}^{J} \mu_j^{(2)} \boldsymbol{X}_j^{2(R\&D)} \leq \boldsymbol{X}_k^{2(R\&D)}$$

$$\sum_{j=1}^{J} \mu_j^{(2)} \boldsymbol{X}_j^{2(R\&D)O} \leq \boldsymbol{\delta}_{X_{2O}} \boldsymbol{X}_k^{2(R\&D)O}$$

$$\sum_{j=1}^{J} \mu_j^{(2R\&D)} \boldsymbol{Y}_j^{2T} \geq \beta_k^{R\&D} \boldsymbol{Y}_k^{2T}$$

$$\sum_{j=1}^{J} \mu_j^{(1R\&D)} = 1$$

$$\sum_{j=1}^{J} \mu_j^{(2R\&D)} = 1$$

$$\boldsymbol{\mu}^{(1R\&D)}, \boldsymbol{\mu}^{(2R\&D)}, \boldsymbol{\delta}_{X_{1O}}, \boldsymbol{\delta}_{X_{2O}} \geq \mathbf{0};\ \beta_k^{RD}, \overset{\sim(R\&D)}{\boldsymbol{Z}_k} \quad \text{unrestricted}$$

$\boldsymbol{\delta}_{X_{1O}}, \boldsymbol{\delta}_{X_{2O}}$ are introduced in model (5) to relax the bound on other inputs. The third and the fourth sets of constraints ensure that intermediate measures do not affect risk and disbursement efficiency. The optimal objective function value of model (5), denoted by $\beta_k^{R\&D*}$ computes risk and disbursement efficiency. Risk and disbursement efficiency score of DMU_k ($R\&DE_k^*$) is given by $1/\beta_k^{R\&D*}$. $\beta_k^{R\&D*}=$ 1 implies DMU_k is risk and disbursement efficient. That is, at the current level of risk and disbursements, DMU_k produces maximum possible outputs and therefore is *risk and disbursement efficient*. On the other hand, $\beta_k^{R\&D*} > 1$ implies that DMU_k is risk and disbursement inefficient. Risk and disbursement efficiency is essentially a measure of efficiency of DMU_k in the management of its risk and disbursements. The constraints associated with the other inputs, $\boldsymbol{X}_k^{1O}$ and $\boldsymbol{X}_k^{1O}$ have no influence on

$\beta_k^{R\&D*}$ calculation and hence, in this case, they are effectively unrestricted inputs. If DMU_k is risk and disbursement inefficient, then it can become risk and disbursement efficient by increasing its outputs to $\beta_k^{R\&D*}\boldsymbol{Y}_k^2$

Disbursement Efficiency Evaluation

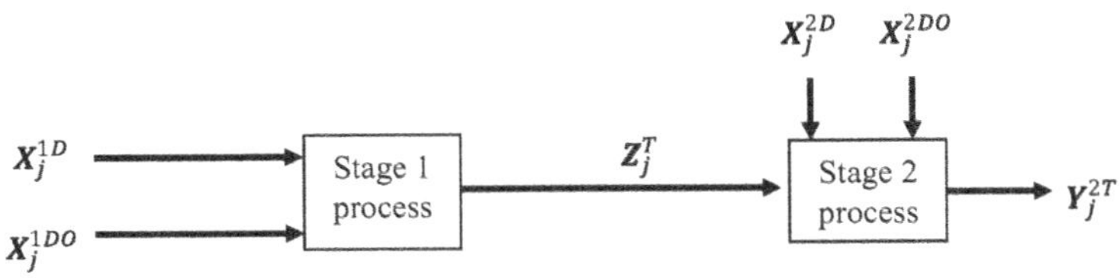

Fig. 8 Variable configuration for disbursement efficiency evaluation of DMU_j

Now let, $\boldsymbol{X}_j^{1D} = \left(X_{1j}^{1D}, X_{2j}^{1D}, \ldots, X_{ND_1j}^{1D}\right) \subset \boldsymbol{X}_j^{1(R\&D)}$ and $\boldsymbol{X}_j^{2D} = \left(X_{1j}^{2D}, X_{2j}^{2D}, \ldots, X_{N_{D2}j}^{2D}\right) \subset \boldsymbol{X}_j^{2(R\&D)}$ denote the subsets of disbursements at stage 1 inputs and stage 2 inputs respectively and $\boldsymbol{X}_j^{1DO} = \left(X_{1j}^{1DO}, X_{2j}^{1DO}, \ldots, X_{N_{DO1}j}^{1DO}\right) = \boldsymbol{X}_j^{1(R\&D)} - \boldsymbol{X}_j^{1D}$ and $\boldsymbol{X}_j^{2DO} = \left(X_{1j}^{2DO}, X_{2j}^{2DO}, \ldots, X_{N_{DO2}j}^{2DO}\right) = \boldsymbol{X}_j^{2(R\&D)} - \boldsymbol{X}_j^{2D}$ denote the subsets of other inputs at stage 1 and stage 2 respectively of DMU_j.[13] Then, an output-oriented network DEA model that computes the risk and disbursement efficiency score of DMU_j under the VRS assumption is

$$\max \beta_k^D \tag{6}$$

$$\text{Subject to} \sum_{j=1}^{J} \mu_j^{(1D)} \boldsymbol{X}_j^{1D} \leq \boldsymbol{X}_k^{1D}$$

$$\sum_{j=1}^{J} \mu_j^{(1D)} \boldsymbol{X}_j^{1DO} \leq \boldsymbol{\gamma}_{X_{1O}} \boldsymbol{X}_k^{1DO}$$

$$\sum_{j=1}^{J} \mu_j^{(1D)} \boldsymbol{Z}_j^T \geq \tilde{\boldsymbol{Z}}_k^D$$

[13] $\boldsymbol{X}_j^1 = \boldsymbol{X}_j^{1D} \cup \boldsymbol{X}_j^{1DO}$ and $\boldsymbol{X}_j^2 = \boldsymbol{X}_j^{2D} \cup \boldsymbol{X}_j^{2DO}$. $N_{D1} + N_{DO1} = N_{X1}$ and $N_{D2} + N_{DO2} = N_{X2}$.

$$\sum_{j=1}^{J} \mu_j^{(2D)} \mathbf{Z}_j^T \leq \overset{\sim D}{\mathbf{Z}_k}$$

$$\sum_{j=1}^{J} \mu_j^{(2D)} \boldsymbol{X}_j^{2D} \leq \boldsymbol{X}_k^{2D}$$

$$\sum_{j=1}^{J} \mu_j^{(2D)} \boldsymbol{X}_j^{2DO} \leq \boldsymbol{\gamma}_{X_{2O}} \boldsymbol{X}_k^{2DO}$$

$$\sum_{j=1}^{J} \mu_j^{(2D)} \mathbf{Y}_j^{2T} \geq \beta_k^D \mathbf{Y}_k^{2T}$$

$$\sum_{j=1}^{J} \mu_j^{(1D)} = 1$$

$$\sum_{j=1}^{J} \mu_j^{(2D)} = 1$$

$$\boldsymbol{\mu}^{(1D)}, \boldsymbol{\mu}^{(2D)}, \boldsymbol{\gamma}_{X_{1O}}, \boldsymbol{\gamma}_{X_{2O}} \geq \mathbf{0};\ \beta_k^D, \overset{\sim D}{\mathbf{Z}_k} \text{ unrestricted}$$

$\boldsymbol{\gamma}_{X_{1O}}, \boldsymbol{\gamma}_{X_{2O}}$ are introduced in model (6) to relax the bound on other inputs. The third and the fourth sets of constraints ensure that intermediate measures do not affect disbursement efficiency. The objective function value of model (6), denoted by β_k^{D*} computes disbursement efficiency. Disbursement efficiency score of DMU_k (DE_k^*) is given by $1/\beta_k^{D*}$. $\beta_k^{D*} = 1$ implies DMU_k is disbursement efficient. That is, at the current level of disbursements, DMU_k produces maximum possible outputs and therefore is *disbursement efficient*. On the other hand, $\beta_k^{D*} > 1$ implies that DMU_k is disbursement inefficient. Disbursement efficiency is essentially a measure of overall efficiency of DMU_k in the use of its disbursements. The constraints associated with the other inputs, $\boldsymbol{X}_k^{1DO}$ and $\boldsymbol{X}_k^{2DO}$ have no influence on β_k^{D*} calculation and hence, in this case, they are effectively unrestricted inputs. If DMU_k is disbursement inefficient, then it can become disbursement efficient by increasing its outputs to $\beta_k^{D*} \boldsymbol{Y}_k^2$.

Remark B1 According to our modelling framework, if DMU_k is disbursement efficient $\left(1/\beta_k^{D*} = 1\right)$, then it is risk and disbursement efficient $\left(1/\beta_k^{R\&D*} = 1\right)$ as well as technical efficient $\left(1/\beta_k^{T*} = 1\right)$. The converse may not hold.

This follows, as the constraint set of model (3) is a subset of the constraint set of model (2), the constraint set of model (2) is a subset of the constraint set of model (1), and they have similar objective functions (scalar). Therefore, the optimal value of model (3) must be at least as that of model (2) and the optimal value of model (5) must be at least as that of model (1). That is $\beta_k^{D*} \geq \beta_k^{R\&D*} \geq \beta_k^{T*}$. Now since β_k^{D*}, $\beta_k^{R\&D*}$ and β_k^{T*} are ≥ 1, $\beta_k^{D*} = 1$ (DMU_k is disbursement efficient) implies $\beta_k^{R\&D*} = 1$ (DMU_k is risk and disbursement efficient) and $\beta_k^{T*} = 1$ (DMU_k is technical efficient). When $\beta_k^{T*} = 1$, we have that $\beta_k^{R\&D*} \geq 1$ and hence the converse may not hold.

Remark B2 According to our modelling framework, disbursement efficiency score of DMU_k is less than or equal to its risk and disbursement efficiency score and in turn, risk and disbursement efficiency score is less than or equal to its technical efficiency score.

This follows from $\beta_k^{D*} \geq \beta_k^{R\&D*} \geq \beta_k^{T*}$ as disbursement efficiency score, risk and disbursement efficiency score and technical efficiency score of DMU_k is computed as $1/\beta_k^{D*}$, $1/\beta_k^{R\&D*}$ and $1/\beta_k^{T*}$ respectively.

References

Badiezadeh, T., Saen, R. F., & Samavati, T. (2018). Assessing sustainability of supply chains by double frontier network DEA: A big data approach. *Computers and Operations Research, 98*, 284–290.

Basso, A., & Funari, S. (2001). A data envelopment analysis approach to measure the mutual fund performance. *European Journal of Operational Research, 135*, 477–492.

Basso, A., & Funari, S. (2016). DEA performance assessment of mutual funds. In J. Zhu (Ed.), *Data envelopment analysis, international series in operations research and management science*. Springer.

Basso, A., & Funari, S. (2017). The role of fund size in the performance of mutual funds assessed with DEA models. *The European Journal of Finance, 23*, 457–473.

Berkowitz, M. K., & Qiu, J. (2003). Ownership, risk and performance of mutual fund management companies. *Journal of Economics and Business, 55*, 109–134.

Choi, Y. K., & Murthy, P. B. S. (2001). Relative performance evaluation of mutual funds: A non-parametric approach. *Journal of Business Finance and Accounting, 28*, 853–876.

Choi, T.-M., Wallace, S. W., & Wang, Y. (2017). Big data analytics in operations management. *Production and Operations Management, 27*, 1868–1883.

Färe, R. (1984). On the existence of plant capacity. *International Economic Review, 25*, 209–213.

Färe, R., & Grosskopf, S. (2000). Network DEA. *Socio-Economic Planning Sciences, 34*, 35–49.

Färe, R., Grosskopf, S., & Kokkelenberg, E. C. (1989). Measuring plant capacity, utilisation and technical change: A nonparametric approach. *International Economic Review, 30*, 655–666.

Färe, R., Grosskopf, S., & Lovell, C. A. K. (1994). *Production frontiers*. Cambridge University Press.

Galagedera, D. U. A., Watson, J., Premachandra, I. M., & Chen, Y. (2016). Modelling leakage in two-stage DEA models: An application to US mutual fund families. *Omega, 61*, 62–77.

Galagedera, D. U. A., Roshdi, I., Fukuyama, H., & Zhu, J. (2018). A new network DEA model for mutual fund performance appraisal: An application to US equity mutual funds. *Omega, 77*, 168–179.

Galagedera, D. U. A., Fukuyama, H., Watson, J., & Tan, K. A. (2020). Do mutual fund managers earn their fees? New measures for performance appraisal. *European Journal of Operational Research, 287*, 653–667.

Holland, D. S., & Lee, S. T. (2002). Impacts of random noise and specification on estimates of capacity derived from data envelopment analysis. *European Journal of Operational Research, 137*, 10–21.

Johansen, L. (1968). Production functions and the concept of capacity, Recherches recentes sur la fonction de production, collection. *Economie Mathématique et Econométrie, 2*, 49–72.

Kang, S.-M., & Kim, M.-H. (2015). Short-run cost minimization and capacity utilization of regional public hospitals in South Korea. *Modern Economy, 6*, 21–29.

Kao, C. (2014). Network data envelopment analysis: A review. *European Journal of Operational Research, 239*, 1–16.

Khezrimotlagh, D., Zhu, J., Cook, W.D. & Toloo, M. (2019). Data envelopment analysis and big data. *European Journal of Operational Research, 274*, 1047–1054.

Kirkley, J., Paul, C.J.M. & Squires, D. (2002). Capacity and capacity utilization in common-pool resource industries. *Environmental and Resource Economics, 22*, 71–97.

Kerstens, K., Sadeghi, J., & Van de Woestyne, I. (2019). Convex and nonconvex input-oriented technical and economic capacity measures: An empirical comparison. *European Journal of Operational Research, 276*, 699–709.

Mahajan, J. (1991). A data envelopment analytic model for assessing the relative efficiency of the selling function. *European Journal of Operational Research, 53*, 189–205.

McCune, D. (2019). *U.S. mutual funds: Lower expenses and the changing competitive landscape.* https://www.broadridge.com/_assets/pdf/broadridge-mutual-fund-expense-trends.pdf

McMullen, P. R., & Strong, R. A. (1998). Selection of mutual funds using data envelopment analysis. *The Journal of Business and Economic Studies, 4*, 1–12.

Morey, M. R., & Morey, R. C. (1999). Mutual fund performance appraisals: A multi-horizon perspective with endogenous benchmarking. *Omega, 27*(2), 241–258.

Morningstar Inc. (2019). *The Morningstar active/passive barometer.* https://www.morningstar.com/lp/active-passive-barometer

Murthi, B. P. S., Choi, Y. K., & Desai, P. (1997). Efficiency of mutual funds and port- folio measurement: A non-parametric approach. *European Journal of Operational Research, 98*, 408–418.

Premachandra, I. M., Zhu, J., Watson, J., & Galagedera, D. U. A. (2012). Best performing US mutual fund families from 1993 to 2006: Evidence from a novel two-stage DEA model for efficiency decomposition. *Journal of Banking and Finance, 36*, 3123–3482.

Premachandra, I. M., Zhu, J., Watson, J., & Galagedera, D. U. A. (2016). Mutual fund industry performance: A network data envelopment analysis approach. In J. Zhu (Ed.), *Data envelopment analysis, international series in operations research and management science*. Springer.

Sánchez-González, C., Luis Sarto, J., & Vicente, L. (2017). The efficiency of mutual fund companies: Evidence from an innovative network SBM approach. *Omega, 71*, 114–128.

Sharpe, W. F. (1966). Mutual fund performance. *Journal of Business, 39*, 119–138.

Treynor, J. (1965). How to rate management of investment funds. *Harvard Business Review, 43*, 63–76.

U.S. Securities and Investment Commission. (2000). *Division of investment management: report on mutual fund fees and expenses*. http://www.sec.gov/news/studies/feestudy.htm

Yang, G.-L., & Fukuyama, H. (2018). Measuring the Chinese regional production potential using a generalized capacity utilisation indicator. *Omega, 76*, 112–127.

Zhu, J. (2020). DEA under big data: Data enabled analytics and network data envelopment analysis. *Annals of Operations Research*. in press.

Zhu, Q., Wu, J., & Song, M. (2018). Efficiency evaluation based on data envelopment analysis in the big data context. *Computers and Operations Research, 98*, 291–300.

Measuring Chinese Bank Performance with Undesirable Outputs: A Slack-Based Two-Stage Network DEA Approach

Ya Chen, Mengyuan Wang, and Jingyu Yang

1 Introduction

In order to offset the huge negative impact of the financial crisis in 2008 and enhance the ability immune to systemic risks of the world economic system, global financial development are under great pressure from structural reforms and transformation of growth patterns. Due to successfully coping with the financial crisis in 2008, China has played an active role in the world economic and financial affairs. As the second largest economy in the world, China will assume more responsibilities on global financial issues such as financial security governance and financial value chain restructuring. Undoubtedly, banks are always the core for financial reforms and development.

As reported by the *Financial Times*, the Chinese banking system has surpassed the euro area and has become the largest banking system in 2019. At the end of 2018, the total assets of Chinese banking industry reached 268.2 trillion yuan, increasing by 6.27%. And the assets of commercial banks accounted for 77.96% of Chinese banking industry. According to statistics from the China Banking Regulatory Commission, as of the end of June 2019, China has a total of 6 large state-owned commercial banks (SOCBs), 12 joint-stock commercial banks (JSCBs),

Y. Chen (✉)
School of Economics and Center for Industrial information and Economy, Hefei University of Technology, Hefei, P.R. China
e-mail: ychen@hfut.edu.cn

M. Wang
School of Economics, Hefei University of Technology, Hefei, P.R. China

J. Yang
School of Management, Hefei University of Technology, Hefei, P.R. China

J. Zhu, V. Charles (eds.), *Data-Enabled Analytics*, International Series in Operations Research & Management Science 312, https://doi.org/10.1007/978-3-030-75162-3_11

134 city commercial banks and rural cooperative banks as well as other corporation structures.[1]

Commercial banks are the basis for Chinese financial regulation. Through the study of operating performance, they can improve the management level and further promote the international competitiveness of Chinese economy. In fact, the efficiency of the banking system is germane to the performance of the entire economy because only an efficient system guarantees the smooth functioning of nation's payment system and effective implementation of the monetary policy (Kumar and Gulati 2014). Moreover, improving bank efficiency is a key step for the continued injecting vitality of the development of commercial banks (An et al. 2015). So, it is important to study how to accurately measure and improve the efficiency of commercial banks. In this paper we investigate Chinese bank efficiency from 2009 to 2018 based on a two-stage network data envelopment analysis (DEA) framework (Cook et al. 2010; Kao 2014), which is important for the Chinese financial regulation and high-quality development.

Looking back on the reform process of Chinese interest rate liberalization, the marketization of loan interest rates is significantly faster than the marketization of deposit interest rates, that is, the artificially suppressed ceiling of deposit interest rates, while relaxing the floating range of loan interest rates. Through the above operations, it provided stable spread protection for the domestic banking industry, which was maintained at about 2%. However, some scholars have shown that interest rate liberalization may have various effects on Chinese commercial banks under different ownership systems (Ba et al. 2013). In this paper, we also examine the impact of Chinese interest rate liberalization on bank efficiency and efficiency difference in terms of different bank types.

Data envelopment analysis (DEA), which was coined by Charnes et al. (1978), is a popular method for evaluating the relative efficiency of decision-making units (DMUs) with multi-inputs and multi-outputs. DEA is evolving into data enabled analytics for big data analysis, and network DEA (Cook et al. 2010; Kao 2014) is actually related to the value dimension of big data (Zhu 2020). For example, the operation process of banks is gradually viewed as a two-stage network system, which includes more valuable and accurate information for performance improvemnet of banks. In this paper, we measure Chinese bank performance by network DEA.

Particularly, we propose an undesirable variable intermediate measures SBM (UVSBM) model to calculate both overall and sub-stage efficiencies of each bank. We show that the overall inefficiency is equal to the sum of the sub-stage inefficiencies. Our empirical results reveal that there is a sharp downward trend for the overall or fund-raising efficiency of Chinese listed banks in the short term

[1]SOBs are state-controlled joint-stock commercial banks listed in the Shanghai and Hong Kong Stock Exchanges. JSCBs could be state-owned or private enterprises. And the main role of CCBs is to provide financial support for small-medium enterprises and promote local economic development.

(2009–2011) and an upward trend then except for 2015. However, the efficiency change during the fund-using stage has been relatively stable. Our results also show that the average efficiency of Chinese commercial banks has improved with the acceleration of interest rate liberalization. But there is no significant difference in the efficiency distribution of these banks before and after the acceleration of interest rate liberalization. For different types of listed banks, our results find that the average efficiency of SOCBs is highest, followed by CCBs and JSCBs. And the efficiency difference is mainly reflected in the fund-raising stage.

The rest of the paper is organized as follows: Sect. 2 briefly reviews Chinese bank efficiency literature closely related to the paper. Section 3 proposes a slack-based two-stage DEA approach for measuring performance of Chinese commercial banks and analyzes the empirical results. Conclusions and directions for future research are given in Sect. 4.

2 Literature Review

In a recent survey on DEA applications by Liu et al. (2013), the banking industry was one of the top five industries. Next we review the existing literature on Chinese bank efficiency.

2.1 Chinese Bank Efficiency in a Single Process Framework

To the best of our knowledge, Chen et al. (2005) initially applied DEA to study the performance of Chinese banks and investigated the efficiency change of 43 Chinese banks before and after the deregulation initiated by the government in 1995. It showed that the financial deregulation improved the efficiency of Chinese banks, and the large state-owned banks and small banks were more efficient than medium sized Chinese banks. Following them, many studies further examine the performance of Chinese banks based on DEA (Barros et al. 2011). For example, Yao et al. (2008) evaluated whether Chinese state-owned commercial banks reacted positively and successfully to the new reforms over the period 1998–2005, while Luo and Yao (2010) examined whether initial public offering (IPO) was effective in enhancing bank performance during 1999–2008. Du et al. (2018) used truncated regression and double bootstrapping methods to correct the traditional DEA bias, and further studied the changes in the efficiency of banks in China over time and its influencing factors from 2006 to 2011.

Focusing on economy efficiency, Laurenceson and Qin (2008) investigated the effect of introducing foreign investment on the cost efficiency of China's banks during 2001–2006. And Ariff and Can (2008) measured the cost and profit efficiency of China's 28 banks during 1995–2004.

2.2 Bank Efficiency in a Two-Stage Process Framework

The above studies in Sect. 2.1 viewed a bank as a "black box" and considered only the initial inputs and final outputs of the black system. Namely, it evaluated each bank's efficiency using traditional DEA models, without considering the production process of intermediate products. In recent years, the internal production or operation process of banks is often considered. Two-stage network DEA, which is a special case of Network DEA, is widely used in bank efficiency evaluation (Cook et al. 2010; Kao 2014). As pointed out by Lozano (2016), existing literature on bank efficiency mostly considers two-stage systems in series although in some cases three or even four processes are considered. As a result, this paper evaluates efficiency of Chinese commercial banks under a two-stage system.

In the research of bank efficiency evaluation under two-stage production process, Wang et al. (1997) and Seiford and Zhu (1999) considered the bank production system as a two-stage production process earlier. And they treat intermediate measures as both input and output indicators, but the efficient system is not equivalent to efficient sub-stages under the treatment. Following these two work, there have been a lot of studies evaluating bank efficiency based on two-stage DEA. Avkiran (2009) used the network slacks-based measure (SBM) model to evaluate the profit efficiency of the UAE banking industry.

With respect to Chinese bank efficiency, Wang et al. (2014) evaluated the system and stage efficiencies of 16 major Chinese commercial banks during the third round of the Chinese banking reform period (2003–2011) using two-stage network DEA. Simimarly, An et al. (2015) measured the input-output efficiency and substage efficiencies of Chinese listed banks during 2008–2012 by using a two-stage DEA model.

2.3 Undesirable Outputs and Modelling in Two-Stage or Network Bank Efficiency

When analyzing a bank's efficieny, undesirable (or bad) outputs such as non-performing loan are important indicators and by-products of the two-stage banking production process. By including bad outputs, Fukuyama and Weber (2010) measured the slacks-based inefficiency of Japanese banks under a two-stage system based on directional distance function (DDF) under an assumption of weak disposability for bad outputs (Färe et al. 1989; Färe and Grosskopf 2003, 2009). Akther et al. (2013) considered the inter-temporal impact of bad loans based on the model proposed by Fukuyama and Weber (2010), and investigated the efficiency of Bangladesh banking sector. Fukuyama and Weber (2015) extended the model proposed by Akther et al. (2013) to multi-periods, allowing banks to reduce current loans and securities investment, thereby increasing future production efficiency. More recent extension see Fukuyama and Weber (2017).

The above studies based on DDF need to assign a directional vector when calculating the banks efficiency. Different from DDF, Wang et al. (2014) incorporated undesirable outputs into the additive two-stage DEA model proposed by Chen et al. (2009). Based on the work of Tone and Tsutsui (2009) and Fukuyama and Mirdehghan (2012), An et al. (2015) measured two-stage bank efficiency based on slacks-based measure (SBM) (Tone 2001) under an assumption of strong disposability for undesirable outputs. However, weak disposability for undesirable outputs is usually assumed and seems more reasonable (see also Pham and Zelenyuk 2018; Liu et al. 2015; Epure and Lafuente 2015).

Except for efficiency evaluation, performance improvement (i.e. frontier projection) for commercial banks is also an important issue. But it is usually neglected by existing literature above. To evaluate DMUs' efficiency and obtain their frontier projections of classical two-stage processes, Chen et al. (2016) proposed the variable intermediate measures SBM (VSBM) model and illustrated it by a bank application.

2.4 The Contributions of this Paper

Compared to existing studies, this paper makes contributions to the literature in two aspects. First, this paper investigates the latest Chinese bank performance during 2009–2018, which naturally complements existing literature on Chinese bank efficiency such as Wang et al. (2014) and An et al. (2015). Our analysis provides more accurate efficiency results and so new (latest) insights for Chinese bank regulators. One one hand, we compare the distributions of estimated efficiency using the adapted Li test (Simar and Zelenyuk 2006; Kenjegalieva et al. 2009). One the other hand, we study bank efficiency before and after the acceleration of interest rate liberalization, and also examine the efficiency difference due to different types of banks. Second, this paper extends Chen et al. (2016)'s research to measure bank efficiency in the presence of undesirable outputs. Difference from Wang et al. (2014) and An et al. (2015) and similar to Pham and Zelenyuk (2018), we model undesirable outputs under the assumption of weak disposability. Moreover, our proposed approach successfully measures bank efficiency without assigning a weight of each stage or a directional vector, and could also provide frontier projection or optimal targets for each bank. Thus, it provides an alternative to existing approaches (Wang et al. 2014; An et al. 2015; Fukuyama and Weber 2010; Fukuyama and Matousek 2017).

3 The Slack-Based Two-Stage DEA Approach with Undesirable Outputs

3.1 The Proposed UVSBM Model

Consider the two-stage production process in Fig. 1. The first stage uses m inputs $x_{ij}, i = 1, 2, ..., m$ to produce D intermediate measures (regarded as outputs) $z_{dj}, d = 1, 2..., D$. The second stage treats these intermediate measures as inputs to generate s_1 non-joint desirable outputs $y^1_{r_1,j}, r_1 = 1, 2, ..., s_1$, and s_2 joint desirable outputs $y^1_{r_2,j}, r_2 = 1, 2, ..., s_2$ and undesirable outputs $u_{kj}, k = 1, ..., p$. In this paper, we assume a weak disposability for joint desirable outputs and undesirable outputs, which implies an opportunity cost of reducing undesirable outputs.

Based on the VSBM model in Chen et al. (2016) and the weak disposability for joint desirable outputs and undesirable outputs, we develop a slack-based model to measure the system efficiency of DMU_0:

$$Min\, e_0 = \frac{1 - \frac{1}{m}\sum_{i=1}^{m} \frac{S_i^-}{x_{io}}}{1 + \frac{1}{s_1+s_2+p}(\sum_{r_1=1}^{s_1} \frac{s_{r_1}^+}{y_{r_10}} + \sum_{r_2=1}^{s_2} \frac{s_{r_2}^+}{y_{r_20}} + \sum_{k=1}^{p} \frac{s_k^{'-}}{u_{k0}})} \tag{1}$$

$$s.t. \sum_{d=1}^{D} w_d^1 z_{dj} - \sum_{i=1}^{m} v_i x_{ij} + A \leq 0,\ j = 1, ..., n \ldots\ldots\ldots\ldots(1a)$$

$$\sum_{j=1}^{n} \lambda_j^1 z_{dj} = \tilde{z}_{d0},\ d = 1, ..., D \ldots\ldots\ldots\ldots(1b)$$

$$\sum_{j=1}^{n} \lambda_j^2 z_{dj} = \tilde{z}_{do},\ d = 1, ..., D \ldots\ldots\ldots\ldots(1c)$$

$$\sum_{j=1}^{n} \lambda_j^2 y_{r_1 j} - s_{r_1}^+ = y_{r_1 0},\ r_1 = 1, ..., s_1 \ldots\ldots\ldots\ldots(1d)$$

$$\theta \sum_{j=1}^{n} \lambda_j^2 y_{r_2 j} - s_{r_2}^+ = y_{r_2 0},\ r_2 = 1, ..., s_2 \ldots\ldots\ldots\ldots(1e)$$

$$\theta \sum_{j=1}^{n} \lambda_j^2 u_{kj} + s_k^{'-} = u_{ko},\ k = 1, ..., p \ldots\ldots\ldots\ldots(1f)$$

$$\sum_{j=1}^{n} \lambda_j^1 = \sum_{j=1}^{n} \lambda_j^2 = 1 \ldots\ldots\ldots\ldots(1g)$$

$$0 < \theta \leq 1 \ldots\ldots\ldots\ldots(1h)$$

$$s_i^-, s_r^+, s_k^{'-}, \lambda_j^1, \lambda_j^2 \geq 0$$

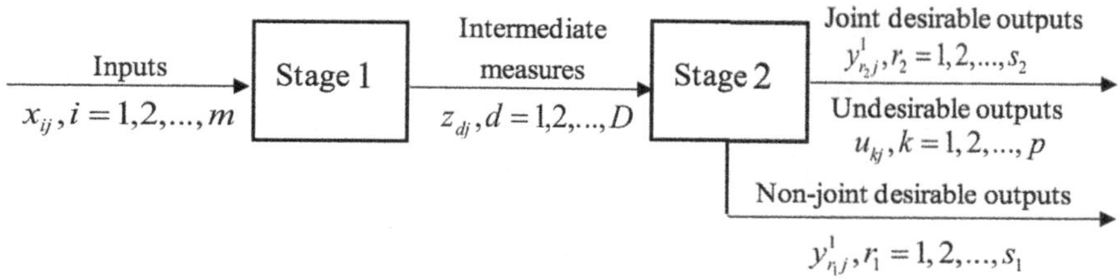

Fig. 1 The two-stage production process

where λ_j^1 and λ_j^2 are intensity variables, and s_i^-, s_i^+ and $s_k^{'-}$ are input, desirable and undesirable output slacks, respectively. We call model (1) as the ***undesirable VSBM (UVSBM) model.***

Remark In model (1), constraints (1b) and (1c) model the intermediate measures, i.e. the link between Stage 1 and Stage 2. Note that $\tilde{z}_{do}$ is a variable, which could increase or decrease for connecting two sub-stages. As pointed by Fukuyama and Mirdehghan (2012), the equality constraints on intermediate products in Tone and Tsutsui (2009) may be inappropriate to identify the efficiency status of DMUs. Slacks for the intermediate products should be considered and inequalities should be employed. As $\tilde{z}_{do}$ is a variable, we actually take the slacks on the intermediate products into account though the constraints on intermediate products are in equalities in model (1). In fact, we use a different way to measure system (stage) efficiency hereinafter. Following most of existing literature, constraints (1e), (1f) and (1h) model the joint desirable outputs and undesirable outputs under the assumption of weak disposability. And constraint (1g) shows that model (1) is under variable returns to scale (VRS), a common assumption for banking efficiency evaluation.

Denote e_0^* as the optimal efficiency of DMU_0, we have:

Definition 1 (UVSBM-Efficient) DMU_0 is UVSBM-efficient if and only if e_0^*=1.

Note that model (1) is a non-linear programming. Next, we transform it into equivalent linear programming. Let $\phi_j^2 = \theta\lambda_j^2$, $\varphi_j^2 = (1-\theta)\lambda_j^2$, model (1) can be transformed into:

$$Min\ e_0 = \frac{1 - \frac{1}{m}\sum_{i=1}^{m}\frac{S_i^-}{x_{io}}}{1 + \frac{1}{s_1+s_2+p}\left(\sum_{r_1=1}^{s_1}\frac{s_{r_1}^+}{y_{r_10}} + \sum_{r_2=1}^{s_2}\frac{s_{r_2}^+}{y_{r_20}} + \sum_{k=1}^{p}\frac{s_k^{'-}}{u_{k0}}\right)} \tag{2}$$

$$s.t. \sum \lambda_j^1 x_{ij} + s_i^- = x_{io}, i = 1,\ldots,m$$

$$\sum_{j=1}^{n} \lambda_j^1 z_{dj} = \tilde{z}_{d0}, d = 1,\ldots,D$$

$$\sum_{j=1}^{n}(\phi_j^2+\varphi_j^2)z_{dj}=\tilde{z}_{do}, d=1,\ldots,D$$

$$\sum_{j=1}^{n}(\phi_j^2+\varphi_j^2)y_{r_1j}-s_{r_1}^+=y_{r_10}, r_1=1,\ldots,s_1$$

$$\sum_{j=1}^{n}\phi_j^2 y_{r_2j}-s_{r_2}^+=y_{r_20}, r_2=1,\ldots,s_2$$

$$\sum_{j=1}^{n}\phi_j^2 u_{kj}+s_k^{\prime-}=u_{ko}, k=1,\ldots,p$$

$$\sum_{j=1}^{n}\lambda_j^1=\sum_{j=1}^{n}(\phi_j^2+\varphi_j^2)=1$$

$$s_i^-, s_r^+, s_k^{\prime-}, \lambda_j^1, \phi_j^1, \varphi_j^2\geq 0$$

Considering ϕ_j^2, φ_j^2 is self-constrained in model (2), the constraint $(1h)$ in model (1) doesn't add any more information and we exclude it from model (2). Using Charnes-Cooper transformation (Charnes and Cooper 1962), we let:

$$t=\frac{1}{1+\frac{1}{s_1+s_2}(\sum_{r_1=1}^{s_1}\frac{S_{r_1}^+}{y_{r_10}}+\sum_{r_2=1}^{s_2}\frac{S_{r_2}^+}{y_{r_20}})} \tag{3}$$

$$S_i^-=ts_i^-, S_{r_1}^+=ts_{r_1}^+, S_{r_2}^+=ts_{r_2}^+, S_k^{'-}=ts_k^{'-}$$

$$\alpha_j^1=t\lambda_j^1, \hat{\phi}_j^2=t\phi_j^2, \hat{\varphi}_j^2=t\varphi_j^2, z_{d0}^{'}=t\tilde{z}_{d0}$$

then model (2) is rewritten as:

$$Min\ e_0=t-\frac{1}{m}\sum_{i=1}^{m}\frac{S_i^-}{x_{io}} \tag{4}$$

$$s.t.\ t+\frac{1}{s_1+s_2+p}(\sum_{r_1=1}^{s_1}\frac{S_{r_1}^+}{y_{r_10}}+\sum_{r_2=1}^{s_2}\frac{S_{r_2}^+}{y_{r_20}}+\sum_{k=1}^{p}\frac{S_k^{'-}}{u_{ko}})=1$$

$$\sum_{j=1}^{n}\alpha_j^1x_{ij}+S_i^-=tx_{io}, i=1,\ldots,m$$

$$\sum_{j=1}^{n} \alpha_j^1 z_{dj} = z'_{d0}, d = 1, \ldots, D$$

$$\sum_{j=1}^{n} (\hat{\phi}_j^2 + \varphi_j^2) z_{dj} = z'_{do}, d = 1, \ldots, D$$

$$\sum_{j=1}^{n} (\hat{\phi}_j^2 + \varphi_j^2) y_{r_1 j} - s_{r_1}^{+} = t y_{r_1 0}, r_1 = 1, \ldots, s_1$$

$$\sum_{j=1}^{n} \hat{\phi}_j^2 y_{r_2 j} - s_{r_2}^{+} = y_{r_2 0}, r_2 = 1, \ldots, s_2$$

$$\sum_{j=1}^{n} \hat{\phi}_j^2 u_{kj} + s_k^{'-} = t u_{ko}, k = 1, \ldots, p$$

$$\sum_{j=1}^{n} \alpha_j^1 = \sum_{j=1}^{n} (\hat{\phi}_j^2 + \hat{\varphi}_j^2) = t$$

$$S_i^{-}, S_r^{+}, S_k^{'-}, \alpha_j^1, \hat{\phi}_j^2, \hat{\varphi}_j^2 \geq 0, t\ free$$

3.2 Properties of the UVSBM *Model*

In this section we propose some properties of the UVSBM model. We examine the DEA frontier and potential conflicts of intermediate measures between the two stages that arise from the dual role of intermediate measures as both inputs and outputs. Namely, the first stage may have to increase its intermediate measures (as outputs) to improve its efficiency while the second stage may have to decrease its intermediate measures (as inputs) to improve its efficiency.

Denote the optimal solution of model (2) as ($s_i^{-*}, s_{r_1}^{+*}, s_{r_2}^{+*}, s_k^{'-*}, \lambda_j^{1*}, \varphi_j^{2*}, \phi_j^{2*}, \tilde{z}_{d0}^{*}$), we have the following theorem to obtain an efficient frontier projection for any DMU_0 under evaluation:

Theorem 1 *For any DMU_0 under evaluation, the projection ($x_{io} - s^{-*}, \tilde{z}_{d0}^{*}, y_{r_1 0} + s_{r_1}^{+*}, y_{r_2 0} + s_{r_2}^{+*}, u_{k0} - s_k^{'-*}$) based on the UVSBM model is UVSBM-efficient.*

Proof: See Appendix

Next, we study the proposed UVSBM model on potential conflicts arising from the intermediate measures between the two sub-stages. To address this conflict, we first transform model (4) into its multiplier form. The dual of model (4) is:

$$
\begin{aligned}
& Max\ \alpha \\
s.t.\ & v_i x_{io} \geq \frac{1}{m}, i = 1, ..., m \\
& \mu_{r_1} y_{r_1 0} \geq \frac{\alpha}{s_1 + s_2 + p}, r_1 = 1, ..., s_1 \\
& \mu_{r_2} y_{r_2 0} \geq \frac{\alpha}{s_1 + s_2 + p}, r_2 = 1, ..., s_2 \\
& \beta_k u_{k0} \geq \frac{\alpha}{s_1 + s_2 + p}, k = 1, ..., p \\
& \alpha + \sum_{i=1}^{m} v_i x_{io} + \sum_{k=1}^{p} \beta_k u_{k0} - \sum_{r_1=1}^{s_1} \mu_{r_1} y_{r_1 0} - \sum_{r_2=1}^{s_2} \mu_{r_2} y_{r_2 0} - A - B = 1 \\
& \sum_{r_1=1}^{s_1} \mu_{r_1} y_{r_1 j} + \sum_{r_2=1}^{s_2} \mu_{r_2} y_{r_2 j} - \sum_{k=1}^{p} \beta_k u_{kj} - \sum_{d=1}^{D} w_d^2 z_{dj} + B \leq 0, j = 1, ..., n \\
& \sum_{d=1}^{D} w_d^1 z_{dj} - \sum_{i=1}^{m} v_i x_{ij} + A \leq 0, j = 1, ..., n \\
& \sum_{r_1=1}^{s_1} \mu_{r_1} y_{r_1 j} - \sum_{d=1}^{D} w_d^2 z_{dj} + B \leq 0, j = 1, ..., n \\
& - w_d^1 + w_d^2 = 0, d = 1, 2, ..., D \\
& \alpha, v_i, \mu_{r_1}, \mu_{r_2}, \beta_k, w_d^1, w_d^2, A, B\ free
\end{aligned}
\tag{5}
$$

The constraints $-w_d^1 + w_d^2 = 0, d = 1, \ldots, D$ in model (5) show that the two weights are the same for intermediate measures. So, the UVSBM model successfully handles potential conflicts of intermediate measures. Note that the two weights of intermediate measures may not be the same if $\tilde{z}_{d0}$ is constant in model (1).

3.3 System and Stage Efficiencies of the Two-Stage System

Based on model (5), the system efficiency of DMU_0 is:

$$
e_0 = \alpha = 1 - (\sum_{i=1}^{m} v_i x_{io} + \sum_{k=1}^{p} \beta_k u_{ko} - \sum_{r_1=1}^{s_1} \mu_{r_1} y_{r_1 o} - \sum_{r_2=1}^{s_2} \mu_{r_2} y_{r_2 o} - A - B) \tag{6}
$$

Let $s_0 = \sum_{i=1}^{m} v_i x_{io} + \sum_{k=1}^{p} \beta_k u_{ko} - \sum_{r_1=1}^{s_1} \mu_{r_1} y_{r_1 o} - \sum_{r_2=1}^{s_2} \mu_{r_2} y_{r_2 o} - A - B$, then we have $e_0 = 1 - s_0$. Here s_0 is equal to the system inefficiency. Similarly, we have the first and second stage efficiencies as $e_0^{Dec1} = 1 - s_0^1$ and $e_0^{Dec2} = 1 - s_0^2$, where $s_0^1 = \sum_{i=1}^{m} v_i x_{io} - \sum_{d=1}^{D} w_d^1 z_{do} - A$ and $s_0^2 = \sum_{k=1}^{p} \beta_k u_{ko} + \sum_{d=1}^{D} w_d^2 z_{do} - \sum_{r_1=1}^{s_1} \mu_{r_1} y_{r_1 o} - \sum_{r_2=1}^{s_2} \mu_{r_2} y_{r_2 o} - B$. The fifth and sixth rows of constraints in model (5) ensure $s_0^1 \geq 0, s_0^2 \geq 0$. Hence, s_0^1 and s_0^1 equal inefficiencies of the first and second stages, respectively.

In model (5), the equality $w_d^1 = w_d^2$ gives

$$s_0 = s_0^1 + s_0^2, s_0, s_0^1, s_0^2 \geq 0. \tag{7}$$

Thus, the system inefficiency equals the sum of inefficiencies of its two stages. Chen et al. (2016) claim that the statement is important because the system efficiency can be improved by reducing its stage inefficiencies. Due to Eq. (6), the overall efficiency of the system is no larger than stage efficiencies of its two stages.

Next, we could further obtain the efficiency decomposition of the system, i.e. the stage efficiency scores (e_0^{Dec1}, e_0^{Dec2}) of each stage. Based on Eq. (6) we have

$$e_0 = e_0^{Dec1} + e_0^{Dec2} - 1, e_0, e_0^{Dec1}, e_0^{Dec2} \leq 1. \tag{8}$$

Note that the stage efficiency scores (e_0^{Dec1}, e_0^{Dec2}) may not be unique. Following Liang et al. (2008) and Chen et al. (2016), we could easily obtain an efficiency interval for the stage efficiencies when the system efficiency keeps unchanged. And we neglect the detailed analysis in this paper.

3.4 Variables and Data

According to the inputs and outputs selected by existing studies (Fukuyama and Weber 2010, 2017; Holod and Lewis 2011; Akther et al. 2013; An et al. 2015), the production process of banks was divided into fund-raising stage and fund-using stage. In this paper, we focus on the same production process and take the deposit as intermediate products to evaluate the efficiency of banks within a two stage production process. The detailed information of the variables used in this paper is shown in Table 1. And the two-stage production process of banks is shown in Fig. 2.

Considering that the governance structure of listed banks is more reasonable and its information disclosure is more strictly, we investigate the performance of Chinese listed banks in this paper. And we focus on the performance of Chinese listed banks after the global financial crisis in 2008. Considering the availability of data, we eliminate 4 rural commercial banks among the 24 listed banks, and select the remaining 20 stock-listed commercial banks in China as the sample. They are divided into 3 types, namely 5 SOCBs, 8 JSBs and 7 CCBs. The SOCBs include Industrial and Commercial Bank of China (ICBC), Bank of China (BOC),

Table 1 Defined inputs, intermediate products and outputs

	Variables	Meaning
Inputs	Fixed assets (X1)	The net fixed assets at the end of the year
	Employees (X2)	The number of employees at the end of the year
	Equity (X3)	The sum of capital invested in the establishment of the business and capital created in the operation
Intermediate products	Total funding (Z)	Total customer deposits and other funding
Outputs	Loans (Y1)	The grant loans and bank advances
	Securities investment (Y2)	The Trading Securities and at FV through Income, Derivatives, Available for Sale Securities, Held to Maturity Securities, At-equity Investments in Associates and Other Securities.
	Non-performing loans (NPLs) (U)	An undesirable output

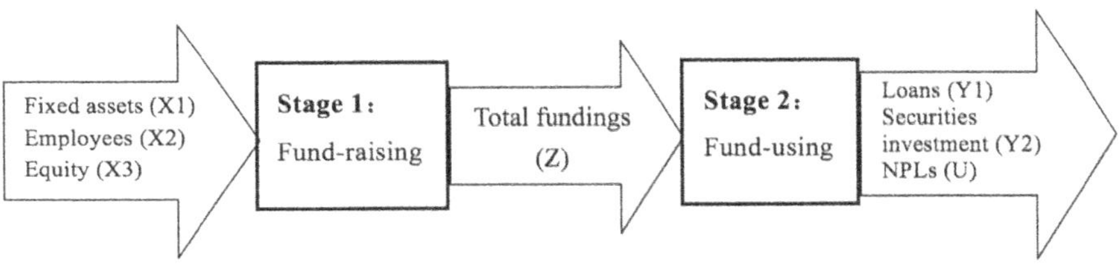

Fig. 2 Two-stage production process of commercial banks

China Construction Bank (CCB), Agricultural Bank of China (ABC) and Bank of Communications (BOCOM). The NJCBs include China CITIC Bank (CNCB), China Minsheng Bank (CMBC), China Merchants Bank (CMB), Industrial Bank (IB), Shanghai Pudong Development Bank (SPDB), Ping An Bank (PAB), Hua Xia Bank (HXB) and China Everbright Bank (CEB). The CCBs include Bank of Beijing (BOBJ), Bank of Nanjing (BONJ), Bank of Ningbo (BONB), Bank of Shanghai (BOSH), Bank of Hangzhou (BOHZ), Bank of Jiangsu (BOJS) and Bank of Guiyang (BOGY). The original data of Chinese commercial banks mainly come from the annual reports of 20 Chinese listed banks from 2009–2018, Orbis Bank Focus database and Wind database.

Table 2 provides the descriptive statistics on the inputs, intermediate products, and outputs of 20 Chinese listed banks for the years 2009 and 2018.

It can be seen from Table 2 that all indicators of listed commercial banks showed an upward trend between 2009 and 2018. Among them, equity has increased by about 4 times, loan has increased by about 3 times, total funding and securities

Table 2 Descriptive statistics of 20 Chinese listed banks from 2009 to 2018

Variable	Mean	Std. Dev.	Max	Min
2009 (20 observations)				
Fixed assets (X1)	24,646.39	38,842.83	111,973.00	388.78
Employees (X2)	84,163.70	140,619.88	441,144.00	2192.00
Equity (X3)	145,748.87	209,378.08	678,934.00	2683.50
Total funding (Z)	2,546,057.08	3,452,336.47	10,909,047.00	46,320.69
Loans (Y1)	1,396,311.89	1,805,741.25	5,583,174.00	24,277.69
Securities investment (Y2)	708,320.21	1,061,988.24	3,621,938.00	14,019.01
NPLs (U)	22,493.98	35,718.26	120,241.00	644.51
2018 (20 observations)				
Fixed assets (X1)	60,361.37	79,265.02	253,525.00	1417.38
Employees (X2)	108,793.75	148,073.76	473,691.00	6474.00
Equity (X3)	616,763.31	692,143.26	2,344,883.00	35,843.27
Total funding (Z)	6,231,083.31	7,045,998.41	23,690,800.00	345,419.57
Loans (Y1)	4,140,276.39	4,578,950.26	15,046,132.00	164,169.85
Securities investment (Y2)	1,348,774.78	810,261.54	3,008,167.00	263,125.33
NPLs (U)	66,049.85	70,358.30	235,084.00	2306.00
Pooled data (200 observations)				
Fixed assets (X1)	42,222.80	62,224.56	253,525.00	388.78
Employees (X2)	101,793.70	148,924.05	503,082.00	2192.00
Equity (X3)	365,299.09	488,106.51	2,344,883.00	2683.50
Total funding (Z)	4,575,171.96	5,567,888.37	23,690,800.00	46,320.69
Loans (Y1)	2,658,104.07	3,378,944.95	15,046,132.00	24,277.69
Securities investment (Y2)	1,307,769.80	1,456,980.59	6,832,348.00	14,019.01
NPLs (U)	37,785.01	53,445.69	235,084.00	279.21

Note: The unit of all financial measures is Million Yuan

investment have increased by about 2 times. Specifically, in fund-raising stage, listed banks raise the average funds of 4,575,171.95 million Yuan. In fund-using stage, banks use the funds raised in fund-raising stage to offer loans and investment. They have offered the average loans of 2,658,104.07 million Yuan and securities investment of 1,307,769.80 million Yuan, at the same time have produced the average NPLs of 37,785.01 million Yuan in the study period.

Furthermore, the change ratio of variables from 2009 to 2018 is shown in Fig. 3. We can find that the inputs and outputs scale of Chinese listed banks increased significantly. In fund-raising stage, the investment in human capital, physical capital, and financial capital increased by 29.26%, 144.91%, and 323.17%, respectively. Compared with physical capital and financial capital, the investment in human capital has been relatively reduced. This is consistent with the phenomenon of personnel redundancy that is common in Chinese state-owned banks. And the total number of funding increases by 144.73%. In fund-using stage, the number of loans and securities investment increases by 196.52% and 90.42%, respectively. But the

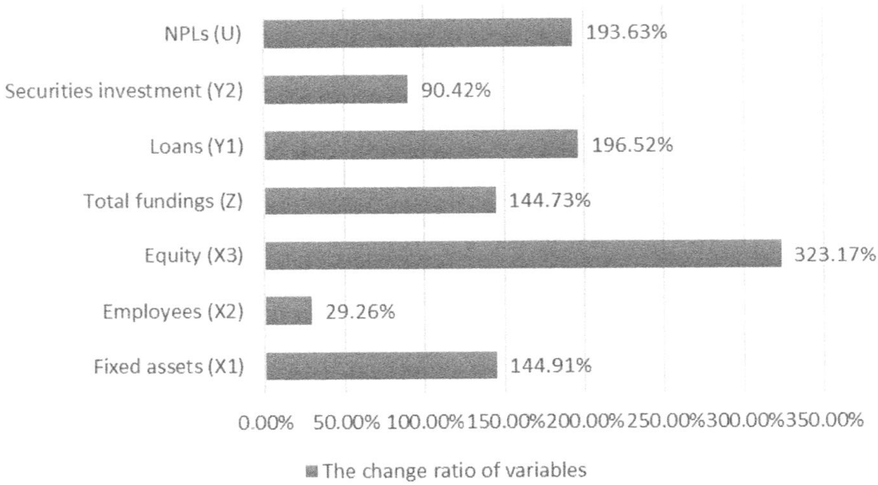

Fig. 3 The change ratio of variables from 2009 to 2018

number of NPLs also increases by 193.63%, which may have a negative impact on the efficiency of fund-using stage.

Most notably, according to the Table 2, the total amount of non-performing loans is increased from 22,493.98 in year 2009–66,049.85 in 2018. According to data publicly released by the China Banking Regulatory Commission, compared with the year before 2010, the non-performing loan balance of Chinese commercial banks increased from 0.43 trillion yuan to 2.37 trillion yuan in September 2019, a 451.2% increase over the 10-year period. The loan ratio also rose from 1.1% in 2010 to 1.86%, an increase of about 72.7%. Since the implementation of the industrial structure upgrade policy in 2013, unfair competition has intensified, which is reflected in the financial industry as the scale of non-performing loans has increased year by year, and the rate of non-performing loans has continued to rise. Therefore, 2013–2018 is the most "anxious" 5 years for Chinese banks. There are two main reasons: first, the rise of internet finance has triggered a technology-driven financial revolution, which has caused internal and external shocks to the banking industry; Second, the supply-side structural reform at the economic level has made the overall operation of commercial banks in MPA, liquidity management and capital management tend to be passive and cautious. The banking industry is facing a crisis of development with the disappearance of dividends and the dilemma of innovation. Thus, in the face of increasingly large commercial banks' non-performing financial assets, the traditional non-performing financial asset disposal model has become inadequate. Therefore, how to innovate the non-performing asset disposal model and improve the bank's operating efficiency is a very necessary issue to consider.

Table 3 Overall efficiencies of 20 Chinese listed banks from 2009 to 2018

Bank	2009	2010	2011	2012	2013	2014	2015	2016	2017	2018	Mean	Rank
ICBC	1.00	1.00	1.00	1.00	1.00	1.00	1.00	1.00	1.00	1.00	1.00	1
ABC	0.80	0.68	0.67	0.74	0.77	0.81	0.86	1.00	1.00	0.71	0.80	12
CCB	0.96	0.96	0.91	0.89	0.96	1.00	1.00	1.00	1.00	1.00	0.97	3
BOC	1.00	1.00	1.00	0.81	0.83	1.00	0.98	0.84	0.89	1.00	0.94	5
BOCOM	0.90	1.00	0.84	0.80	0.75	0.80	0.73	0.85	1.00	1.00	0.87	8
CNCB	0.79	0.73	0.59	0.68	0.72	0.88	0.91	0.95	0.80	1.00	0.82	11
CMBC	0.58	0.68	0.65	0.77	0.60	0.64	0.58	0.74	0.73	0.72	0.67	18
CMB	0.85	0.86	0.77	0.71	0.69	0.75	0.67	0.71	0.66	0.70	0.75	14
IB	0.89	1.00	1.00	1.00	1.00	1.00	1.00	1.00	1.00	1.00	0.99	2
SPDB	0.85	1.00	0.97	1.00	1.00	1.00	1.00	1.00	0.89	0.85	0.96	4
PAB	1.00	0.83	0.56	0.72	0.76	0.79	0.68	0.76	0.82	0.83	0.80	13
HXB	1.00	0.45	0.43	0.54	0.77	0.77	0.58	0.66	0.61	0.58	0.64	19
CEB	0.61	0.65	0.63	1.00	0.65	0.72	0.69	0.82	0.76	0.76	0.74	16
BOBJ	0.88	0.85	0.80	0.97	0.93	0.93	0.70	0.80	0.67	0.67	0.85	9
BONJ	1.00	1.00	0.71	1.00	1.00	1.00	0.72	0.82	1.00	0.78	0.90	6
BONB	0.78	0.74	0.49	0.68	0.75	0.68	0.60	0.71	0.83	1.00	0.74	15
BOHZ	1.00	0.85	0.84	0.67	0.84	0.86	0.80	1.00	0.89	1.00	0.89	7
BOSH	0.66	0.68	0.57	0.77	0.85	0.83	0.81	1.00	1.00	1.00	0.83	10
BOJS	0.41	0.44	0.43	0.59	0.58	0.80	0.90	0.89	0.86	0.95	0.69	17
BOGY	1.00	1.00	1.00	1.00	1.00	1.00	1.00	1.00	1.00	1.00	1.00	1
Mean	0.85	0.82	0.74	0.82	0.82	0.86	0.81	0.88	0.87	0.88	0.84	

3.5 *Overall Efficiency and Sub-stage Efficiencies for Chinese Listed Banks*

According to the optimal solution of model (4), we can get the overall efficiency and sub-stage efficiencies for Chinese listed banks, which are organized in Tables 3, 4 and 5. And the penultimate column in Tables 3, 4 and 5 shows the average efficiency of each listed bank. The last column indicates the rank of each listed bank based on the average efficiency.

According to Table 3, we can see most of banks improved their efficiencies during 2009–2018, such as ABC, CCB, BOC, CNCB, IB, SPDB, BONJ, BOHZ, BOJS. At the same time, the overall efficiency of some banks is showing a trend of continued inefficiency, such as BOCOM, CMBC, CMB, PAB, CEB, BONB, that is, the efficiency improvement is not obvious during the sample period. In addition, there are a few banks showing a decline in efficiency, such as BOBJ. Among them we can find that ICBC and BOGY perform best, and they are efficient from 2009 to 2018 continuously. And HXB performs worst, with an average efficiency score of 0.64. The average efficiency of the above 20 listed banks showed a slight upward trend from 2009 to 2018, but it was not obvious.

Table 4 Fund-raising efficiencies of Chinese listed banks from 2009 to 2018

Bank	2009	2010	2011	2012	2013	2014	2015	2016	2017	2018	Mean	Rank
ICBC	1.00	1.00	1.00	1.00	1.00	1.00	1.00	1.00	1.00	1.00	1.00	1
ABC	1.00	1.00	1.00	1.00	1.00	1.00	1.00	1.00	1.00	1.00	1.00	1
CCB	1.00	0.96	0.91	0.92	0.96	1.00	1.00	1.00	1.00	1.00	0.97	4
BOC	1.00	1.00	1.00	0.81	0.89	1.00	1.00	0.86	0.89	1.00	0.94	5
BOCOM	1.00	1.00	0.89	0.80	0.84	0.92	0.89	1.00	1.00	1.00	0.93	7
CNCB	1.00	0.74	0.92	0.72	0.87	1.00	1.00	1.00	0.86	1.00	0.91	10
CMBC	0.71	0.71	0.71	0.83	0.63	0.70	0.65	0.83	0.73	0.72	0.71	17
CMB	0.93	0.86	0.77	0.76	0.75	0.87	0.78	0.75	0.66	0.70	0.78	15
IB	0.89	1.00	1.00	1.00	1.00	1.00	1.00	1.00	1.00	1.00	0.99	2
SPDB	1.00	1.00	1.00	1.00	1.00	1.00	1.00	1.00	1.00	0.85	0.99	3
PAB	1.00	1.00	0.63	1.00	1.00	1.00	0.86	0.87	0.95	0.92	0.92	9
HXB	1.00	1.00	0.73	0.86	1.00	1.00	0.76	0.66	0.61	0.58	0.82	13
CEB	0.80	0.72	0.68	1.00	0.71	0.79	0.74	0.88	0.77	0.79	0.79	14
BOBJ	1.00	1.00	1.00	1.00	1.00	1.00	0.83	0.81	0.80	0.83	0.93	8
BONJ	1.00	1.00	0.71	1.00	1.00	1.00	0.72	0.82	1.00	0.78	0.90	11
BONB	0.82	0.74	0.61	0.73	0.91	0.73	0.68	0.71	0.83	1.00	0.78	16
BOHZ	1.00	1.00	0.87	1.00	1.00	1.00	1.00	1.00	1.00	1.00	0.99	2
BOSH	0.88	0.81	0.71	1.00	1.00	1.00	1.00	1.00	1.00	1.00	0.94	6
BOJS	0.65	0.70	0.59	0.71	0.67	1.00	1.00	1.00	0.94	0.95	0.82	12
BOGY	1.00	1.00	1.00	1.00	1.00	1.00	1.00	1.00	1.00	1.00	1.00	1

From the internal ranks of each type of listed bank, it can be found that among the state-owned listed banks, ICBC performs best. Judging from the actual performance, ICBC does present certain advantages in many aspects. First, it is the largest banks in China, which ranks first both in market value and the scale of deposits and loans. According to data released by Forbes' Fortune Magazine in 2018, ICBC ranked 26th, the highest ranking of all banks, with operating income of 153 billion dollars. Second, it has strong profitability in five state-owned banks and even some World Banks. According to the list of the top 1000 global banks in 2019, ICBC's annual net profit was 43.6 billion dollars, equivalent to 305.1 billion yuan. In addition, by comparing the non-performing loan ratios of the five state-owned banks, ICBC's credit risk control ability is relatively strong. In the first half of 2019, the non-performing loan ratio was 1.48%, showing a downward trend for 10 consecutive quarters, while ABC is the worst performer. The relative inefficiency of ABC may be related to its later implementation of the shareholding system reform. By the end of 2006, BOC, CCB, BOCOM and ICBC have completed joint-stock reforms and successfully listed on the stock market. Their internal control mechanism and management ability are constantly enhanced, and the risk regulation system is constantly improved. However, ABC starts the reform of shareholding system in 2008 and completes the public listing in 2010. The improvement of its corporate governance structure and the construction of its risk control system lag behind.

Table 5 Fund-using efficiencies of 20 Chinese listed banks from 2009 to 2018

Bank	2009	2010	2011	2012	2013	2014	2015	2016	2017	2018	Mean	Rank
ICBC	1.00	1.00	1.00	1.00	1.00	1.00	1.00	1.00	1.00	1.00	1.00	1
ABC	0.80	0.68	0.67	0.74	0.77	0.81	0.86	1.00	1.00	0.71	0.80	16
CCB	0.96	1.00	0.99	0.97	1.00	1.00	1.00	1.00	1.00	1.00	1.00	2
BOC	1.00	1.00	1.00	1.00	0.94	1.00	0.98	0.99	1.00	1.00	1.00	1
BOCOM	0.90	1.00	0.95	1.00	0.91	0.88	0.84	0.85	1.00	1.00	0.93	8
CNCB	0.79	0.99	0.67	0.96	0.85	0.88	0.91	0.95	0.94	1.00	0.90	10
CMBC	0.87	0.96	0.94	0.93	0.97	0.94	0.93	0.91	1.00	1.00	0.96	6
CMB	0.92	1.00	1.00	0.95	0.95	0.87	0.89	0.96	1.00	1.00	0.97	4
IB	1.00	1.00	1.00	1.00	1.00	1.00	1.00	1.00	1.00	1.00	1.00	1
SPDB	0.85	1.00	0.97	1.00	1.00	1.00	1.00	1.00	0.89	1.00	0.98	3
PAB	1.00	0.83	0.93	0.72	0.76	0.79	0.81	0.89	0.87	0.91	0.87	13
HXB	1.00	0.45	0.70	0.68	0.77	0.77	0.82	1.00	1.00	1.00	0.82	15
CEB	0.81	0.93	0.94	1.00	0.94	0.93	0.95	0.93	1.00	0.97	0.95	7
BOBJ	0.88	0.85	0.80	0.97	0.93	0.93	0.87	0.99	0.86	0.85	0.92	9
BONJ	1.00	1.00	1.00	1.00	1.00	1.00	1.00	1.00	1.00	1.00	1.00	1
BONB	0.96	1.00	0.88	0.94	0.85	0.95	0.92	1.00	1.00	1.00	0.96	5
BOHZ	1.00	0.85	0.97	0.67	0.84	0.86	0.80	1.00	0.89	1.00	0.90	11
BOSH	0.78	0.87	0.87	0.77	0.85	0.83	0.81	1.00	1.00	1.00	0.89	12
BOJS	0.77	0.75	0.84	0.88	0.91	0.80	0.90	0.89	0.93	1.00	0.87	14
BOGY	1.00	1.00	1.00	1.00	1.00	1.00	1.00	1.00	1.00	1.00	1.00	1

Among the national joint-stock listed banks, IB performs best with an average efficiency of 0.99, and HXB and CMBC are relatively low. The higher efficiency of IB may be related to its obvious advantages in the management of credit risk and operating costs. Specifically, IB has the lowest non-performing loan ratio, with an average of only 0.77% and its operating cost is 1.33%, which is only slightly higher than that of SPDB. Among the city commercial banks, BOGY performs best and BOJS performs worst. Despite the small size of BOGY, it has developed rapidly in recent years. First, its net profit growth rate is much higher than the average growth rate of the 20 banks, which is the highest among them. Second, from the perspective of the growth rate of deposits, loans, and total assets, the development speed of BOGY ranks among the top in listed city commercial banks. Based on the overall efficiency of all banks, our paper continues to summarize the efficiency results of the above-mentioned bank sub-stages.

According to Tables 4 and 5, taking the analysis of the ABC in 2018 as an example, we can further find that the overall inefficiency of ABC is primarily caused by the inefficiency of fund-using stage. Specifically, in fund-using stage, ABC issues a total of 11,461,542 million yuan in loans and generates 190,002 million yuan of NPLs based on 18,852,849 million yuan of total funding. At this time, the loan disbursement rate in ABC reached only 60%, while non-performing loans accounted for 1.1%. Compared with the best-performing ICBC, ABC held fewer loans and a higher number of non-performing loans. Therefore, ABC's sources of invalidity mainly exist in the fund-using stage.

However, the overall inefficiency of CMBC is mainly caused by the inefficiency of fund-raising stage. For example, the investment scale of CMBC and SPDB in 2018 are similar, and the investment in physical, human, and financial capital of CMBC is slightly higher, but the funds raised by SPDB were 271,212 million yuan more than that of CMBC. However, HXB and BOJS perform poorly both in the fund-raising stage and fund-using stage.

Combining Tables 3, 4 and 5, we can see that the overall efficiency of each bank is not greater than its sub-stage efficiency. And the relationship between the overall efficiency and the sub-stage efficiency satisfies Eq. (6). That is, the system inefficiency is equal to the sum of the sub-stage inefficiencies. Moreover, each bank's inefficiency may result from the fund-raising stage and/or the fund-using stage.

3.6 The Efficiency Changes for Chinese Listed Banks Over Time

Our paper further considers the trend differences in stage efficiency and overall efficiency, and analyzes the sources of inefficiency in the banking industry during the sample period. Figure 4 depicts the annual average efficiency of Chinese listed banks from 2009 to 2018. We can find that the trend of fund-raising efficiency is similar to the trend of overall efficiency. From 2009 to 2011, the fund-raising efficiency shows a downward trend, falling by 10.85%. The possible reason may come from the impact of the global financial crisis. Namely, the financial crisis has caused depositors to decline their expectations of bank credit, and the fund-raising is directly related to the bank's credit. In 2011–2018, there is an overall upward trend except for 2015. During this period, the slowdown in economic growth and the penetration of internet finance into the core business of banking may have a negative impact on the fund-raising efficiency of banks. However, the promotion of marketization of deposit interest rate has brought opportunities for commercial banks to improve their fund-raising efficiency. With the application of block chain, artificial intelligence and other new technologies in the banking industry in recent years, more and more banks are slowing down the growth rate of traditional physical and human capital investment. And it also helps improve the efficiency of fund-raising. In 2015, the decline in the fund-raising efficiency of banks may be related to the central bank's five consecutive reductions in benchmark interest rates for deposits and loans. This is reflected in the fact that the net interest margin yield of banks has dropped significantly.

However, during the research period, the fund-using efficiency does not change too much. Specifically, in 2009–2011, it slightly decreases by 2.26% while it gradually increases in 2011–2018. During 2009–2011, Chinese macro economy has shown an upward trend during this period, and the investment command has been

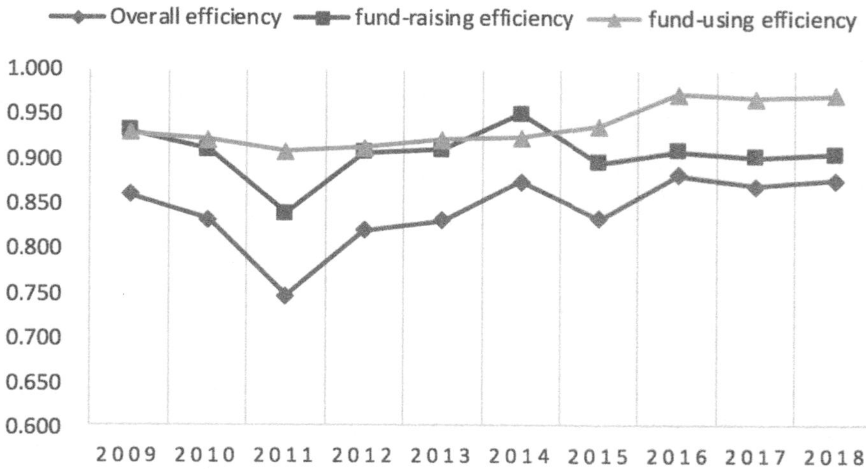

Fig. 4 The average overall efficiency and sub-stage efficiencies of banks

booming, which is conductive to the banking industry to maintain a high fund-using efficiency. Nevertheless, the slowdown in the growth of bank credit scale and the increase in supervision by China Banking Regulatory Commission (CBRC) may have a negative impact on the fund-using of banks, which leads to a slightly decrease for the fund-using efficiency in 2009–2011. In 2011–2018, the continuous increase in NPLs caused by the slowdown in economic growth has a certain negative impact on the fund-using efficiency of the banking industry. However, the advancement of the marketization of loan interest rate is conducive to the improvement of fund-using efficiency and results in a mild increase for the fund-using efficiency. In 2016, Chinese commercial banks began to make strategic adjustments. On the one hand, the growth rate of profit has rebounded. On the other hand, exposure to credit risk has slowed down. According to the disclosure by CBRC, the growth rate of NPLs of commercial banks has decreased by 32.7 percentage points, and the positive effect of interest rate marketization has gradually emerged. Therefore, it can explain the significant increase in the fund-using efficiency of Chinese banks in 2016 to some extent.

Furthermore, we can see from Fig. 5a and b that 70% of the banks are effective at the fund-raising stage and 60% of the banks are efficient at the fund-using stage. However, it is worth noting that the cabinets in the fund-using stage are obviously on the upper side, which are between 0.82 and 1.00, that is, in the overall sense, the efficiency of the fund-using stage is significantly higher than the fund-raising stage. It can be seen from Fig. 4 that the overall efficiency and fund-raising efficiency of Chinese listed banks show the same trend.

We further compile the average efficiency scores before and after the acceleration of interest rate liberalization, as shown in Fig. 6. We find that with the acceleration of the full marketization of interest rates, the overall efficiency of the banking

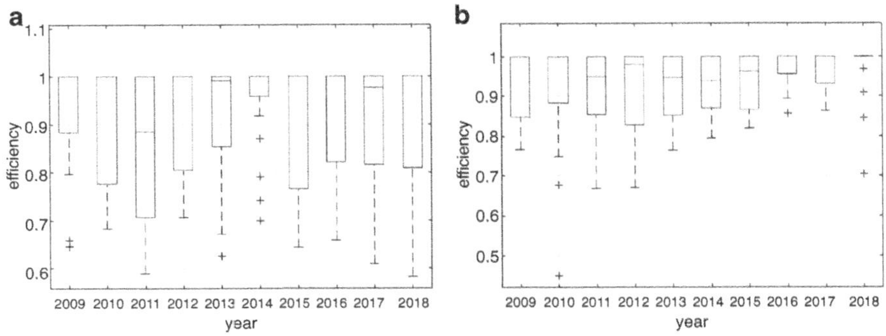

Fig. 5 The boxplot of efficiency distribution in sub-stage. (**a**) The boxplot of fund-raising efficiency. (**b**) The boxplot of fund-using efficiency

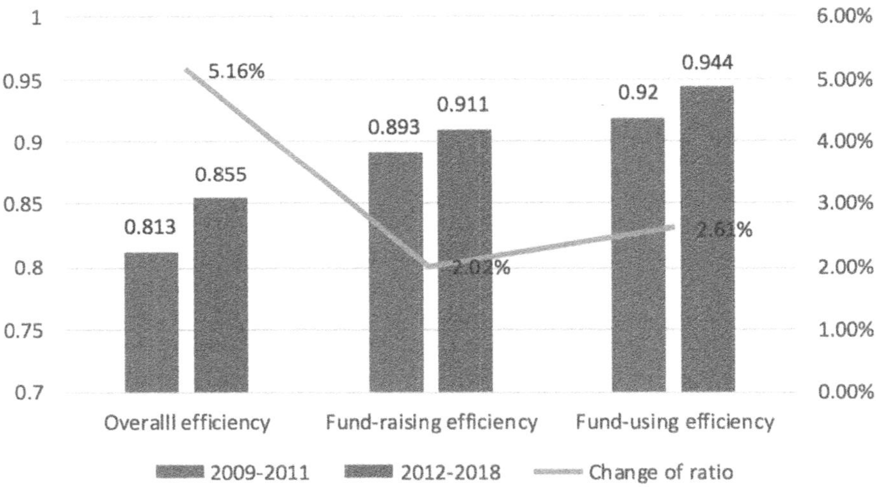

Fig. 6 The efficiency comparison of Chinese listed banks before and after the acceleration of interest rate liberalization

industry is lightly affected, with a change of 5.16% before and after. In particular, the impact of marketization of interest rates on the fund-using stage was greater than the impact on the fund-raising stage. The reason for this phenomenon is that after the marketization of interest rates on deposits and loans, banks can obtain more interest income on loans while reducing the generation of non-performing loans, and can absorb idle funds from customers effectively and improve capital utilization efficiency.

In order to further analyze whether there is a significant difference in the efficiency distribution of Chinese listed banks before and after the acceleration of interest rate marketization, we use the *bootstrap-based Simar-Zelenyuk-adapted-*

Table 6 The results of bootstrap-based Simar-Zelenyuk adapted Li-test

	2009–2011	2012–2018	Change of ratio	Significance
Overall efficiency	0.813	0.855	5.16%	INSIG(0.25)
Fund-raising efficiency	0.893	0.911	2.02%	INSIG(0.64)
Fund-using efficiency	0.920	0.944	2.61%	INSIG(0.87)

Note: INSIG represents non-significant under adapted-Li test at the significance level of 10%, and the tested p-values are shown in parentheses

Li test to test the significance of the distribution of overall efficiency, fund-raising efficiency and fund-using efficiency. And the test results are provided in the last column of Table 6.

However, by the testing p-value, we unexpectedly found that although the marketization of interest rates has a certain impact on bank efficiency, the impact is not significant. The possible reasons are: (1) although China has completely liberalized the lower limit of loan interest rates in 2013, the marketization of deposit interest rates has not made substantial progress, so the overall efficiency improvement effect brought by the marketization of interest rates is not obvious. It is precisely for this reason that our paper considers that the effect of interest rate marketization on the efficiency of the fund-using stage is greater than the effect on the efficiency of the fund-raising stage; (2) The marketization of interest rates and the resulting price competition effect will lead to a reduction in the return on net assets and an increase in the risk of non-performing loans, which is not conducive to the improvement of the overall efficiency and stage efficiency of banks; (3) There is a certain lag in policies. In the short term, banks have not established an appropriate forecasting and evaluation system for interest rate risk. In order to further understand the impact of interest rate marketization on the efficiency of banks of different ownership types, we propose the analysis in the next section.

3.7 *The Efficiency Comparisons for Different Types of Listed Banks*

Like most studies on bank efficiency, we further study the performance of listed banks with different ownerships. The average efficiencies of SOBs, NJBs and CBs are illustrated in Fig. 7. The results show that, on average, the efficiency of SOBs is the highest, followed by CBs and NJBs. Although, this finding is different from the common sense that the SOBs should be the least inefficient banks as the employees may work to rule and the large scale of banks may also cause a decrease in performance. But it is reasonable as what we measure in this paper is technical (input-output) efficiency of banks. And the results are similar to those in Chen et al. (2005) who find that SOBs banks show a relatively higher mean efficiency score, and large banks and small banks are the most efficient. It is interesting because this

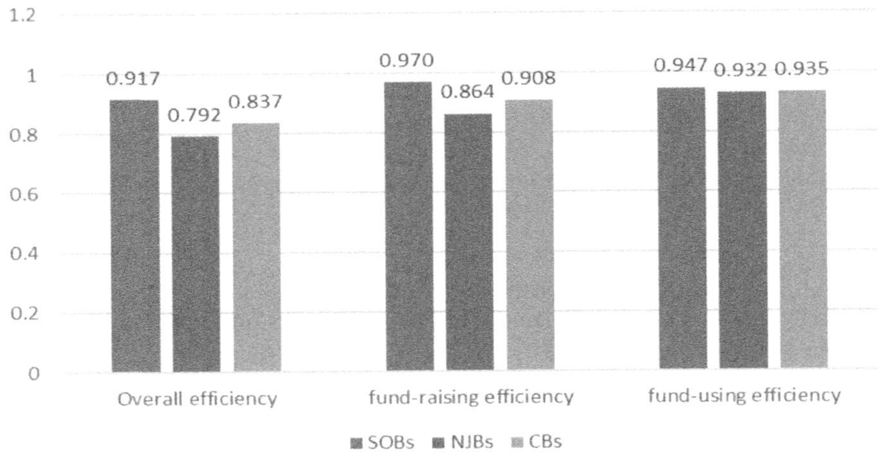

Fig. 7 The average efficiency for three types of listed banks

Table 7 Nonparametric tests for the efficiency distribution of SOBs, NJBs and CBs

Null hypothesis	Overall		Fund-raising		Fund-using	
	Test statistic	p-value	Test statistic	p-value	Test statistic	p-value
f (SOBs)=f (NJBs)	5.748	0.000***	5.001	0.000***	2.245	0.018**
f (SOBs)=f (CBs)	1.724	0.035**	3.337	0.003***	1.165	0.110
f (CBs)=f (NJBs)	1.613	0.056*	1.813	0.040**	−0.267	0.684

Notes: All calculations are done by authors in MATLAB using, after adopting from programs used for the work in Simar and Zelenyuk (2006). The ***, ** and * indicate significance at 1%, 5% and 10% level under adapted-Li-DEA-test respectively

paper measures banking efficiency by using different models, indicators and data set. Comparing the efficiency of fund-raising stage and fund-using stage, we can also find that the differences in the efficiency of the three types of listed banks are mainly reflected in the fund-raising stage.

To make the above conclusions more convincing, we use the adapted Li-test to further analyze the efficiency distribution between different ownership banks. The p-value results are shown in Table 7. All types of banks reject the null hypothesis within a certain confidence interval and believe that there is a significant difference in the efficiency distribution of SOBs, NJBs, and CBs during the fund-raising stage. Taking the capital raising stage as an example, it can be seen from Fig. 8a, b, and c that the three types of banks have significantly different efficiency distributions, that is, SOBs are better than NJBs, followed by CBs. And this difference is mainly reflected in the fund-raising stage. In addition, the efficiency performance of the three types of banks during the fund-using stage is very similar.

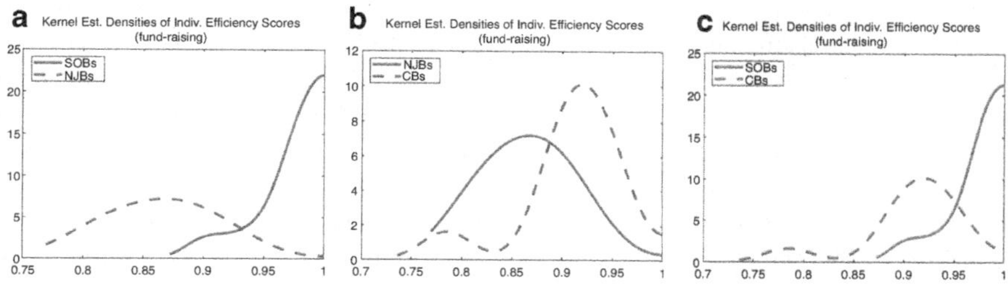

Fig. 8 Estimated Densities of the efficiency scores in fund-raising stage. (**a**) SOBs and NJBs. (**b**) NJBs and CBs. (**c**) SOBs and CBs

Table 8 The frontier projection of Chinese listed banks in 2017

DMU	Fixed Assets	Employees	Equity	Total Funding	Loans	Total Securities	NPLs
ICBC	0.00	0.00	0.00	0.00	0.00	0.00	0.00
ABC	0.00	0.00	0.00	0.00	0.00	0.00	0.00
CCB	0.00	0.00	0.00	0.00	0.00	0.00	0.00
BOC	62,107.89	7709.36	12,997.89	0.00	0.00	0.00	0.00
BOCOM	0.00	0.00	0.00	0.00	0.00	0.00	0.00
CNCB	7458.25	0.00	16,011.65	42,304.37	0.00	377,979.67	1399.09
CMBC	32,409.49	2599.17	1120.72	114,818.59	0.00	0.00	3892.37
CMB	34,609.92	12,380.69	68,598.13	0.00	0.00	0.00	0.00
IB	0.00	0.00	0.00	0.00	0.00	0.00	0.00
SPDB	2947.15	0.00	11,959.09	107,104.63	0.00	0.00	15,313.98
PAB	1403.48	4404.53	25,803.93	240,546.78	0.00	119,791.69	0.00
HXB	7733.24	20,657.35	14,520.98	0.00	0.00	0.00	0.00
CEB	5678.93	9978.78	30,432.69	2890.42	0.00	0.00	0.00
BOBJ	6558.38	1210.51	12,299.07	77,299.19	0.00	0.00	12,714.45
BONJ	0.00	0.00	0.00	0.01	0.00	0.00	0.00
BONB	1335.48	2948.66	0.00	0.00	0.00	0.00	0.00
BOHZ	0.00	0.00	0.00	0.00	0.00	3336.64	1639.37
BOSH	0.00	0.00	0.00	0.00	0.00	0.00	0.00
BOJS	1917.21	0.00	0.00	66,413.94	0.00	0.00	711.68
BOGY	0.00	0.00	0.00	0.00	0.00	0.00	0.00

3.8 *The Frontier Projection of Two-Stage Processes*

Based on model (3), we can get the projections of inputs and outputs for each listed bank. Furthermore, we also provide the benchmark for the banks to increase the efficiencies like the results obtained by the An et al. (2015). Taking 2017 as an example, we get the results shown in Table 8. We can see that banks with overall efficiency equal to unity do not need to change the value of inputs or outputs, nor do they need to change the value of sub-stage such as ICBC, ABC, CCB, BOCOM, IB, BONJ, BOSH and BOGY. And for banks with efficient fund-using stage and inefficient fund-raising stage, they do not need to change the values of intermediate measure and final outputs, such as BOC, CMB, HXB, BONB. For other banks, they have to adjust initial inputs, intermediate indicators, and final outputs to reach an effective status. For example, for CNCB, its overall efficiency in 2017 is 0.799. In order to be efficient, CNCB needs to reduce the investment in fixed assets of 7458.25 million yuan, financial capital of 16,011.65 million yuan at the same time and its bad loans by 1399.09 million yuan while increasing its total securities by 377,979.67 million yuan and total funding by 42,304.37 million yuan. The corresponding optimization scales of other banks are available in the same way.

4 Conclusions

The big data era brings a great opportunity for network DEA, which could be used to gain valuable information in data enable analytics. In this paper, we develop a slack-based two-stage DEA model with undesirable outputs under VRS (the *UVSBM* model) to measure banking efficiency. The advantage of the proposed method is that the overall and sub-stage efficiency can be measured at the same time, and an efficient frontier projection can be obtained to provide a reasonable optimization plan for banks. Then we evaluate the efficiencies of 20 Chinese listed banks during the period 2009–2018 by using the *UVSBM* model.

The model results show that the overall inefficiency is proved to be equal to the sum of the sub-stage inefficiencies, that is the overall efficiency is no greater than its sub-stage efficiencies. Therefore, it helps to figure out the source of inefficiency in bank operations by considering the internal production process in bank efficiency evaluation. Our empirical results are following that: (1) from 2009 to 2018, the trend of fund-raising efficiency and overall efficiency was similar, and only showed signs of decline in 2011, after which they were all increasing, and the efficiency of the fund-using stage was almost unchanged. This important finding suggests that the banking industry should focus on improving the efficiency of the fund-raising stage. (2) The overall efficiency and capital raising efficiency of SOBs banks are significantly better than CBs, followed by NJBs. Testing through the *adapted-Li test* shows that after the marketization of interest rates, bank performance has improved, but the increase is not obvious. The average efficiency between 2012 and 2018 is slightly higher than the average efficiency from 2009 to 2011. (3) By calculating the frontier projection, we have provided reasonable optimization solutions for all banks, showing that the method proposed in this paper has an important role in guiding the development of the banking industry, which is mainly reflected in adjusting the input and output of indicators and reducing waste of resources. The specific suggestion is that according to the efficiencies results, banks must improve the efficiency of the corresponding stage in a targeted manner. For example, SOB should further enhance the efficiency of the fund-using stage, that is, enhance the ability of fund management. For CBs and NJBs, they should focus on improving the efficiency of the fund-raising stage. They can appropriately expand the scale to achieve economies of scale and provide more banking services to local locally-operated small and medium-sized enterprises through cooperation with local governments.

As previously pointed out, we do not evaluate the efficiency of rural commercial banks. But they play an increasing important role in Chinese banking industry and the non-performing loan ratio of Chinese rural commercial banks is generally higher. So, we can extend the performance analysis of this paper to efficiency evaluation of Chinese rural commercial banks. Besides, this paper mainly evaluates the efficiency of commercial banks from the perspective of banks' traditional deposit and loan business, without considering intermediate businesses. Therefore, we could further investigate the performance of Chinese commercial banks when intermediate business is included.

Although we develop a slack-based two-stage DEA model with undesirable outputs and investigate the efficiencies of 20 Chinese listed banks during 2009–2018, the results may be totally different when cost or profit efficiency for banks is analyzed. For future research it could be further examined to conduct cost or profit efficiency analysis for Chinese banks. Moreover, even this paper finds a clear performance difference in Chinese banking industry. But the reasons are just given with qualitative analysis and descriptive explanations. The determinants of the performance difference in Chinese banking industry still need to be figure out with a quantitative analysis for future research.

Acknowledgments Ya Chen is grateful to Professors Zhimin Huang and Valentin Zelenyuk for their suggestions and comments on an earlier version of the paper. This research was supported by National Natural Science Foundation of China [grant numbers 71601064, 72071067, 71871081], and the Major Program of National Fund of Philosophy and Social Science of China [grant number 18ZDA064].

Appendix: Proof of Theorem

Theorem 1 *For any DMU_0 under evaluation, the projection $(x_{i0}-s_i^{-*}, \tilde{z}_{d0}^*, y_{r_1 0}+s_{r_1}^{+*}, y_{r_2 0}+s_{r_2}^{+*}, u_{k0}-s_k^{'-*})$ based on the UVSBM model is UVSBM-efficient.*

Proof Denote the projection for DMU_0 as $(x_{i0}^*, \tilde{z}_{d0}^*, y_{r_1 0}^*, y_{r_2 0}^*, u_{k0}^*)$. Denote the optimal solution for the projected DMU as $(\overline{s}_i^{-*}, \overline{s}_{r_1}^{+*}, \overline{s}_{r_2}^{+*}, \overline{s}_k^{'-*}, \overline{\lambda}_j^{1*}, \overline{\phi}_j^{2*}, \overline{\varphi}_j^{2*}, \overline{\tilde{z}}_{d0}^*)$.

Then we have:

$$\sum_{j=1}^{n} \overline{\lambda}_j^{1*} x_{ij} + \overline{s}_i^{-*} = x_{io}^*; \sum_{j=1}^{n} (\overline{\phi}_j^{2*} + \overline{\varphi}_j^{2*}) y_{r_1 j} - \overline{s}_{r_1}^{+*} = y_{r_1 0}^*$$

$$\sum_{j=1}^{n} \overline{\varphi}_j^{2*} y_{r_2 j} - \overline{s}_{r_2}^{+*} = y_{r_2 o}^*; \sum_{j=1}^{n} \overline{\varphi}_j^{2*} u_{kj} + \overline{s}_k^{'-*} = u_{k0}^*$$

According to model (2), the optimal solution is $(s_i^{-*}, s_{r_1}^{+*}, s_{r_2}^{+*}, s_k^{'-*}, \lambda_j^{1*}, \phi_j^{2*}, \varphi_j^{2*}, \tilde{z}_{d0}^*)$.

Then we have:

$$\sum_{j=1}^{n} \lambda_j^{1*} x_{ij} = x_{i0} - s_i^{-*} = x_{io}^*; \sum_{j=1}^{n} (\phi_j^{2*} + \varphi_j^{2*}) y_{r_1 j} = y_{r_1 0} + s_{r_1}^{+*} = y_{r_1 0}^*$$

$$\sum_{j=1}^{n} \varphi_j^{2*} y_{r_2 j} = y_{r_2 0} + s_{r_2}^{+*} = y_{r_2 o}^*; \sum_{j=1}^{n} \varphi_j^{2*} u_{kj} = u_{k0} - s_k^{'-*} = u_{k0}^*$$

According to (6) and (7), we obtain:

$$\sum_{j=1}^{n} \overline{\lambda}_j^{1*} x_{ij} = x_{io} - \overline{s}_i^{-*} - \overline{s}_i^{-*}; \sum_{j=1}^{n} (\overline{\phi}_j^{2*} + \overline{\varphi}_j^{2*}) y_{r_1 j} = y_{r_1 0} + s_{r_1}^{+*} + \overline{s}_{r_1}^{+*}$$

$$\sum_{j=1}^{n} \overline{\varphi}_j^{2*} y_{r_2 j} = y_{r_2 0} + s_{r_2}^{+*} + \overline{s}_{r_2}^{+*}; \sum_{j=1}^{n} \overline{\varphi}_j^{2*} u_{kj} = u_{k0} - s_k^{'-*} - \overline{s}_k^{'-*}$$

Obviously, (8) is feasible for model (2), i.e. the optimal overall efficiency of model (2) is not larger than:

$$\overline{e}_0^* = \frac{1 - \frac{1}{m} \sum_{i=1}^{m} \frac{S_i^{-*} + \overline{s}_i^{-*}}{x_{io}}}{1 + \frac{1}{s_1 + s_2 + p} \left(\sum_{r_1=1}^{s_1} \frac{s_{r_1}^{+*} + \overline{s}_{r_1}^{+*}}{y_{r_1 0}} + \sum_{r_2=1}^{s_2} \frac{s_{r_2}^{+*} + \overline{s}_{r_2}^{+*}}{y_{r_2 0}} + \sum_{k=1}^{p} \frac{s_k^{'-*} + \overline{s}_k^{'-*}}{u_{k0}} \right)}$$

On the other hand, the optimal overall efficiency of model (2) is e_0^*. If any of $\overline{s}_i^{-*}, \overline{s}_{r_1}^{+*}, \overline{s}_{r_2}^{+*}, \overline{s}_k^{'-*}$ is larger than zero, then $\overline{e}_0^* < e_0^*$, a contradiction. It indicates that $\overline{s}_i^{-*} = \overline{s}_{r_1}^{+*} = \overline{s}_{r_2}^{+*} = \overline{s}_k^{'-*} = 0$. Hence, the projected DMU is efficient. It completes the proof of Theorem 1. Q.E.D.

References

Akther, S., Fukuyama, H., & Weber, W. L. (2013). Estimating two-stage network slacks-based inefficiency: An application to Bangladesh banking. *Omega, 41*, 88–96.

An, Q., Chen, H., Wu, J., & Liang, L. (2015). Measuring slacks-based efficiency for commercialv banks in China by using a two-stage DEA model with undesirable output. *Annals of Operations Research, 235*, 13–35.

Ariff, M., & Can, L. (2008). Cost and profit efficiency of Chinese banks: A non-parametric analysis. *China Economic Review, 19*(2), 260–273.

Avkiran, N. K. (2009). Opening the black box of efficiency analysis: an illustration with UAE banks. *Omega, 37*, 930–941.

Ba, S., Yan, M., & Wang, Y. (2013). Analysis of the Influence of Interest Rate Liberalization on Commercial Banks. *Journal of Huazhong Normal University (Humanities and Social Sciences), 52*(4), 27–37.

Barros, C. P., Chen, Z., Liang, Q. B., & Peypoch, N. (2011). Technical efficiency in the Chinese banking sector. *Economic Modelling, 28*(5), 2083–2089.

Charnes, A., & Cooper, W. W. (1962). Programming with linear fractional functions. *Naval Research Logistics Quarterly, 9*, 181–185.

Charnes, A., Cooper, W. W., & Rhodes, E. (1978). Measuring the efficiency of decision-making units. *European Journal of Operational Research, 2*, 429–444.

Chen, X., Skully, M., & Brown, K. (2005). Banking efficiency in China: Application of DEA to pre-and post-deregulation eras: 1993–2000. *China Economic Review, 16*(3), 229–245.

Chen, Y., Cook, W. D., Li, N., & Zhu, J. (2009). Additive efficiency decomposition in two-stage DEA. *European Journal of Operational Research, 196*, 1170–1176.

Chen, Y., Li, Y.J., Liang, L., Salo, A., & Wu, H.Q. (2016). Frontier Projection and Efficiency Decomposition in Two-stage Processes with Slacks-based Measures. *European Journal of Operational Research, 250*, 543–554.

Cook, W. D., Liang, L., & Zhu, J. (2010). Measuring performance of two-stage network structures by DEA: a review and future perspective. *Omega, 38*, 423–430.

Du, K., Worthington, A. C., & Zelenyuk, V. (2018). Data envelopment analysis, truncated regression and double-bootstrap for panel data with application to Chinese banking. *European Journal of Operational Research, 265*(2), 748–764.

Epure, M., & Lafuente, E. (2015). Monitoring bank performance in the presence of risk. *Journal of Productivity Analysis, 44*(3), 265–281.

Färe, R., & Grosskopf, S. (2003). Nonparametric productivity analysis with undesirable outputs: Comment. *American Journal of Agricultural Economics, 85*(4), 1070–1074.

Färe, R., & Grosskopf, S. (2009). A comment on weak disposability in nonparametric production analysis. *American Journal of Agricultural Economics, 91*(2), 535–538.

Färe, R., Grosskopf, S., Lovell, C. K., & Pasurka, C. (1989). Multilateral productivity comparisons when some outputs are undesirable: a nonparametric approach. *The Review of Economics and Statistics, 71*, 90–98.

Fukuyama, H., & Matousek, R. (2017). Modelling bank performance: A network DEA approach. *European Journal of Operational Research, 259*, 721–732.

Fukuyama, H., & Mirdehghan, S. M. (2012). Identifying the efficiency status in network DEA. *European Journal of Operational Research, 220*, 85–92.

Fukuyama, H., & Weber, W. L. (2010). A slacks-based inefficiency measure for a two-stage system with bad outputs. *Omega, 38*, 398–409.

Fukuyama, H., & Weber, W. L. (2015). Measuring Japanese bank performance: a dynamic network DEA approach. *Journal of Productivity Analysis, 44*, 249–264.

Fukuyama, H., & Weber, W. L. (2017). Measuring bank performance with a dynamic network Luenberger indicator. *Annals of Operations Research, 250*(1), 85–104.

Holod, D., & Lewis, H. F. (2011). Resolving the deposit dilemma: A new DEA bank efficiency model. *Journal of Banking & Finance, 35*, 2801–2810.

Kao, C. (2014). Network data envelopment analysis: A review. European *Journal of Operational Research, 239*, 1–16.

Kenjegalieva, K., Weyman-Jones, T., Simper, R., & Zelenyuk, V. (2009). Comparative analysis of banking production frameworks in eastern European financial markets. *European Journal of Operational Research, 198*, 326–340.

Kumar, S., & Gulati, R. (2014). *Deregulation and efficiency of Indian Banks*. India: Springer.

Laurenceson, J., & Qin, F. (2008). Has minority foreign investment in China's banks improved their cost efficiency? *China & World Economy, 16*(3), 57–74.

Liang, L., Cook, W. D., & Zhu, J. (2008). DEA models for two-stage processes: Game approach and efficiency decomposition. *Naval Research Logistics, 55*, 643–653.

Liu, J. S., Lu, L. Y. Y., Lu, W. M., & Lin, B. J. (2013). A survey of DEA applications. *Omega, 41*, 893–902.

Liu, W., Zhou, Z., Ma, C., Liu, D., & Shen, W. (2015). Two-stage DEA models with undesirable input-intermediate-outputs. *Omega, 56*, 74–87.

Lozano, S. (2016). Slacks-based inefficiency approach for general networks with bad outputs: An application to the banking sector. *Omega, 60*, 73–84.

Luo, D., & Yao, S. (2010). World financial crisis and the rise of Chinese commercial banks: an efficiency analysis using DEA. *Applied Financial Economics, 20*(19), 1515–1530.

Pham, M. D., & Zelenyuk, V. (2018). Slack-based directional distance function in the presence of bad outputs: Theory and application to Vietnamese banking. *Empirical Economics, 54*(1), 153–187.

Seiford, L. M., & Zhu, J. (1999). Profitability and marketability of the top 55 US commercial banks. *Management Science, 45*, 1270–1288.

Simar, L., & Zelenyuk, V. (2006). On testing equality of distributions of technical efficiency scores. *Econometric Reviews, 25*(4), 497–522.

Tone, K. (2001). A slacks-based measure of efficiency in data envelopment analysis. *European Journal of Operational Research, 130*, 498–509.

Tone, K., & Tsutsui, M. (2009). Network DEA: A slacks-based measure approach. *European Journal of Operational Research, 197*, 243–252.

Wang, C. H., Gopal, R. D., & Zionts, S. (1997). Use of data envelopment analysis in assessing information technology impact on firm performance. *Annals of Operations Research, 73*, 191–213.

Wang, K., Huang, W., Wu, J., & Liu, Y. N. (2014). Efficiency measures of the Chinese commercial banking system using an additive two-stage DEA. *Omega, 44*, 5–20.

Yao, S., Han, Z., & Feng, G. (2008). Ownership Reform, Foreign Competition and Efficiency of Chinese Commercial Banks: A Non-parametric Approach. *The World Economy, 31*(10), 1310–1326.

Zhu, J. (2020). DEA under big data: Data enabled analytics and network data envelopment analysis. *Annals of Operations Research*. https://doi.org/10.1007/s10479-020-03668-8

Using Network DEA and Grey Prediction Model for Big Data Analysis: An Application in the Global Airline Efficiency

Wen-Min Lu, Qian Long Kweh, Mohammad Nourani, and Hsiu-Fei Wang

1 Introduction

Carbon emissions of commercial flights have been increasing more quickly than originally predicted.[1] According to data published by the Air Transport Action Group, the global airline industry accounts for 2% of human-induced carbon dioxide (CO_2) emissions.[2] A forecast by the United Nations International Civil Aviation Organization indicates that if this situation continues, then CO_2 emissions from air transport will triple by 2050.[3] Advanced aircrafts are much more fuel efficient than they were in the 1960s; nevertheless, the CO_2 emissions of the airline industry have attracted considerable attention. Smith (1977) discussed trends in energy use and fuel efficiency in the U.S. commercial airline industry since the 1973 oil embargo.

[1] https://www.theguardian.com/business/2019/sep/19/airlines-co2-emissions-rising-up-to-70-faster-than-predicted

[2] https://www.atag.org/facts-figures.html

[3] https://www.nytimes.com/2019/09/19/climate/air-travel-emissions.html

W.-M. Lu (✉)
Department of International Business Administration, Chinese Culture University, Taipei, Taiwan

Q. L. Kweh
Faculty of Management, Canadian University Dubai, Dubai, United Arab Emirates

M. Nourani
The University of Waikato Joint Institute at Zhejiang University City College, University of Waikato, Hangzhou, China

H.-F. Wang
Graduate Institute of Educational Information and Measurement, National Taichung University of Education, Taichung, Taiwan

J. Zhu, V. Charles (eds.), *Data-Enabled Analytics*, International Series in Operations Research & Management Science 312, https://doi.org/10.1007/978-3-030-75162-3_12

Findings suggest the important role of sustainable growth in future economic growth and development.

Overall, the future global trends in growth and the associated environmental implications carbon dioxide emissions from air transport need to be understood. Simply put, energy consumption forecasting is an important issue in the study of CO_2 emissions. In response to this matter, several studies have examined airline energy efficiency from a dynamic viewpoint (for examples, Cui & Li, 2017; Cui, Li, et al., 2016; Li et al., 2016) and network perspective (for examples, Cui & Jin, 2020; Li & Cui, 2017a, Li & Cui, 2017b). More recent studies include those by Cui and Li (2017) and Li and Cui (2017b), who conducted forecasting of related data through the 2020s.

An airline's efforts to control its CO_2 emissions need to be taken into account when estimating airline efficiency. This detail is determined by transforming the minimum possible inputs into the maximum possible outputs. CO_2 emissions are considered an undesirable output that needs to be minimized if an airline is to achieve high efficiency. Unlike prior studies, this study explores airline efficiency in terms of two aspects: energy efficiency and revenue efficiency. Estimations are performed by integrating multiplicative efficiency aggregation (MEA) in data envelopment analysis (DEA) along with second-order cone programming (SOCP).

The conventional DEA models are not capable of characterizing the accurate benchmarking of the airline operation due to the complexity of its various data measures and performance metrics which can be used for a particular purpose of the analysis. Such complexity cannot be addressed in a single dimensional DEA model. As such, network DEA is able to transform the diverse data into useful information. This process involves utilizing the "value" dimension of big data as it positions in the apex of big data pyramid (Zhu, 2020). Hence, network DEA, which is different from DEA, has become a data enabled analytics tool by extracting the value dimension of big data analytics.

Two-stage network DEA models (Cook & Zhu, 2014) have been used in previous studies to examine the network and internal structures of decision-making units (DMUs). MEA can cope with scenarios in which outputs from the first stage are not inputted in the second stage with assumed variable returns to scale (VRS) (Chen & Zhu, 2017, 2019). In this study, the undesirable output of carbon dioxide emissions leaves stage 1. Moreover, MEA does not require any predetermined weights, thereby enabling the free flow of disposability of CO_2 emissions. Because of the nonlinearity of the network DEA model using MEA and its computational difficulty, there is a need to apply optimization techniques to solve the nonlinear network DEA problem in a big data environment. Hence, this study applies the SOCP technique to solve MEA for its nonlinearity problem by following the technique of Chen and Zhu (2017, 2019). Another unique feature of the present study is that it forecasts airline efficiency by using the grey model (Mao & Chirwa, 2005).

A key detail to keep in mind is that the past performance of an airline is not indicative of future results. Given its use of the grey prediction technique to estimate the future efficiency scores of each airline by using past and present data, this study is able to provide useful insights for managers when formulating a competitive strategy to improve long-term resource allocation, especially to reduce CO_2 emissions. In short, this study addresses the impacts of all ratio combinations of inputs, intermediates, and outputs on overall efficiency.

The remainder of this study is organized as follows. A literature review is provided in the second chapter. The methodology and data are given in the third chapter. The findings and discussions are presented in the fourth chapter. This study is concluded in the fifth chapter.

2 Literature Review

Airline frontier analysis research employs different scientific techniques; one such technique is DEA, which is considered the most popular efficiency evaluation approach (Cui, 2020). Schefczyk (1993) pioneered the use of DEA in the airline industry, applying it on a sample of 15 international airlines during 1989 to 1992. The use of conventional DEA has continued until recent years (for example, b; Alam & Sickles, 1998; Bhadra, 2009; Capobianco & Fernandes, 2004; Cui & Li, 2015a; Good et al., 1995; Hong & Zhang, 2010; Ouellette et al., 2010; Ray & Mukherjee, 1996; Seufert et al., 2017; Wang et al., 2011). In another strand of the literature, DEA analysis is combined with other techniques to measure airline efficiency. Such techniques include the Malmquist productivity index in Distexhe and Perelman (1994), the stochastic frontier approach in Good et al. (1995), the Fisher productivity index in Ray and Mukherjee (1996), free disposal hull in Alam and Sickles (1998), full disposal hull in Semenick Alam and Sickles (2000), Tobit analysis in Chiou and Chen (2006), bootstrapped Tobit regression in Barros and Peypoch (2009), and balanced scorecard in Wu and Liao (2014).

However, previous studies ignored the inner process of an airline operating system; therefore, they examined airline performance by using a set of selected inputs and outputs (Xu & Cui, 2017). The connecting divisions in the airline operation are not considered in this so-called black-box process. Most, if not all, airline divisions conjoin for the overall operations (Li et al., 2015). Each division uses its inputs to produce outputs, and linking activities connect the divisions. This issue was overcome by recent studies by applying the concept of a network structure into the airline production process. As a result, network DEA models have become popular among airline efficiency researchers in recent years.

Chiou and Chen (2006) decomposed airline performance in terms of air routes into cost efficiency and service effectiveness. They found that cost efficiency was more prevalent among 15 routes operated by a Taiwanese domestic airline. Zhu (2011) developed a centralized efficiency model in a two-stage framework to measure the cost efficiency and revenue efficiency of 21 airlines between 2007 to 2008. Gramani (2012) studied Brazilian and American airlines, segregating the efficiency performance into operational performance and financial performance. Findings indicate that airlines have better operational efficiency than financial efficiency in emergent markets; however, this condition does not lead to improved financial efficiency. Lu et al. (2012) divided airline efficiency into production efficiency and marketing efficiency among 30 U.S. airlines, and they found that low-cost airlines are more operationally efficient than full-service airlines but less efficient in marketing activities. Lozano and Gutiérrez (2014) used the slack-based measure (SBM) model to estimate the efficiency of 16 European airlines divided into production and sales divisions. The network approach has a higher discriminatory power and obtains more valid efficiency scores and rankings than those of conventional DEA models. Tavassoli et al. (2014) used the SBM model and explored the technical efficiency and service effectiveness of 11 Iranian airlines. An innovative approach was adopted by Mallikarjun (2015), who proposed a three-stage efficiency model that was used to evaluate the operations, services, and sales efficiencies of 27 U.S. domestic airlines. Li et al. (2015) and Li and Cui (2017a, 2017b) introduced a similar framework in a three-stage setting.

Given the differences in the production process in the abovementioned studies, efficiency stages depend on the research objective. Undoubtedly, with the vast and increasing research on network efficiency in the airline industry, the importance of this field is evident. Thus, the ongoing research provides a satisfactory foundation for this particular study. As a result of this recent development in the field, fuel efficiency or energy efficiency has recently been the subject of increased attention, with the unfavorable characteristic of CO2 emissions receiving less focus (Cui & Li, 2016, 2017; Cui, Li, et al., 2016; Ko et al., 2017). Babikian et al. (2002) examined the fuel efficiency of various types of airlines in the US and found that differences in aircraft operations can be a factor in the differences in fuel efficiency. Zou et al. (2014) investigated the fuel efficiency of 15 U.S. jet operators, and they found that up to one billion dollars in cost savings could be achieved from improved efficiency beginning in 2010. Cui and Li (2015b, 2017) measured the energy efficiency of airlines by using virtual frontier benevolent DEA and dynamic epsilon-based measure model. They also highlighted that the global financial crisis had a highly negative effect on energy efficiency.

The treatment of undesirable outputs is a critical topic in energy efficiency research. These outputs exit from the airlines' production system and are therefore not transferred over to the next functional division. Instead of being considered an input in succeeding divisions, CO_2 should be considered an undesirable output during the estimation of the energy efficiency of airlines. Chen et al. (2017) explored the energy efficiency of 13 major Chinese airlines by using stochastic network DEA to account for undesirable outputs in their two-stage framework; these two stages are network efficiency and flight efficiency. Much more work is needed even with the increased attention to the exploration of undesirable outputs in measuring energy efficiency. To the best of our knowledge, no airline efficiency study has considered energy efficiency and revenue efficiency while taking CO2 emissions as an undesirable output. We specifically integrated MEA into DEA because MEA can handle the scenario in which an output from the first stage (energy efficiency) is not inputted to the second stage (revenue efficiency). In addition, MEA does not require a predetermined assumption on weights. MEA is nonlinear, which is why we utilized the SOCP technique for transformation into a linear model (Chen & Zhu, 2017, 2019).

The prediction of inputs and outputs is another important area to which the airline efficiency literature has not paid sufficient attention (Cui & Li, 2017; Li & Cui, 2017a). The estimation of efficiency reflects the past performance of airlines, and this performance does not guarantee future outcomes. Hence, forecasting future efficiency may provide an intuitive roadmap for airline operators. Grey system theory is a prediction approach that can most effectively deal with uncertain and poor or incomplete information (Huang et al., 2008; Zhang & Chen, 2004). The grey model (GM), as the core of grey system theory, gathers existing data to achieve internal regularity without any predetermined assumptions (Mao & Chirwa, 2005). One innovation of the present study is the use of the GM prediction technique to predict energy efficiency and revenue efficiency, thus potentially assisting policymakers in implementing necessary CO2 emission reduction measures.

3 Research Design

3.1 Definition of the Two-Stage Production Process for Airlines

In the big data context, a two-stage production process is used in this study to uncover hidden information that are useful for decision making; as mentioned previously, the two stages are the energy efficiency and revenue efficiency of airlines. The first stage (energy efficiency) includes fuel expense and depreciation and amortization to produce CO_2 emissions, available seat miles (ASM), and revenue passenger miles (RPM). In this stage, CO_2 emission is an unintended output item. The second stage (revenue efficiency) includes ASM, RPM, operating expense, employees, and fixed assets to produce market value, and sales. The framework of the two-stage production process for airlines is shown below.

In line with prior studies (for example, Zhu, 2020), we employ network DEA as a big data analytics approach by first gaining insights from historical data. The variables (inputs, mediates and outputs) used in this study are summarized according to the references. Table 1 presents the definitions of relevant variables and sources of the cited reference.

3.2 Checking the Model Validity

This study utilizes the approaches of the two-stage DEA model as a data enabled analytics tool on a sample of 23 global airlines, which have a Standard Industrial Code of 451 or 452, from the Compustat database for the period of 2013–2017.

Table 1 Definition of inputs, intermediates, and outputs in this study

Variables	Definition	Reference
Inputs		
Fuel expense	Fuel expense is incurred during the current period and is not available for banks or utility companies.	Chiu et al. (2016), Zhang et al. (2019), and Yu et al. (2017)
Depreciation and amortization	Depreciation involves distributing the actual cost or other basic value of tangible capital assets over their estimated useful life in a systematic manner. Amortization involves cost allocation for intangible assets.	Wang et al. (2014)
Intermediates		
CO_2 emission	The amount of CO_2 emitted during the combustion process of the aviation fuel used in the current period.	Cui et al. (2016), Arjomandi et al. (2018) and Chen et al. (2017)
Available seat miles	This item refers to the total seat miles available to passengers (i.e., aircraft miles flown multiplied by the number of seats that are available for revenue passenger use).	Kuljanin et al. (2019), Arjomandi et al. (2018), and Yu et al. (2017)
Revenue passenger miles	This item refers to the scheduled revenue miles flown by passengers (i.e., revenue passengers carried multiplied by the distance flown).	Kuljanin et al. (2019), Zhang et al. (2019), and Yu et al. (2017)
Operating expense	This item represents the sum of all operations-related expenses.	Zhang et al. (2019), Kottas and Madas (2018) and Wang et al. (2017)
Employees	This item represents the number of people employed by the company and its consolidated subsidiaries in thousands.	Kuljanin et al. (2019), Zhang et al. (2019) and Arjomandi et al. (2018)
Fixed assets	Gross property, plant, and equipment minus the accumulated reserves for depreciation, depletion, and amortization.	Tone et al. (2019)
Outputs		
Market value	Market value for companies' outstanding common shares multiplied by the month-end price that corresponds to the period end date.	Zhang et al. (2019) and Wang et al. (2017)
Sales	This item represents gross sales and other operating revenue less discounts, returns, and allowances.	Cui, Li, et al. (2016) and Zhang et al. (2019)

Upon the removal of airlines with missing and unavailable data that are required for fitting the proposed airline efficiency evaluation model, the final balanced panel dataset of 115 firm-year observations is derived. The median values of DEA indicators used in the evaluation model are summarized in Table 2.

The summary statistics reveal interesting information related to the DEA indicators used and highlighted in this study. First, the median values of the DEA indicators are reported, given that the Kolmogorov–Smirnov (K–S) test statistics are all significant at the 1% level. This finding indicates that the DEA indicators are not normally distributed. With non-normally distributed samples, median values are an improved description of the central tendency (Farnum, 1996). This finding also necessitates the use of DEA, which is a nonparametric approach.

Our study focuses only on airlines, thereby ensuring the homogeny of DMUs, with each unique airline having the same objectives in operations and wealth creation. This criterion is suggested by Golany and Roll (1989). The 23 DMUs number more than twice the 10 DEA indicators used in the present study, thereby meeting the minimum required ratio of two, as stipulated in Golany and Roll (1989). Therefore, the construct validity of the proposed airline DEA evaluation model is stable and reliable. The validity of our DEA analyses is verified by running a correlation analysis of the DEA indicators. The results in Table 3 corroborate the significant and positive correlation among the input and output indicators in each stage. The results indicate the isotonicity in the current study's dataset (Golany & Roll, 1989).

3.3 Multiplicative Efficiency Aggregation with SOCP

The idea of big data context lies in the ability to transform data into useful information. When evaluating the complex production structure of airlines, we need to apply network DEA as the big data analytical tool to uncover not only the black box but also the complicated interrelationships of inputs, intermediates, and outputs. Three methods can be applied to two-stage network DEA structures: the decomposition method additive efficiency decomposition (AED), multiplicative efficiency decomposition (MED), and the aggregation method MEA to extract values from big data.

With its additive form, AED decomposes the overall efficiency as a weight average of the two-stage efficiency scores (Chen et al., 2009). The a priori weights need to be set to combine the two-stage efficiencies; simply put, overall efficiency depends on the predetermined weights (Guo et al., 2017). With its multiplicative form, MED decomposes the overall efficiency as a product of the two-stage efficiency scores (Kao & Hwang, 2011).

The decomposition method enables the objective function in two-stage network DEA to remain as a single linear fractional form that can be solved by using linear models through the Charnes-Cooper transformation (Charnes & Cooper, 1962). Thus, in accordance with classic conventional DEA theories, typical issues such

Table 2 Descriptive statistics for 23 global airlines from 2013 to 2017

Variables	Unit	Sample	Mean	Median	Interquartile Range	Std.Dev.	K-S test
Fuel expense	Millions	115	3408.44	2139.72	4527.10	3291.27	p = 0.00***
Depreciation and amortization	Millions	115	838.29	486.69	1446.90	745.31	p = 0.00***
CO2 emission	Thousands of gallons	115	11,588.52	4906.89	13,412.30	12,373.69	p = 0.00***
Available seat miles	Millions of miles	115	96,310.30	73,889.00	136,328.40	80,987.72	p = 0.00***
Revenue passenger miles	Millions of miles	115	79,634.59	61,616.00	108,723.30	67,215.71	p = 0.00***
Operating expense	Millions	115	11,857.68	4142.00	20,666.80	13,605.12	p = 0.00***
Employees	Thousands	115	42.74	21.15	70.40	38.59	p = 0.00***
Fixed assets	Millions	115	13,787.24	7912.00	21,028.10	12,768.20	p = 0.00***
Market value	Millions	115	9218.32	3819.40	11,891.90	10,927.18	p = 0.00***
Sales	Millions	115	21,584.78	10,576.02	21,397.10	27,196.08	p = 0.00***

Note:* is the $p < 0.1$; ** is the $p < 0.05$; *** is the $p < 0.01$.

Table 3 Correlation coefficients among input, intermediate, and output variables

Variables		X1	X2	UZ	Z1	Z2	ZI1	ZI2	ZI3	Y1	Y2
X1:	Fuel Expense	1.0000			Stage	1					
		p = ···									
X2:	Depreciation and amortization	0.8381	1.0000								
		p = 0.00***	p = ···								
UZ:	CO2 emission	0.7733	0.7130	1.0000							
		p = 0.00***	p = 0.00***	p = ···							
Z1:	Available seat miles	0.8863	0.9206	0.8538	1.0000					Stage	2
		p = 0.00***	p = 0.00***	p = 0.00***	p = ···						
Z2:	Revenue passenger miles	0.8836	0.9146	0.8608	0.9993	1.0000					
		p = 0.00***	p = 0.00***	p = 0.00***	p = 0.00***	p = ···					
ZI1:	Operating expense	0.8646	0.7956	0.6324	0.8400	0.8380	1.0000				
		p = 0.00***	p = 0.00***	p = 0.00***	p = 0.00***	p = 0.00***	p = ···				
ZI2:	Employees	0.8399	0.9455	0.7009	0.9066	0.8965	0.8155	1.0000			
		p = 0.00***	p = 0.00***	p = 0.00***	p = 0.00***	p = 0.00***	p = 0.00***	p = ···			
ZI3:	Fixed assets	0.8358	0.9281	0.8345	0.9522	0.9493	0.7547	0.9083	1.0000		
		p = 0.00***	p = 0.00***	p = 0.00***	p = 0.00***	p = 0.00***	p = 0.00***	p = 0.00***	p = ···		
Y1:	Market value	0.5770	0.4867	0.6134	0.6840	0.6971	0.6065	0.4267	0.6217	1.0000	
		p = 0.00***	p = 0.00***	p = 0.00***	p = 0.00***	p = 0.00***	p = 0.00***	p = 0.00***	p = 0.00***	p = ···	
Y2:	Sales	0.4580	0.6954	0.4893	0.5620	0.5515	0.1770	0.6398	0.6689	0.0822	1.0000
		p = 0.00***	p = 0.00***	p = 0.00***	p = 0.00***	p = 0.00***	p = 0.06*	p = 0.00***	p = 0.00***	p = 0.38	p = ···

Note: * is $p < 0.1$; ** is $p < 0.05$; *** is $p < 0.01$.

as scale efficiency and frontier projection can also be studied in two-stage network DEA. However, the decomposition method has two main disadvantages. First, the requirement of a priori weights is a major limitation of AED during the testing process of all possible values of weights, thus resulting in a tedious and cumbersome process of obtaining the optimal solution to overall efficiency (Guo et al., 2017). Second, MED (Kao & Hwang, 2011) is limited to basic two-stage structures, in which only the stage 1 outputs (a.k.a. links or intermediate measures) are the only inputs to stage 2. For the objective function to remain as a single fractional form, MED cannot be applied to general two-stage network structures, in which some outputs from stage 1 may be removed from the production system and/or additional inputs may enter stage 2. This limitation is derived from its definition of overall efficiency, which considers only the inputs and outputs of a production system but neglects the relationship between overall efficiency and stage efficiencies (Chen & Zhu, 2019). Therefore, the overall system efficiency of two-stage DEA based on MED and the weighted average of the divisional efficiencies may not be necessarily equal (Kao, 2017).

With such issues, the aggregation method (i.e., MEA) appears to be a viable way to overcome those issues. MEA, which is also in a multiplicative form as MED, defines the overall efficiency as a geometric mean of the individual stage efficiencies. However, measuring efficiencies by using MEA remains challenging even though it does not require a priori weights and is applicable to both basic and general two-stage DEA structures under the assumption of either CRS or VRS. The problem is due to MEA in a two-stage network DEA model being a highly nonlinear optimization problem and thus beyond the application scope of the Charnes-Cooper transformation, thereby making it different from AED or MED. Chen and Zhu (2019) avoided this difficulty by combining SOCP with MEA to solve two-stage network DEA models with the use of a nonheuristic algorithm (i.e., interior point method). SOCP, which is a form of generalized linear programming, is a mature convex optimization technique (Boyd & Vandenberghe, 2004). Moreover, SOCP is considered as effective and accurate as linear programming, and it can also deal with complex nonlinear convex problems (Chen & Zhu, 2019).

In its investigation of the two-stage production process of airlines, this study uses balanced panel data and integrates MEA two-stage network DEA with SOCP to estimate the energy and revenue efficiencies. The following section explains the two-stage production process and the adopted MEA two-stage network DEA model.

The two-stage network production process investigated in this study (Fig. 1) contains n DMUs during T terms. In stage 1, m inputs x_{ij}^{t} $(i = 1, \ldots, m;\ j = 1, \ldots, n;\ t = 1, \ldots, T)$ at time t enter the production system and produce d outputs ez_{pj}^{t} $(p = 1, \ldots, d;\ j = 1, \ldots, n;\ t = 1, \ldots, T)$ that leave the system and h intermediate outputs z_{vj}^{t} $(v = 1, \ldots, h;\ j = 1, \ldots, n;\ t = 1, \ldots, T)$ (a.k.a. links) together with additional k inputs ax_{qj}^{t} $(q = 1, \ldots, k;\ j = 1, \ldots, n;\ t = 1, \ldots, T)$, which enter stage 2 to generate s final outputsy_{rj}^{t} $(r = 1, \ldots, s;\ j = 1, \ldots, n;\ t = 1, \ldots, T)$.

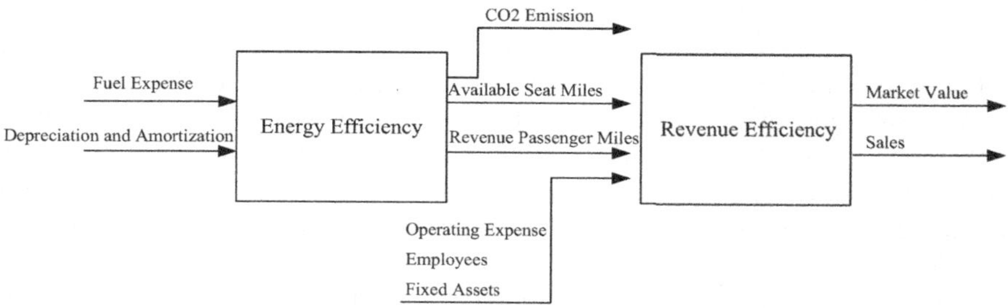

Fig. 1 Framework of the two-stage production process for airlines

The aforementioned production process is a general two-stage structure, in which some outputs from stage 1 leave the production system and additional inputs enter stage 2. Therefore, MEA is adopted in this study for two reasons. First, MEA is capable of handling a general two-stage network structure, which MED cannot handle. Second, unlike AED, MEA does not require a priori weights, thereby circumventing the tedious process of testing all possible weights for the best solution. Below, the corresponding MEA two-stage network DEA model under VRS assumption to attain the efficiency of observed DMU ($o = 1, \ldots, n$) is shown.

$$\begin{aligned}
&\text{Max } \frac{\left(\sum_{p=1}^{d} w_p^1 ez_{po}^t + \sum_{h=1}^{b} w_h^2 z_{ho}^t + u_1\right)}{\sum_{i=1}^{m} v_i^1 x_{io}^t} \cdot \frac{\left(\sum_{r=1}^{s} \mu_r y_{ro}^t + u_2\right)}{\left(\sum_{h=1}^{b} w_h^2 z_{ho}^t + \sum_{q=1}^{k} v_q^2 ax_{qo}^t\right)} \\
&S.T. \\
&\left(\sum_{p=1}^{d} w_p^1 ez_{pj}^t + \sum_{h=1}^{b} w_h^2 z_{hj}^t + u_1\right) / \sum_{i=1}^{m} v_i^1 x_{ij}^t \leq 1, (j = 1, \ldots, n; t = 1, \ldots, T) \\
&\left(\sum_{r=1}^{s} \mu_r y_{rj}^t + u_2\right) / \left(\sum_{h=1}^{b} w_h^2 z_{hj}^t + \sum_{q=1}^{k} v_q^2 ax_{qj}^t\right) \leq 1, (j = 1, \ldots, n; t = 1, \ldots, T) \\
&v_i^1, v_q^2, w_p^1, w_h^2, \mu_r \geq \varepsilon, \forall i, q, p, h, r \\
&u_1, u_2 \text{ free in sign,}
\end{aligned} \tag{1}$$

where $v_i^1, v_q^2, w_p^1, w_h^2, \mu_r$ are assumed-positive weights by integrating the small non-Archimedean in the model; and u_1, u_2 are free variables under the VRS assumption. The efficiencies of stages 1 and 2 are defined as $\frac{\left(\sum_{p=1}^{d} w_p^1 ez_{po}^t + \sum_{h=1}^{b} w_h^2 z_{ho}^t + u_1\right)}{\sum_{i=1}^{m} v_i^1 x_{io}^t}$ and $\frac{\left(\sum_{r=1}^{s} \mu_r y_{ro}^t + u_2\right)}{\left(\sum_{h=1}^{b} w_h^2 z_{ho}^t + \sum_{q=1}^{k} v_q^2 ax_{qo}^t\right)}$, respectively. w_h^2 in both stages are weights of the links (a.k.a. intermediate products/measures) and are assumed equal.

According to this MEA two-stage network DEA model, the overall efficiency is defined as $\frac{\left(\sum_{p=1}^{d} w_p^1 ez_{po}^t + \sum_{h=1}^{b} w_h^2 z_{ho}^t + u_1\right)}{\sum_{i=1}^{m} v_i^1 x_{io}^t} \cdot \frac{\left(\sum_{r=1}^{s} \mu_r y_{ro}^t + u_2\right)}{\left(\sum_{h=1}^{b} w_h^2 z_{ho}^t + \sum_{q=1}^{k} v_q^2 ax_{qo}^t\right)}$.

Evidently, the overall efficiency defined as a product of two fractional terms is highly nonlinear; thus it cannot be changed into a linear problem by using Charnes-Cooper transformation (Charnes & Cooper, 1962). This difficulty is addressed by using SOCP with an interior point method, because SOCP, with such a nonheuristic algorithm, is close to polynominal complexity, thereby enabling the global optimal solution to nonlinear convex problems to be solved effectively and efficiently (Boyd & Vandenberghe, 2004). Therefore, in line with Chen and Zhu (2019), this study transforms the multiplicative-aggregation objective function in model (1) into a cone structure. Thus, the global optimal solution can be solved nonheuristically by a sequence of feasibility problems of SOCP. The transformation process is explicated as follows:

First, model (1) is changed into a minimization problem of model (2) by taking a reciprocal of the objective function and modifying the constraint functions.

$$
\begin{aligned}
&\text{Min } \frac{\sum_{i=1}^{m} v_i x_{io}^t}{\left(\sum_{p=1}^{d} w_p^1 ez_{po}^t + \sum_{h=1}^{b} w_h^2 z_{ho}^t + u_1\right)} \cdot \frac{\left(\sum_{h=1}^{b} w_h^2 z_{ho}^t + \sum_{q=1}^{k} v_q^2 ax_{qo}^t\right)}{\left(\sum_{r=1}^{s} \mu_r y_{ro}^t + u_2\right)} \\
&S.T. \\
&\left(\sum_{p=1}^{d} w_p^1 ez_{pj}^t + \sum_{h=1}^{b} w_h^2 z_{hj}^t + u_1\right) \le \sum_{i=1}^{m} v_i x_{ij}^t, \ (j = 1, \ldots, n; t = 1, \ldots, T) \\
&\left(\sum_{r=1}^{s} \mu_r y_{rj}^t + u_2\right) \le \left(\sum_{h=1}^{b} w_h^2 z_{hj}^t + \sum_{q=1}^{k} v_q^2 ax_{qj}^t\right), \ (j = 1, \ldots, n; t = 1, \ldots, T) \\
&v_i^1, v_q^2, w_p^1, w_h^2, \mu_r \ge \varepsilon, \forall i, q, p, h, r \\
&u_1, u_2 \ free \ in \text{ sign.}
\end{aligned} \tag{2}
$$

Second, model (2) is converted to model (3) by applying an epigraph transformation.

$$
\begin{aligned}
&\text{Min } \theta_o^t \\
&S.T. \\
&\frac{\sum_{i=1}^{m} v_i x_{io}^t}{\left(\sum_{p=1}^{d} w_p^1 ez_{po}^t + \sum_{h=1}^{b} w_h^2 z_{ho}^t + u_1\right)} \cdot \frac{\left(\sum_{h=1}^{b} w_h^2 z_{ho}^t + \sum_{q=1}^{k} v_q^2 ax_{qo}^t\right)}{\left(\sum_{r=1}^{s} \mu_r y_{ro}^t + u_2\right)} \le \theta_o^t, \\
&\left(\sum_{p=1}^{d} w_p^1 ez_{pj}^t + \sum_{h=1}^{b} w_h^2 z_{hj}^t + u_1\right) \le \sum_{i=1}^{m} v_i x_{ij}^t, \ (j = 1, \ldots, n; t = 1, \ldots, T) \\
&\left(\sum_{r=1}^{s} \mu_r y_{rj}^t + u_2\right) \le \left(\sum_{h=1}^{b} w_h^2 z_{hj}^t + \sum_{q=1}^{k} v_q^2 ax_{qj}^t\right), \ (j = 1, \ldots, n; t = 1, \ldots, T) \\
&\theta_o^t \ge 1, \\
&v_i^1, v_q^2, w_p^1, w_h^2, \mu_r \ge \varepsilon, \forall i, q, p, h, r \\
&u_1, u_2 \text{ free in sign}.
\end{aligned} \tag{3}
$$

$v_i^1, v_q^2, w_p^1, w_h^2, \mu_r \ge 0$ and θ_o^t is bounded above. The efficiency scores of stages 1 and 2 range between 0 and 1. Therefore, $\theta_o^t \ge 1$ $(o = 1, \ldots, n; t = 1, \ldots, T)$.

Third, model (3) is further transformed into a form of SOCP by applying algebraic manipulations. The first nonlinear constraint in model (3), $\frac{\sum_{i=1}^{m} v_i x_{io}^t}{\left(\sum_{p=1}^{d} w_p^1 ez_{po}^t + \sum_{h=1}^{b} w_h^2 z_{ho}^t + u_1\right)} \cdot \frac{\left(\sum_{h=1}^{b} w_h^2 z_{ho}^t + \sum_{q=1}^{k} v_q^2 ax_{qo}^t\right)}{\left(\sum_{r=1}^{s} \mu_r y_{ro}^t + u_2\right)} \le \theta_o^t$, can be equivalently displayed in matrix form

$$M = \begin{bmatrix} \theta_o^t \cdot \left(\sum_{p=1}^d w_p^1 ez_{po}^t + \sum_{h=1}^b w_h^2 z_{ho}^t + u_1\right) & \left(\sum_{h=1}^b w_h^2 z_{ho}^t + \sum_{q=1}^k v_q^2 ax_{qo}^t\right) \\ \sum_{i=1}^m v_i x_{io}^t & \left(\sum_{r=1}^s \mu_r y_{ro}^t + u_2\right) \end{bmatrix} \succ= 0, \tag{4}$$

where '$\succ=$' indicates that M is positive semidefinite because all elements in the matrix are positive. Moreover, given that M is a square matrix, M $\succ= 0$ also indicates $(1/2)(M + M^T) \succ = 0$. As such, formula (4) can be equivalently expressed as follows:

$$\begin{bmatrix} \theta_o^t \cdot \left(\sum_{p=1}^d w_p^1 ez_{po}^t + \sum_{h=1}^b w_h^2 z_{ho}^t + u_1\right) & \frac{1}{2}\left(\sum_{i=1}^m v_i x_{io}^t + \sum_{h=1}^b w_h^2 z_{ho}^t + \sum_{q=1}^k v_q^2 ax_{qo}^t\right) \\ \frac{1}{2}\left(\sum_{i=1}^m v_i x_{io}^t + \sum_{h=1}^b w_h^2 z_{ho}^t + \sum_{q=1}^k v_q^2 ax_{qo}^t\right) & \left(\sum_{r=1}^s \mu_r y_{ro}^t + u_2\right) \end{bmatrix} \succ= 0 \tag{5}$$

The positive semi-definite property of the square matrix in formula (5) implies that the determinant of this matrix is also non-negative. As such, let $a = \theta_o^t \cdot \left(\sum_{p=1}^d w_p^1 ez_{po}^t + \sum_{h=1}^b w_h^2 z_{ho}^t + u_1\right)$, $b = \frac{1}{2}\left(\sum_{i=1}^m v_i x_{io}^t + \sum_{h=1}^b w_h^2 z_{ho}^t + \sum_{q=1}^k v_q^2 ax_{qo}^t\right)$, and $c = \left(\sum_{r=1}^s \mu_r y_{ro}^t + u_2\right)$. Then, formula (5) can be expressed and denoted as inequality (6).

$$ac - b^2 \geq 0 \tag{6}$$

Furthermore, with the use of some math operations, inequality (6) can be converted to

$$\sqrt{b^2 + \left[\frac{1}{2}(a-c)\right]^2} \leq \frac{1}{2}(a+c) \tag{7}$$

Let $N = \begin{bmatrix} b^2 \\ \frac{1}{2}(a-c) \end{bmatrix}$ and the Euclidean norm or ℓ_2 norm of vector N, denoted by $\|N\|_2$, is defined as the length or magnitude of the vector and can be calculated using the formula $\|N\|_2 = \sqrt{b^2 + \left[\frac{1}{2}(a-c)\right]^2}$. Therefore, inequality (7) can also be expressed as $\left\| \begin{matrix} b^2 \\ \frac{1}{2}(a-c) \end{matrix} \right\|_2 \leq \frac{1}{2}(a+c)$ and then represented as follows after $\frac{1}{2}$ is canceled out simultaneously from both sides of the inequality:

$$\begin{gathered} \left\| \begin{matrix} \left(\sum_{i=1}^m v_i x_{io}^t + \sum_{h=1}^b w_h^2 z_{ho}^t + \sum_{q=1}^k v_q^2 ax_{qo}^t\right) \\ \left(\theta_o^t \sum_{p=1}^d w_p^1 ez_{po}^t + \theta_o^t \sum_{h=1}^b w_h^2 z_{ho}^t + \theta_o^t u_1 - \sum_{r=1}^s \mu_r y_{ro}^t - u_2\right) \end{matrix} \right\|_2 \\ \leq \left(\theta_o^t \sum_{p=1}^d w_p^1 ez_{po}^t + \theta_o^t \sum_{h=1}^b w_h^2 z_{ho}^t + \theta_o^t u_1 + \sum_{r=1}^s \mu_r y_{ro}^t + u_2\right) \end{gathered} \tag{8}$$

Finally, with the use of the aforementioned mathematic transformation processes (2)–(8), model (3) is equivalently transformed to model (9) with a conical constraint.

$$
\begin{aligned}
&\text{Min } \theta_o^t \\
&S.T. \\
&\left\| \begin{matrix} \left(\sum_{i=1}^{m} v_i x_{io}^t + \sum_{h=1}^{b} w_h^2 z_{ho}^t + \sum_{q=1}^{k} v_q^2 a x_{qo}^t\right) \\ \left(\theta_o^t \sum_{p=1}^{d} w_p^1 e z_{po}^t + \theta_o^t \sum_{h=1}^{b} w_h^2 z_{ho}^t + \theta_o^t u_1 - \sum_{r=1}^{s} \mu_r y_{ro}^t - u_2\right) \end{matrix} \right\|_2 \\
&\leq \left(\theta_o^t \sum_{p=1}^{d} w_p^1 e z_{po}^t + \theta_o^t \sum_{h=1}^{b} w_h^2 z_{ho}^t + \theta_o^t u_1 + \sum_{r=1}^{s} \mu_r y_{ro}^t + u_2\right) \\
&\left(\sum_{p=1}^{d} w_p^1 e z_{pj}^t + \sum_{h=1}^{b} w_h^2 z_{hj}^t + u_1\right) \leq \sum_{i=1}^{m} v_i x_{ij}^t, \; (j = 1, \ldots, n; t = 1, \ldots, T) \\
&\left(\sum_{r=1}^{s} \mu_r y_{rj}^t + u_2\right) \leq \left(\sum_{h=1}^{b} w_h^2 z_{hj}^t + \sum_{q=1}^{k} v_q^2 a x_{qj}^t\right), \; (j = 1, \ldots, n; t = 1, \ldots, T) \\
&\theta_o^t \geq 1, \\
&v_i^1, v_q^2, w_p^1, w_h^2, \mu_r \geq \varepsilon, \forall i, \text{q}, p, h, r \\
&u_1, u_2 \text{ free in sign}.
\end{aligned} \tag{9}
$$

When θ_o^t is given, model (9) becomes a feasibility problem in SOCP form in which $v_i^1, v_q^2, w_p^1, w_h^2, \mu_r \geq 0$ and θ_o^t is bounded above. With the use of a nonheuristic algorithm (i.e., interior point method), model (9) can then be effectively and efficiently solved in a convergent manner by searching the optimal value of θ_o^t, denoted by θ_o^{t*}, in its domain. Chen and Zhu (2019) provide additional details related to this SOCP-based transformation technique.

Once model (9) is solved, the overall efficiency $\left(OE_o^t\right)$ and the efficiencies of stage 1 $\left(TE_{1o}^t\right)$ and stage 2 $\left(TE_{2o}^t\right)$ at time t for a specific DMU under evaluation are determined and can be expressed as

$$
\begin{aligned}
TE_{1o}^t &= \left(\sum_{p=1}^{d} w_p^{1*} e z_{po}^t + \sum_{h=1}^{b} w_h^{2*} z_{ho}^t + u_1^*\right) / \sum_{i=1}^{m} v_i^* x_{i0}^t, \\
TE_{2o}^t &= \left(\sum_{r=1}^{s} \mu_r^* y_{ro}^t + u_2^*\right) / \left(\sum_{h=1}^{b} w_h^2 z_{ho}^t + \sum_{q=1}^{k} v_q^2 a x_{qo}^t\right), \\
OE_o^t &= TE_{1o}^t \times TE_{2o}^t,
\end{aligned} \tag{10}
$$

where an asterisk indicates the optional solution from model (9).

3.4 GM (1, N) Model – Grey Relational Model

In addition to the network structure, DEA can be used as a predictive analytics tool to better forecast the performance of DMUs. With regard to information, traditional DEA usually regards the production process of a DMU as analogous to a black box, in which the information inside the box is ignored. Here, a two-stage network DEA model is adopted to open the black box by examining the key subprocesses, as discussed in Section "Multiplicative Efficiency Aggregation with SOCP". To further understand the information contained within the box, the relationships between overall efficiency and essential ratio combinations of inputs, intermediates, and outputs are explored by using a GM (1, N) grey model. The GM (1, 1) grey model is also used to predict the overall efficiency of airlines on the basis of past and present data in the big data context. Both types of grey models are expounded as follows.

In grey system theory, GM (1, N) is a multivariable grey model that can be used to evaluate the relationships between a main behavior and its key factors in a system; 1 refers to the first-order derivative, and N denotes the number of variables. Suppose $c_1^{(0)} = \left\{c_1^{(0)}(1), c_1^{(0)}(2), \ldots, c_1^{(0)}(m)\right\}, k = 1, \ldots, m$ is the sequence of a main behavior for observed m DMUs. Likewise,

$$c_p^{(0)} = \left\{\begin{array}{l} c_2^{(0)}(1), c_2^{(0)}(2), \ldots, c_2^{(0)}(m) \\ c_3^{(0)}(1), c_3^{(0)}(2), \ldots, c_3^{(0)}(m) \\ \ldots \\ c_N^{(0)}(1), c_N^{(0)}(2), \ldots, c_N^{(0)}(m) \end{array}\right\} = \left\{c_p^{(0)}(1), c_p^{(0)}(2), \ldots, c_p^{(0)}(k)\right\},$$

$$\mathrm{p} = 2, \ldots, \mathrm{N};\ \mathrm{k} = 1, \ldots, \mathrm{m}. \tag{11}$$

where $c_p^{(0)}$ represents the sequences of (N-1) key factors that influence the main behavior. Then, the first accumulated generating operation (AGO) of $c^{(0)}$ is defined as

$$\text{AGO}\left\{c^{(0)}\right\} = c^{(1)} = \left(\sum\nolimits_{k=1}^{1} c^{(0)}(k), \sum\nolimits_{k=1}^{2} c^{(0)}(k), \ldots, \sum\nolimits_{k=1}^{m} c^{(0)}(k)\right). \tag{12}$$

With the use of this operation, $c^{(0)}$ is transformed to $c^{(1)}$, which becomes a monotonic increasing sequence. On the basis of $c^{(1)}$, the first-order grey differential equation of $GM(1, N)$ is defined as follows:

$$\frac{dc_1^{(1)}}{dt} + ac_1^{(1)}(k) = \sum\nolimits_{p=2}^{N} b_p c_p^{(1)}(k), k = 1, \ldots, \mathrm{m}, \tag{13}$$

where a and b_p are the parameters to be estimated; a is called the developing coefficient; b_p is the influence coefficient (a.k.a. the grey input); k is the number of observed entries; and p is the number of key factors. $c_1^{(1)}$ is the first AGO of $c_1^{(0)}$ (i.e., the original sequence of a main behavior), and $c_p^{(1)}$ is the first AGO of $c_p^{(0)}$ (i.e., the original sequences of key factors).

With $c_1^{(1)}$ being a discrete sequence, $\frac{dc_1^{(1)}}{dt}$, the grey derivative in Eq. (2) is therefore approximately evaluated by $c_1^{(1)}(k) - c_1^{(1)}(k-1)$, which is also equal to $c_1^{(0)}(k)$ in accordance with the definition of AGO. Therefore, the differential equation of GM (1, N) can be approximately converted to the discrete equation of GM (1, N) as follows (Deng, 1982):

$$c_1^{(0)}(k) + az_1^{(1)}(k) = \sum_{p=2}^{N} b_p c_p^{(1)}(k), k = 2, \ldots, m, \tag{14}$$

where $z_1^{(1)}(k)$ represents the background value of $\frac{dc_1^{(1)}}{dt}$ and is defined as

$$z_1^{(1)}(k) = 0.5c_1^{(1)}(k) + 0.5c_1^{(1)}(k-1), k \geq 2. \tag{15}$$

This study uses adopts GM (1, N) to investigate the relationship between overall efficiency (i.e., the main behavior) and ratio combinations of inputs, intermediates, and outputs (i.e., the key factors). The step-by-step procedure is described as follows:

Step 1: The original sequence $c_i^{(0)}$, $(i = 1, \ldots, N)$ is established.

$$\begin{aligned} c_1^{(0)} &= \left\{c_1^{(0)}(1), c_1^{(0)}(2), \ldots, c_1^{(0)}(k)\right\} \\ c_2^{(0)} &= \left\{c_2^{(0)}(1), c_2^{(0)}(2), \ldots, c_2^{(0)}(k)\right\} \\ &\cdots\cdots\cdots \\ c_{\mathrm{N}}^{(0)} &= \left\{c_Q^{(0)}(1), c_Q^{(0)}(2), \ldots, c_Q^{(0)}(k)\right\}, \mathrm{k} = 1, \ldots, \mathrm{m}, \end{aligned} \tag{16}$$

where k is the number of observed DMUs (k = 1, . . . ,m), $c_1^{(0)}$ represents the main-behavior sequence, and $c_2^{(0)}, \ldots, c_N^{(0)}$ are the key-factor sequences.

Step 2: The first AGO sequence $c_i^{(1)}$ is generated.

$$\begin{aligned} c_1^{(1)}(k) &= \left\{c_1^{(1)}(1), c_1^{(1)}(2), \ldots, c_1^{(1)}(m)\right\} \\ c_2^{(1)}(k) &= \left\{c_2^{(1)}(1), c_2^{(1)}(2), \ldots, c_2^{(1)}(m)\right\} \\ &\cdots\cdots\cdots \\ c_N^{(1)}(k) &= \left\{c_N^{(1)}(1), c_N^{(1)}(2), \ldots, c_N^{(1)}(m)\right\}, \end{aligned} \tag{17}$$

where $c_i^{(1)} = \mathrm{AGO}\left\{c_i^{(0)}\right\} = \left(\sum_{k=1}^{1} c_i^{(0)}(k), \sum_{k=1}^{2} c_i^{(0)}(k), \ldots, \sum_{k=1}^{m} c_i^{(0)}(k)\right), i = 1, \ldots, N$.

Step 3: The background value sequence $z_1^{(1)}$ is generated.

$$z_1^{(1)} = \left\{z_1^{(1)}(2), z_1^{(1)}(3), \ldots, z_1^{(1)}(m)\right\}, \tag{18}$$

where $z_1^{(1)}(k) = 0.5c_1^{(1)}(k) + 0.5c_1^{(1)}(k-1), k = 2, \ldots, m.$.

Step 4: The grey discrete equations of GM (1, N) are created.

$$
\begin{array}{l}
c_1^{(0)}(2) + az_1^{(1)}(2) = \sum_{p=2}^{N} b_p c_p^{(1)}(2) \\
c_1^{(0)}(3) + az_1^{(1)}(3) = \sum_{p=2}^{N} b_p c_p^{(1)}(3) \\
\cdots\cdots\cdots\cdots\cdots\cdots\cdots\cdots\cdots \\
c_1^{(0)}(m) + az_1^{(1)}(m) = \sum_{p=2}^{N} b_p c_p^{(1)}(m)
\end{array}
\tag{19}
$$

Step 5: The discrete Eq. (5) is converted into the form of a matrix.

$$
\begin{bmatrix} c_1^{(0)}(2) \\ c_1^{(0)}(3) \\ \vdots \\ c_1^{(0)}(m) \end{bmatrix} = \begin{bmatrix} -z_1^{(1)}(2) & c_2^{(1)}(2) & \cdots & c_N^{(1)}(2) \\ -z_1^{(1)}(3) & c_2^{(1)}(3) & \cdots & c_N^{(1)}(3) \\ \vdots & \vdots & \cdots & \vdots \\ -z_1^{(1)}(m) & c_2^{(1)}(m) & \cdots & c_N^{(1)}(m) \end{bmatrix} \begin{bmatrix} a \\ b_2 \\ \vdots \\ b_N \end{bmatrix}
\tag{20}
$$

Step 6: The parameters are estimated by using the ordinary least square (OLS) method.

Let

$$
Y = \begin{bmatrix} c_1^{(0)}(2) \\ c_1^{(0)}(3) \\ \vdots \\ c_1^{(0)}(m) \end{bmatrix}, \mathrm{B} = \begin{bmatrix} -z_1^{(1)}(2) & c_2^{(1)}(2) & \cdots & c_N^{(1)}(2) \\ -z_1^{(1)}(3) & c_2^{(1)}(3) & \cdots & c_N^{(1)}(3) \\ \vdots & \vdots & \cdots & \vdots \\ -z_1^{(1)}(m) & c_2^{(1)}(m) & \cdots & c_N^{(1)}(m) \end{bmatrix}, \hat{a} = \begin{bmatrix} a \\ b_2 \\ \vdots \\ b_N \end{bmatrix},
\tag{21}
$$

and then the parameters can be estimated by using $\hat{a} = \left(B^T B\right)^{-1} B^T Y$.

Once $\hat{a}$ is obtained, the relationships between the main behavior and the key factors are examined by using b_p, ($\mathrm{p} = 2, \ldots, N$). The corresponding key factors are ranked, and the extent to which those factors influence the main behavior is evaluated with the use of the absolute values of b_p, which are denoted by $|\mathrm{b}_p|$.

3.5 GM (1,1) – Grey Prediction Model

When N = 1, the multivariate GM (1, N) model tends to descend to the univariate GM (1, 1) model, which is used to predict the time series that has only one variable of interest. Similar to Eqs. (2) and (3) of GM (1, N), the first-order grey differential equation and the approximate grey discrete equation of GM (1, 1) are defined as Eqs. (11) and (12), respectively.

$$
\frac{dc_1^{(1)}}{dt} + ac_1^{(1)}(t) = u, \mathrm{t} = 1, \ldots, \mathrm{T},
\tag{22}
$$

and

$$c_1^{(0)}(t) + az_1^{(1)}(t) = u, \mathrm{t} = 2, \ldots, \mathrm{T}, \tag{23}$$

where a and u are the parameters to be estimated; a is the developing coefficient; u is the control coefficient (a.k.a. the grey input); and t reflects the equidistant time period that corresponds to the collected data.

GM (1, 1) is used in this work to forecast the inputs, intermediates, and outputs. The obtained figures are then used to calculate the predicted overall efficiency according to the MEA two-stage DEA model. The forecasting process is described as follows.

Step 1: The original sequence $c^{(0)}$ of the forecasted variable for each DMU is established.

$$c^{(0)} = \left\{c^{(0)}(1), c^{(0)}(2), \ldots, c^{(0)}(t)\right\}, \mathrm{t} = 1, \ldots, \mathrm{T}, \tag{24}$$

where t is the equidistant time period that corresponds to the collected data (i.e., Δt =constant.)

Step 2: The first AGO sequence $c^{(1)}$ is generated.

$$c^{(1)} = \left\{c^{(1)}(1), c^{(1)}(2), \ldots, c^{(1)}(T)\right\} \tag{25}$$

where $c^{(1)} = \text{AGO}\left\{c^{(0)}\right\} = \left(\sum_{t=1}^{1} c^{(0)}(t), \sum_{t=1}^{2} c^{(0)}(t), \ldots, \sum_{t=1}^{T} c^{(0)}(t)\right)$..

Step 3: The background value sequence $z^{(1)}$ is generated.

$$z^{(1)} = \left\{z^{(1)}(2), z^{(1)}(3), \ldots, z^{(1)}(T)\right\}, \tag{26}$$

where $z^{(1)}(t) = 0.5c^{(1)}(t) + 0.5c^{(1)}(t-1)$, t = 2, ..., T..

Step 4: The grey discrete equations of GM (1, 1) are created.

$$\begin{aligned} c^{(0)}(2) + az^{(1)}(2) &= b \\ c^{(0)}(3) + az^{(1)}(3) &= b \\ \cdots\cdots&\cdots\cdots \\ c^{(0)}(T) + az^{(1)}(T) &= b \end{aligned} \tag{27}$$

Step 5: The discrete Eq. (8) is converted into the form of a matrix.

$$\begin{bmatrix} c^{(0)}(2) \\ c^{(0)}(3) \\ \vdots \\ c^{(0)}(T) \end{bmatrix} = \begin{bmatrix} -z^{(1)}(2) & 1 \\ -z^{(1)}(3) & 1 \\ \vdots & \\ -z^{(1)}(T) & 1 \end{bmatrix} \begin{bmatrix} a \\ b \end{bmatrix} \tag{28}$$

Step 6: The parameters are estimated by using the OLS method.
Let

$$Y = \begin{bmatrix} c^{(0)}(2) \\ c^{(0)}(3) \\ \vdots \\ c^{(0)}(T) \end{bmatrix}, \mathrm{B} = \begin{bmatrix} -z^{(1)}(2)\ 1 \\ -z^{(1)}(3)\ 1 \\ \vdots \\ -z^{(1)}(T)\ 1 \end{bmatrix}, \hat{a} = \begin{bmatrix} a \\ b \end{bmatrix}, \tag{29}$$

Then, parameters a and b are estimated by using OLS: $\hat{a} = \left(B^T B\right)^{-1} B^T Y$.

Step 7: The forecasted first AGO sequence $\hat{c}^{(1)}$ is computed.

In GM (1,1), the assumption is that $\hat{c}^{(1)}(1) = c^{(1)}(1) = c^{(0)}(1)$. On the basis of the values of coefficients a and u obtained from step 6, the approximate time response function for the forecasted first AGO value of $c^{(1)}(\mathrm{t}+1)$, denoted by $\hat{c}^{(1)}(\mathrm{t}+1)$, is modeled as follows:

$$\hat{c}^{(1)}(\mathrm{t}+1) = \left(c^{(1)}(1) - \frac{u}{a}\right) e^{-at} + \frac{u}{a}, \mathrm{t} = 1, \ldots, \mathrm{T}, \tag{30}$$

where $\hat{c}^{(1)}(1) = c^{(1)}(1) = c^{(0)}(1)$. Then, Eq. (19) obtains the forecasted first AGO sequence $\hat{c}^{(1)}$, in which

$$\hat{c}^{(1)} = \left\{\hat{c}^{(1)}(1), \hat{c}^{(1)}(2), \ldots, \hat{c}^{(1)}(t), \hat{c}^{(1)}(t+1)\right\}, t=1, \ldots, T, \text{where} \hat{c}^{(1)}(1) = c^{(0)}(1).$$

Step 8: The forecasted sequence $\hat{c}^{(0)}$ is obtained.

The definition of AGO indicates that $c^{(1)} = \mathrm{AGO}\ \left\{c^{(0)}\right\} = \left(\sum_{t=1}^{1} c^{(0)}(t), \sum_{t=1}^{2} c^{(0)}(t), \ldots, \sum_{t=1}^{T} c^{(0)}(t)\right), t = 1, \ldots, T$, and that $c^{(1)}(1) = c^{(0)}(1)$. As such, we can derive $c^{(1)}(t+1) = \ = c^{(1)}(t) + c^{(0)}(t+1)$, and thus $c^{(0)}(t+1) = c^{(1)}(t+1) - c^{(1)}(t)$. By the same token, the forecasted value of $\hat{c}^{(0)}(t+1)$ can be obtained as

$$\hat{c}^{(0)}(t+1) = \hat{c}^{(1)}(t+1) - \hat{c}^{(1)}(t), t = 1, \ldots, \mathrm{T}. \tag{31}$$

Then, given that $\hat{c}^{(1)}(1) = c^{(1)}(1) = c^{(0)}(1)$, the forecasted sequence $\hat{c}^{(0)}$ can be restored by applying Eq. (12) accordingly, in which

$$\hat{c}^{(0)} = \left\{\hat{c}^{(0)}(1), \hat{c}^{(0)}(2), \ldots, \hat{c}^{(0)}(t), \hat{c}^{(0)}(t+1)\right\}, t=1, \ldots, T, \text{where} \hat{c}^{(0)}(1) = c^{(0)}(1).$$

Step 9: The forecasting and modeling accuracy is assessed.

The forecast error at time t is defined and measured by

Table 4 MAPE criteria for model evaluation

MAPE (%)	Model fit and accuracy degree
<10	Excellent
10–20	Good
20–50	Reasonable
>50	Poor

$$e(t) = \left| \frac{c^{(0)}(t) - \hat{c}^{(0)}(t)}{c^{(0)}(t)} \right|, t = 1, \ldots, T, \tag{32}$$

where $c^{(0)}(t)$ indicates the actual value, and $\hat{c}^{(0)}(t)$ denotes the forecasted value at time t. The mean absolute percentage error (MAPE), which has seen wide application for evaluating the accuracy of a fitted time series in statistics, is used to measure the overall model fit and forecast accuracy.

$$MAPE = \frac{1}{T} \sum_{t=1}^{T} |e(t)| = \frac{1}{T} \sum_{t=1}^{T} \left| \frac{c^{(0)}(t) - \hat{c}^{(0)}(t)}{c^{(0)}(t)} \right|, \tag{33}$$

where T is the number of data time periods used to construct the GM (1,1) model.

The MAPE criteria for model fit and forecast accuracy are presented in Table 4 (Lewis, 1982).

4 Empirical Analysis

4.1 Dynamic Efficiency Analysis

The business environment nowadays is becoming increasingly dynamic. Therefore, determining how the ability of competitors to use resources can be understood and preparing countermeasures are a crucial matter. This issue is addressed in this work by evaluating the two-stage production process of airlines by using balanced panel data. The multiyear structure of data needs to be accounted for, and the effects of changes between input, intermediate, and output items in terms of efficiency need to be considered. Therefore, this study integrates two-stage network DEA and SOCP technique as a data enabled analytics approach to estimate energy and revenue efficiencies.

The MEA two-stage network DEA model with SOCP is used to evaluate the energy and revenue efficiencies of 23 airlines from the period of 2013 to 2017. The performances of full-service carriers (FSC) and low-cost carriers (LCC) are compared and tested, with the results presented in Table 5.

Table 5 shows the means, standard deviations, and coefficients of variation for management efficiency (TE1) and revenue efficiency (TE2) scores obtained by

Table 5 Airline dynamic efficiency grouped by service type in 2013–2017

	2013	2014	2015	2016	2017	Overall	Mann-Whitney U test
TE1	Average Technical Efficiency						
FSC	0.510	0.562	0.620	0.664	0.638	0.599	p-value < .025
Std.Dev.	0.213	0.246	0.204	0.196	0.178	0.211	
LCC	0.619	0.626	0.704	0.699	0.639	0.657	
Std.Dev.	0.254	0.252	0.216	0.243	0.233	0.230	
Total average	0.548	0.584	0.649	0.676	0.638	0.619	
Total Std.Dev.	0.229	0.244	0.207	0.209	0.193	0.218	
TE2							
FSC	0.466	0.518	0.461	0.451	0.494	0.478	p-value < .01
Std.Dev.	0.269	0.288	0.279	0.265	0.248	0.264	
LCC	0.678	0.724	0.658	0.620	0.623	0.661	
Std.Dev.	0.348	0.283	0.275	0.232	0.277	0.273	
Total average	0.540	0.590	0.530	0.510	0.539	0.542	
Total Std.Dev.	0.309	0.297	0.288	0.262	0.260	0.280	

Note: *LCC* Low-cost carrier, *FSC* Full-service carrier

FSCs and LCCs across the period of 2013–2017. Efficiency scores are presented in a 0–1 scale. As shown in Table 4, the total average and standard deviation for energy efficiency scores are 0.619 and 0.218, respectively, in stage 1, and the total average and standard deviation for revenue efficiency scores are 0.542 and 0.280, respectively, in stage 2. This result indicates that airlines generally perform better in the energy performance stage with less variation than they do in the revenue performance stage. In other words, energy efficiency and revenue efficiency improved by 38% and 46%, respectively. This finding implies that airlines still have considerable room to reach optimal efficiency.

Energy and revenue effectiveness are also subjected to further examination across years. As shown in Fig. 2, the total average and standard deviation of efficiency scores for energy and revenue stages vary throughout the period of 2013–2017. The trends of central tendency and variability for energy and revenue efficiency scores are also presented. In 2017, the average energy efficiency score exhibited an increasing trend from 0.548 in 2013 to 0.638, whereas the average revenue efficiency score demonstrated a decreasing trend from 0.540 in 2013 to 0.539. Moreover, the energy performance stage constantly outperforms the revenue performance stage with less variation across the period of 2013–2017.

The energy and revenue effectiveness are also compared by different carrier types (i.e., FSC and LCC). Table 5 shows that the overall total averages of energy and revenue efficiency scores for FSCs and LCCs are 0.599 and 0.478, and 0.657 and 0.661, respectively. Mann-Whitney U tests at the 5% significance level are conducted to examine the efficiency differences between FSCs and LCCs. LCCs have significantly better energy and revenue performances than FSCs do. This finding also implies that carrier type plays a key role, which influences airlines' performances.

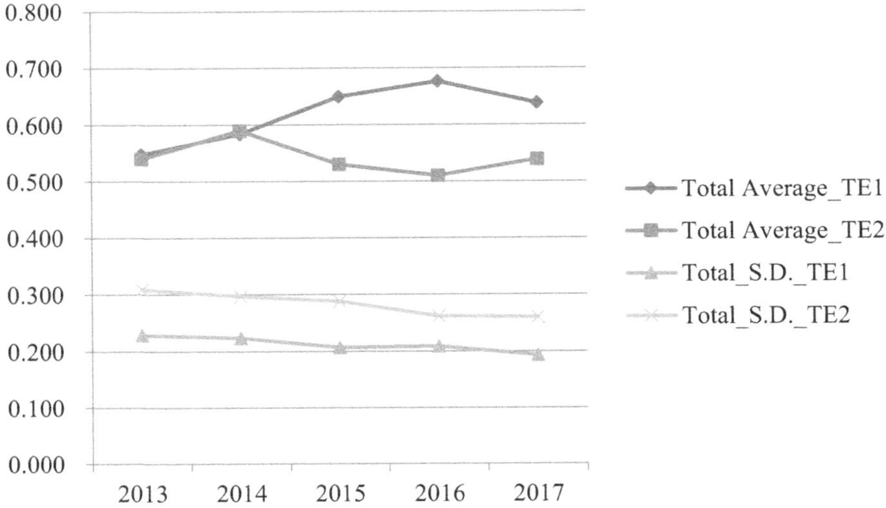

Fig. 2 Total average and standard deviation of energy and revenue efficiency scores

4.2 Opening the Black Box of Overall Efficiency

The previous section showed that network data envelopment analysis uses multiple inputs, intermediaries, and outputs to enable an understanding of the resource use among competitors. However, this method is unable to explain the relationship between overall efficiency and ratio combinations of inputs, intermediates, and outputs. This situation can be considered a black box for network data envelopment analysis. This study overcomes this particular black box problem by first integrating data envelopment and grey relational analyses (Liu & Lin, 2006). Through this approach, the impacts of all ratio combinations of inputs, intermediates, and outputs on overall efficiency are investigated. In other words, with the use of grey relational analysis, this study solves problems that have a complicated interrelationship among multiple factors and variables (Morán et al., 2006).

Tables 6 and 7 report the results. In terms of energy efficiency, Z2 and X1 have the strongest relationship, with a GM (1,N) value of 0.2902, followed by that between Z1 and X1 in energy efficiency. In terms of revenue efficiency, Y1 and Z1 have the strongest relationship, with a GM (1,N) value of 8.5778, followed by that between Y1 and Z2. In both cases, Z1 and Z2 are among the main input–output indicators, that is, airlines have to pay more attention to their available seat miles and revenue passenger miles. This feature is the main advantage of grey modeling, that is, it enables us to understand the relative relationships between the factors and between the factors and the system.

Table 6 Energy efficiency and ratio of input and output in GM (1, N)

AGO (X1) matrix						
X1=						
[3.72	17.34	26.87	123.29	22.26	102.25	0.59
6.9	30.43	50.36	214.47	42.14	179.3	1.133
11.51	46.86	87.61	346.59	73.63	291.18	1.62
16.14	62.29	121.55	458.92	102.25	386.57	2.393
21.01	87.1	152.94	617.84	128.05	517.27	3.35
25.28	106.09	179.83	736.84	149.83	613.22	3.625
30.16	118.22	211.2	806.32	175.1	669.03	4.062
33.34	126.63	246.22	893.84	203.48	739.93	4.623
38.34	190.08	290.82	1459.63	240.83	1212.99	5.616
42.55	210.25	324.6	1621.33	267.49	1339.49	6.026
47.22	230.72	354.67	1752.61	292.99	1450.66	6.862
47.72	232.42	381.95	1839.98	314.83	1520.61	7.365
51.38	257.4	411.26	2039.37	337.34	1673.67	7.772
55.99	280.36	446.87	2216.37	367.02	1820.36	8.275
56.47	282.7	470.53	2326.4	386.19	1909.45	8.878
61.22	304.96	505.26	2489.09	415.54	2046.69	9.485
65.31	319.18	534.61	2590.17	440.07	2130.89	9.984
69.24	337.74	570.68	2752.58	472.15	2273.5	10.826
72.95	340.36	784.82	2890.61	647.82	2386.36	11.751
77.29	358.14	818.02	3025.88	675.42	2498.25	12.505
81.98	393.97	858.59	3332.37	709.83	2760.16	13.123
86.65	414.5	889.74	3468.29	735.71	2873.28	13.973
90.31	426.53	919.83	3567.14	760.3	2953.94	14.254]
Z matrix						
Z = [0.8615, 1.3765, 2.0065, 2.8715, 3.4875, 3.8435, 4.3425, 5.1195, 5.821, 6.444, 7.1135, 7.5685, 8.0235, 8.5765, 9.1815, 9.7345, 10.405, 11.2885, 12.128, 12.814, 13.548, 14.1135]						
GM (1,N)						
GM (1,N) = [0.13, 0.0115, 0.2388, 0.0309, 0.2902, 0.0351]						

Note: EZ/X1 EZ/X2 Z1/X1 Z1/X2 Z2/X1 Z2/X2

4.3 Performance Forecasting

The past performance of an airline is neither indicative of future results nor does it guarantee such results; this issue needs to be overcome. Therefore, we use the grey prediction technique (Mao & Chirwa, 2005), which is a part of grey system theory (Deng, 1982), to employ a forecasting model that estimates the future efficiency scores of each airline with the use of both past and present data. We envision that our proposed forecasting model can assist managers in formulating a competitive

Table 7 Revenue efficiency and ratio of input and output in GM (1, N)

AGO (X1) matrix

X1=

[0.04	0.05	0.17	0.2	0.3	147.73	0.41	1.14	550.68	1.61	0.235
0.06	0.08	0.32	0.38	0.42	193.09	0.62	1.94	830.77	2.97	0.401
0.26	0.32	0.46	0.54	2.39	708.03	2.12	3.27	1187.58	4.02	0.997
0.5	0.6	0.58	0.68	5.31	1581.52	4.89	4.7	1614.09	5.37	1.862
0.59	0.71	0.73	0.87	6.02	1781.96	5.54	5.9	1939.33	6.44	2.299
0.64	0.77	0.89	1.07	6.33	1840.63	5.8	7.03	2148.91	7.35	2.634
0.65	0.78	1.66	2.03	6.9	1855.76	5.84	52.6	3364.66	10.7	3.6
0.66	0.79	2.36	2.9	7.32	1866.73	5.88	100.95	4618.49	14.81	4.579
0.67	0.8	3.64	4.43	8.61	1889.6	6.06	426.93	9765	55.12	5.572
0.89	1.08	3.76	4.58	11.01	2407.22	7.7	428.21	10038.66	55.99	6.194
1.04	1.25	3.92	4.77	12.11	2841.82	8.52	429.44	10525.25	56.91	6.941
1.11	1.33	4.07	4.95	12.38	2936.27	8.94	430.06	10733.41	57.85	7.243
1.12	1.34	4.4	5.38	12.58	2949.24	9.03	441.02	11356.07	62.26	7.583
1.22	1.46	4.53	5.54	13.52	3248.73	9.97	442.26	11767.82	63.55	8.011
1.34	1.6	4.66	5.7	14.23	3544.8	10.88	443.06	12091.43	64.55	8.445
1.46	1.74	4.79	5.85	15.39	3883.22	11.67	444.3	12458.81	65.41	8.826
1.54	1.83	4.92	6.01	16.04	4013.19	12.06	445.4	12675.57	66.07	9.108
1.84	2.17	4.99	6.09	20.04	6191.15	15.19	446.42	13222.21	66.86	9.997
1.88	2.22	5.08	6.2	20.58	6273.43	15.56	447.61	13403.31	67.77	10.258
2.07	2.45	5.22	6.37	22.38	6820.19	17.13	448.88	13793.69	68.9	10.928
2.26	2.68	5.32	6.49	24.71	7610.42	20.59	450.17	14223.85	70.65	11.633
2.34	2.78	5.47	6.67	25.35	7846.51	21.24	451.35	14661.75	71.85	12.013
2.46	2.93	5.62	6.85	26.33	8196.72	22.2	452.58	15095.99	73.01	12.428]

Z matrix

Z = [0.318, 0.699, 1.4295, 2.0805, 2.4665, 3.117, 4.0895, 5.0755, 5.883, 6.5675, 7.092, 7.413, 7.797, 8.228, 8.6355, 8.967, 9.5525, 10.1275, 10.593, 11.2805, 11.823, 12.2205]

GM (1,N)

GM (1,N) = [8.5778, 5.7763, 0.2609, 1.1526, 0.1949, 0.0003, 0.1326, 0.0029, 0.0002, 0.0792]

Note: Y1/Z1 Y1/Z2 Y2/Z1 Y2/Z2 Y1/IZ1 Y1/IZ2 Y1/IZ3 Y2/IZ1 Y2/IZ2 Y2/IZ3

strategy for improving long-term resource allocation. Grey prediction predicts the change of known data and pinpoints the dynamic relationship of the series (Zhang & Chen, 2004). This approach is able to predict future data changes with reduced data uncertainty and is not simultaneously subject to statistical assumptions.

The input, intermediate, and output values and future efficiency scores for airlines are estimated by this study first by integrating grey prediction and DEA. A strong feature of grey modeling is its ability to handle the uncertainty and to perform prediction even with a small sample, as used in this study. Through future prediction, the airline industry is better able to forecast and create plans for its future endeavors. The related results are given in Tables 7 and 8. Airlines can improve their energy efficiency by making more effort to control their CO_2 emissions. On the overall

Table 8 GM (1,1) prediction at r = 1 in fuel expense of Air Canada

X0 = [3534	3747	2823	2606	3339]			
AGO (X1) matrix = [3534	7281	10,104	12,710	16,049]			
Z matrix = [−-	5407.5	8692.5	11,407	14379.5]			
Solution to a,b							
[a b] = [0.0512833277316785	3640.1281128924]						
Ẍ(1) = [3371.698	6574.8429	9617.861	12508.757	15255.1355	17864.2212]		
IAGO Ẍ() = [3371.698	3203.145	3043.0181	2890.896	2746.3785	2609.0856]		
Prediction							
ID	000001	000002	000003	000004	000005	000006	000007
Data	3534.0000	3747.0000	2823.0000	2606.0000	3339.0000		
Pred.	3534.0000	3371.6980	3203.1450	3043.0181	2890.8960	2746.3785	2609.0856
Error	0.0000	10.0161	13.4660	16.7697	13.4203	0.0000	0.0000
Average error = 10.7344%							

Table 9 Average efficiency value and predicted value

Airline	Service Type	Energy efficiency		Revenue efficiency	
		Mean	Predict	Mean	Predict
AIR CANADA	FSC	0.590	0.604	0.235	0.349
AIR FRANCE – KLM	FSC	0.543	0.731	0.166	0.309
ALASKA AIR GROUP INC	FSC	0.487	1.000	0.596	0.410
ALLEGIANT TRAVEL CO	LCC	0.773	0.785	0.865	0.546
AMERICAN AIRLINES GROUP INC	FSC	0.957	0.976	0.437	0.301
AVIANCA HOLDINGS SA	FSC	0.275	0.434	0.335	0.263
CHINA EASTERN AIRLINES CORP	FSC	0.437	0.696	0.966	0.907
CHINA SOUTHERN AIRLINES	FSC	0.561	0.619	0.979	1.000
CONTROLADORA VUELA COMPANIA	LCC	0.993	1.000	0.993	1.000
COPA HOLDINGS SA	FSC	0.410	0.610	0.622	0.860
DELTA AIR LINES INC	FSC	0.836	1.000	0.747	0.611
DEUTSCHE LUFTHANSA AG	FSC	0.503	0.639	0.302	0.623
GOL LINHAS AEREAS INTELIGENT	LCC	0.407	0.581	0.340	0.381
HAWAIIAN HOLDINGS INC	FSC	0.503	0.892	0.428	0.460
INTL CONSOL AIRLINES GROUP	FSC	0.603	1.000	0.434	0.323
JETBLUE AIRWAYS CORP	LCC	0.607	0.834	0.381	0.434
LATAM AIRLINES GROUP SA	FSC	0.499	0.811	0.282	0.344
RYANAIR HOLDINGS PLC	LCC	0.842	1.000	0.889	1.000
SKYWEST INC	FSC	0.925	1.000	0.261	0.706
SOUTHWEST AIRLINES	LCC	0.754	0.852	0.670	1.000
SPIRIT AIRLINES INC	LCC	0.618	0.765	0.705	0.238
UNITED CONTINENTAL HLDGS INC	FSC	0.850	0.914	0.380	0.184
WESTJET AIRLINES LTD	LCC	0.281	0.335	0.415	0.263
Average		0.620	0.786	0.540	0.544

average basis, revenue efficiency will also improve in the future once airlines have improved their energy efficiency. Combining energy efficiency and revenue efficiency is potentially a difficult combinatorial problem. However, the prediction indicates an average prediction error of only 10.73%, that is, the predicted values shown in Table 9 yield more accurate results.

5 Concluding Remarks

This study assesses the energy efficiency and revenue efficiency of global airlines, whereby CO_2 emissions exit the first stage without being inputted to the second stage under a VRS assumption. These considerations are addressed by integrating a two-stage network DEA model in the form of SOCP with the grey prediction model as a data enabled analytics tool to estimate and predict airline efficiency. Aside from addressing the undesirable output of CO_2 emissions, handling the impacts of all ratio combinations of inputs, intermediates, and outputs on overall efficiency provides a reference for how airline efficiency can be measured accurately. That is, we illustrate how to use network DEA to uncover valuable information in a big data context.

This study's implications to the literature have two aspects. First, this study evaluates airline efficiency by using a SOCP-based two-stage network structure. This model can reflect the internal structure of the airline production process and allows undesirable outputs to leave the first stage of efficiency evaluation to optimize the overall efficiency. This study applies an innovative model to estimate the efficiency of 23 global airlines during the period of 2013–2017. Second, this study forecasts airline efficiency with the use of the grey prediction model, thereby determining the relationship between overall efficiency (i.e., the main behavior) and ratio combinations of inputs, intermediates, and outputs (i.e., the key factors) to improve the overall efficiency of airlines.

References

Alam, I. M. S., & Sickles, R. C. (1998). The relationship between stock market returns and technical efficiency innovations: Evidence from the US airline industry. *Journal of Productivity Analysis, 9*(1), 35–51.

Arjomandi, A., Dakpo, K. H., & Seufert, J. H. (2018). Have Asian airlines caught up with European airlines? A by-production efficiency analysis. *Transportation Research Part A: Policy and Practice, 116*, 389–403.

Babikian, R., Lukachko, S. P., & Waitz, I. A. (2002). The historical fuel efficiency characteristics of regional aircraft from technological, operational, and cost perspectives. *Journal of Air Transport Management, 8*(6), 389–400.

Barros, C. P., & Peypoch, N. (2009). An evaluation of European airlines' operational performance. *International Journal of Production Economics, 122*(2), 525–533.

Bhadra, D. (2009). Race to the bottom or swimming upstream: Performance analysis of US airlines. *Journal of Air Transport Management, 15*(5), 227–235.

Boyd, S., & Vandenberghe, L. (2004). *Convex optimization*. Cambridge University Press.

Capobianco, H. M. P., & Fernandes, E. (2004). Capital structure in the world airline industry. *Transportation Research Part A: Policy and Practice, 38*(6), 421–434.

Charnes, A., & Cooper, W. W. (1962). Programming with linear fractional functionals. *Naval Research Logistics Quarterly, 9*(3–4), 181–186.

Chen, K., & Zhu, J. (2017). Second order cone programming approach to two-stage network data envelopment analysis. *European Journal of Operational Research, 262*(1), 231–238.

Chen, K., & Zhu, J. (2019). Scale efficiency in two-stage network DEA. *Journal of the Operational Research Society, 70*(1), 101–110.
Chen, Y., Cook, W. D., Li, N., & Zhu, J. (2009). Additive efficiency decomposition in two-stage DEA. *European Journal of Operational Research, 196*(3), 1170–1176.
Chen, Z., Wanke, P., Antunes, J. J. M., & Zhang, N. (2017). Chinese airline efficiency under CO2 emissions and flight delays: A stochastic network DEA model. *Energy Economics, 68*, 89–108.
Chiou, Y.-C., & Chen, Y.-H. (2006). Route-based performance evaluation of Taiwanese domestic airlines using data envelopment analysis. *Transportation Research Part E: Logistics and Transportation Review, 42*(2), 116–127.
Chiu, C.-R., Chiu, Y.-H., Chen, Y.-C., & Fang, C.-L. (2016). Exploring the source of metafrontier inefficiency for various bank types in the two-stage network system with undesirable output. *Pacific-Basin Finance Journal, 36*, 1–13.
Cook, W. D., & Zhu, J. (2014). *Data envelopment analysis: A handbook of modeling internal structure and network*. Springer.
Cui, Q. (2020). Airline energy efficiency measures using a network range-adjusted measure with unified natural and managerial disposability. *Energy Efficiency, 13*(6), 1195–1211.
Cui, Q., & Jin, Z.-Y. (2020). Airline environmental efficiency measures considering negative data: An application of a modified network modified slacks-based measure model. *Energy, 207*, 118221.
Cui, Q., & Li, Y. (2015a). The change trend and influencing factors of civil aviation safety efficiency: The case of Chinese airline companies. *Safety Science, 75*, 56–63.
Cui, Q., & Li, Y. (2015b). Evaluating energy efficiency for airlines: An application of VFB-DEA. *Journal of Air Transport Management, 44*, 34–41.
Cui, Q., & Li, Y. (2016). Airline energy efficiency measures considering carbon abatement: A new strategic framework. *Transportation Research Part D: Transport and Environment, 49*, 246–258.
Cui, Q., & Li, Y. (2017). Airline efficiency measures under CNG2020 strategy: An application of a dynamic by-production model. *Transportation Research Part A: Policy and Practice, 106*, 130–143.
Cui, Q., Li, Y., Yu, C.-L., & Wei, Y.-M. (2016). Evaluating energy efficiency for airlines: An application of virtual frontier dynamic slacks based measure. *Energy, 113*, 1231–1240.
Cui, Q., Wei, Y.-M., & Li, Y. (2016). Exploring the impacts of the EU ETS emission limits on airline performance via the dynamic environmental DEA approach. *Applied Energy, 183*, 984–994.
Deng, J. (1982). Control problems of grey systems. *Systems and Control Letters, 1*, 288–294.
Distexhe, V., & Perelman, S. (1994). Technical efficiency and productivity growth in an era of deregulation: The case of airlines. *Swiss Journal of Economics and Statistics, 130*(4), 669–689.
Farnum, N. R. (1996). Using JOHNSON curves to describe non-normal ROCESS data. *Quality Engineering, 9*(2), 329–336.
Golany, B., & Roll, Y. (1989). An application procedure for DEA. *Omega, 17*(3), 237–250.
Good, D. H., Röller, L.-H., & Sickles, R. C. (1995). Airline efficiency differences between Europe and the US: Implications for the pace of EC integration and domestic regulation. *European Journal of Operational Research, 80*(3), 508–518.
Gramani, M. C. N. (2012). Efficiency decomposition approach: A cross-country airline analysis. *Expert Systems with Applications, 39*(5), 5815–5819.
Guo, C., Shureshjani, R. A., Foroughi, A. A., & Zhu, J. (2017). Decomposition weights and overall efficiency in two-stage additive network DEA. *European Journal of Operational Research, 257*(3), 896–906.
Hong, S., & Zhang, A. (2010). An efficiency study of airlines and air cargo/passenger divisions: A DEA approach. *World Review of Intermodal Transportation Research, 3*(1–2), 137–149.
Huang, S.-J., Chiu, N.-H., & Chen, L.-W. (2008). Integration of the grey relational analysis with genetic algorithm for software effort estimation. *European Journal of Operational Research, 188*(3), 898–909.

Kao, C. (2017). *Network data envelopment analysis: Foundations and extensions*. Springer International Publishing.

Kao, C., & Hwang, S.-N. (2011). Decomposition of technical and scale efficiencies in two-stage production systems. *European Journal of Operational Research, 211*(3), 515–519.

Ko, Y. D., Jang, Y. J., & Kim, D. Y. (2017). Strategic airline operation considering the carbon constrained air transport industry. *Journal of Air Transport Management, 62*, 1–9.

Kottas, A. T., & Madas, M. A. (2018). Comparative efficiency analysis of major international airlines using data envelopment analysis: Exploring effects of alliance membership and other operational efficiency determinants. *Journal of Air Transport Management, 70*, 1–17.

Kuljanin, J., Kalić, M., Caggiani, L., & Ottomanelli, M. (2019). A comparative efficiency and productivity analysis: Implication to airlines located in Central and South-East Europe. *Journal of Air Transport Management, 78*, 152–163.

Lewis, C. D. (1982). *Industrial and business forecasting methods*. Kent: Butterworth Scientific.

Li, Y., & Cui, Q. (2017a). Airline energy efficiency measures using the Virtual Frontier Network RAM with weak disposability. *Transportation Planning and Technology, 40*(4), 479–504.

Li, Y., & Cui, Q. (2017b). Carbon neutral growth from 2020 strategy and airline environmental inefficiency: A network range adjusted environmental data envelopment analysis. *Applied Energy, 199*, 13–24.

Li, Y., Wang, Y.-Z., & Cui, Q. (2015). Evaluating airline efficiency: An application of Virtual Frontier Network SBM. *Transportation Research Part E: Logistics and Transportation Review, 81*, 1–17.

Li, Y., Wang, Y.-Z., & Cui, Q. (2016). Energy efficiency measures for airlines: An application of virtual frontier dynamic range adjusted measure. *Journal of Renewable and Sustainable Energy, 8*(1), 015901.

Liu, S., & Lin, Y. (2006). *Grey information: Theory and practical applications*. Springer Science & Business Media.

Lozano, S., & Gutiérrez, E. (2014). A slacks-based network DEA efficiency analysis of European airlines. *Transportation Planning and Technology, 37*(7), 623–637.

Lu, W.-M., Wang, W.-K., Hung, S.-W., & Lu, E.-T. (2012). The effects of corporate governance on airline performance: Production and marketing efficiency perspectives. *Transportation Research Part E: Logistics and Transportation Review, 48*(2), 529–544.

Mallikarjun, S. (2015). Efficiency of US airlines: A strategic operating model. *Journal of Air Transport Management, 43*, 46–56.

Mao, M., & Chirwa, E. (2005). Combination of grey model GM (1, 1) with three-point moving average for accurate vehicle fatality risk prediction. *International Journal of Crashworthiness, 10*(6), 635–642.

Morán, J., Granada, E., Míguez, J., & Porteiro, J. (2006). Use of grey relational analysis to assess and optimize small biomass boilers. *Fuel Processing Technology, 87*(2), 123–127.

Ouellette, P., Petit, P., Tessier-Parent, L.-P., & Vigeant, S. (2010). Introducing regulation in the measurement of efficiency, with an application to the Canadian air carriers industry. *European Journal of Operational Research, 200*(1), 216–226.

Ray, S. C., & Mukherjee, K. (1996). Decomposition of the fisher ideal index of productivity: A non-parametric dual analysis of US airlines data. *The Economic Journal, 106*(439), 1659–1678.

Schefczyk, M. (1993). Operational performance of airlines: An extension of traditional measurement paradigms. *Strategic Management Journal, 14*(4), 301–317.

Semenick Alam, I. M., & Sickles, R. C. (2000). Time series analysis of deregulatory dynamics and technical efficiency: The case of the U.S. airline industry. *International Economic Review, 41*(1), 203–218.

Seufert, J. H., Arjomandi, A., & Dakpo, K. H. (2017). Evaluating airline operational performance: A Luenberger-Hicks-Moorsteen productivity indicator. *Transportation Research Part E: Logistics and Transportation Review, 104*, 52–68.

Smith, J. B. (1977). Trends in energy use and fuel efficiency in the US commercial airline industry. *Transportation Research Record, 870*, 90–97.

Tavassoli, M., Faramarzi, G. R., & Farzipoor Saen, R. (2014). Efficiency and effectiveness in airline performance using a SBM-NDEA model in the presence of shared input. *Journal of Air Transport Management, 34*, 146–153.

Tone, K., Kweh, Q. L., Lu, W.-M., & Ting, I. W. K. (2019). Modeling investments in the dynamic network performance of insurance companies. *Omega, 88*, 237–247.

Wang, W.-K., Lin, F., Ting, I. W. K., Kweh, Q. L., Lu, W.-M., & Chiu, T.-Y. (2017). Does asset-light strategy contribute to the dynamic efficiency of global airlines? *Journal of Air Transport Management, 62*, 99–108.

Wang, W.-K., Lu, W.-M., Kweh, Q. L., & Cheng, I.-T. (2014). Does intellectual capital matter? Assessing the performance of CPA firms based on additive efficiency decomposition DEA. *Knowledge-Based Systems, 65*, 38–49.

Wang, W.-K., Lu, W.-M., & Tsai, C.-J. (2011). The relationship between airline performance and corporate governance amongst US listed companies. *Journal of Air Transport Management, 17*(2), 148–152.

Wu, W.-Y., & Liao, Y.-K. (2014). A balanced scorecard envelopment approach to assess airlines' performance. *Industrial Management & Data Systems, 114*(1), 123–143.

Xu, X., & Cui, Q. (2017). Evaluating airline energy efficiency: An integrated approach with network epsilon-based measure and network slacks-based measure. *Energy, 122*, 274–286.

Yu, M.-M., Chen, L.-H., & Chiang, H. (2017). The effects of alliances and size on airlines' dynamic operational performance. *Transportation Research Part A: Policy and Practice, 106*, 197–214.

Zhang, Q., Koutmos, D., Chen, K., & Zhu, J. (2019). Using operational and stock analytics to measure airline performance: A network DEA approach. *Decision Sciences.*

Zhang, S., & Chen, M. (2004). The characteristics analysis of SCGM(1,1) model and the new grey modeling methods. *Kybernetes, 33*(2), 453–463.

Zhu, J. (2011). Airlines performance via two-stage network DEA approach. *Journal of CENTRUM Cathedra: The Business and Economics Research Journal, 4*(2), 260–269.

Zhu, J. (2020). DEA under big data: Data enabled analytics and network data envelopment analysis. *Annals of Operations Research.* https://doi.org/10.1007/s10479-020-03668-8

Zou, B., Elke, M., Hansen, M., & Kafle, N. (2014). Evaluating air carrier fuel efficiency in the US airline industry. *Transportation Research Part A: Policy and Practice, 59*, 306–330.

Index

J. Zhu, V. Charles (eds.), *Data-Enabled Analytics*, International Series in Operations Research & Management Science 312, https://doi.org/10.1007/978-3-030-75162-3

D

The manufacturer's authorised representative in the EU is Springer Nature Customer Service Centre GmbH, Europaplatz 3, 69115 Heidelberg, Germany. If you have any concerns regarding our products, please contact ProductSafety@springernature.com

Printed and bound by CPI Group (UK) Ltd, Croydon, CR0 4YY
28/09/2026
02224088-0019